NITROGEN ACQUISITION AND
ASSIMILATION IN HIGHER PLANTS

Plant Ecophysiology

Volume 3

Series Editors:

Luit J. De Kok and Ineke Stulen

University of Groningen
The Netherlands

Aims & Scope:

The Kluwer Handbook Series of Plant Ecophysiology comprises a series of books that deals with the impact of biotic and abiotic factors on plant functioning and physiological adaptation to the environment. The aim of the Plant Ecophysiology series is to review and integrate the present knowledge on the impact of the environment on plant functioning and adaptation at various levels of integration: from the molecular, biochemical, physiological to a whole plant level. This Handbook series is of interest to scientists who like to be informed of new developments and insights in plant ecophysiology, and can be used as advanced textbooks for biology students.

The titles published in this series are listed at the end of this volume.

Nitrogen Acquisition and Assimilation in Higher Plants

Edited by

Sara Amâncio
Technical University of Lisbon
Portugal

and

Ineke Stulen
University of Groningen
The Netherlands

KLUWER ACADEMIC PUBLISHERS
DORDRECHT/BOSTON/LONDON

A C.I.P. Catalogue record for this book is available from the Library of Congress.

ISBN 1-4020-2727-3 (HB)
ISBN 1-4020-2728-1 (e-book)

Published by Kluwer Academic Publishers,
P.O. Box 17, 3300 AA Dordrecht, The Netherlands.

Sold and distributed in North, Central and South America
by Kluwer Academic Publishers,
101 Philip Drive, Norwell, MA 02061, U.S.A.

In all other countries, sold and distributed
by Kluwer Academic Publishers,
P.O. Box 322, 3300 AH Dordrecht, The Netherlands.

Printed on acid-free paper

Cover illustrations:
Lower left corner: Part of a model on the localisation and regulation of the processes of uptake of nitrate in the root plasmalemma. A, amino acids; E, nitrate efflux; I, nitrate influx (Margreet ter Steege, University of Groningen, The Netherlands). Middle: An agro-ecosystem near Groningen, The Netherlands (Ineke Stulen). Upper right corner: Part of *in vivo* ^{15}N-NMR spectra of root apices perfused with $^{15}NH_4NO_3$. GLN, glutamine; ALA, alanine; GLU, glutamate (Sara Amâncio, Technical University of Lisbon; Helena Santos, ITQB, Oeiras, Portugal; figure by Belmiro J. Vilela, Lisbon).
Indexing: Daniël J.A. De Kok (Groningen)
Technical editing of the figures: Dick Visser (Groningen)

All Rights Reserved
© 2004 Kluwer Academic Publishers
No part of this work may be reproduced, stored in a retrieval system, or transmitted in any form or by any means, electronic, mechanical, photocopying, microfilming, recording or otherwise, without written permission from the Publisher, with the exception of any material supplied specifically for the purpose of being entered and executed on a computer system, for exclusive use by the purchaser of the work.

Printed in China

Dedicated to our friend and colleague

Wolfram Ullrich

We are grateful to the contributing authors for the time and effort spent in writing their chapters.

Contents

	Contributors	xi
	Preface	xiii
1.	**Nitrate uptake by roots - transporters and root development** Bruno Touraine	1
2.	**Metabolic regulation of ammonium uptake and assimilation** Tomoyuki Yamaya and Ann Oaks	35
3.	**Atmospheric nitrogen - pollutant or fertiliser?** Lucy J. Sheppard and Håkan Wallander	65
4.	**Symbiotic nitrogen fixation** Javier Ramos and Ton Bisseling	99
5.	**Nitrogen metabolism and plant adaptation to the environment - the scope for process-based modelling** Marcel van Oijen and Peter Levy	133
6.	**Light regulation of nitrate uptake, assimilation and metabolism** Cathrine Lillo	149
7.	**Modulation of nitrate reduction - environmental and internal factors involved** Werner M. Kaiser, Elisabeth Planchet, Maria Stoimenova and Matasoshi Sonoda	185
8.	**Integrated molecular analysis of the polyamine pathway in abiotic stress signalling** Alejandro Ferrando, Pedro Carrasco, Juan Cruz Cuevas, Teresa Altabella and Antonio F. Tiburcio	207
9.	**Significance of secondary nitrogen metabolites for food quality** Silvia Haneklaus and Ewald Schnug	231
10.	**Biotechnology of nitrogen acquisition in rice - implications for food security** Dev T. Britto and Herbert J. Kronzucker	261
	Index	283

Contributors

Teresa Altabella
Laboratori de Fisiología Vegetal, Facultat de Farmàcia, Universitat de Barcelona,
Diagonal 643, 08028 Barcelona, Spain

Ton Bisseling
Department of Molecular Biology, University of Wageningen,
Dreijenlaan 3, 6703 HA Wageningen, The Netherlands

Dev T. Britto
Division of Life Sciences, Department of Botany,
1265 Military Trail, Scarborough, Ontario, MiC 1A4 Canada

Pedro Carrasco
Departamento de Bioquímica y Biología Molecular, Universitad de Valencia,
Dr. Moliner 50 Burjassot, 46100 Valencia, Spain

Juan Cruz Cuevas
Laboratori de Fisiología Vegetal, Facultat de Farmàcia, Universitat de Barcelona,
Diagonal 643, 08028 Barcelona, Spain

Alejandro Ferrando
Laboratori de Fisiología Vegetal, Facultat de Farmàcia, Universitat de Barcelona,
Diagonal 643, 08028 Barcelona, Spain

Silvia Haneklaus
Institute of Plant Nutrition and Soil Science, Federal Agricultural Research Centre
(FAL), Bundesallee 50, D-38116 Braunschweig, Germany

Werner M. Kaiser
Julius von Sachs Institute of Biosciences, University of Würzburg,
Julius von Sachs Platz 2, D-97082 Würzburg, Germany

Herbert J. Kronzucker
Division of Life Sciences, Department of Botany,
1265 Military Trail, Scarborough, Ontario, MiC 1A4 Canada

Peter Levy
CEH - Edinburgh, Bush Estate, Penicuik, EH26 0QB, Scotland

Cathrine Lillo
School of Technology and Science, Stavanger University College,
Box 8002, 4068 Stavanger, Norway

Ann Oaks
Suite 1604, 685 Woolwich Street, Guelph, Ontario, N1H 8M6, Canada

Elisabeth Planchet
Julius von Sachs Institute of Biosciences, University of Würzburg,
Julius von Sachs Platz 2, D-97082 Würzburg, Germany

Javier Ramos
Departamento de Nutricíon Vegetal, Consejo Superior de Investigaciones Cientificas,
Apdo 202, 50080 Zaragoza, Spain

Ewald Schnug
Institute of Plant Nutrition and Soil Science, Federal Agricultural Research Centre
(FAL), Bundesallee 50, D-38116 Braunschweig, Germany

Lucy J. Sheppard
CEH - Edinburgh, Bush Estate, Penicuik, EH26 0QB, Scotland

Matasoshi Sonoda
Julius von Sachs Institute of Biosciences, University of Würzburg,
Julius von Sachs Platz 2, D-97082 Würzburg, Germany

Maria Stoimenova
Julius von Sachs Institute of Biosciences, University of Würzburg,
Julius von Sachs Platz 2, D-97082 Würzburg, Germany

Antonio F. Tiburcio
Laboratori de Fisiología Vegetal, Facultat de Farmàcia, Universitat de Barcelona,
Diagonal 643, 08028 Barcelona, Spain

Bruno Touraine
UMR 113, Symbioses Tropicales et Méditerranéennes, Université Montpellier 2,
CC 002, Place E. Bataillon, 34095 Montpellier Cedex 5, France

Marcel van Oijen
CEH - Edinburgh, Bush Estate, Penicuik, EH26 0QB, Scotland

Håkan Wallander
Department of Microbial Ecology, Ecology Building, 223 62 Lund, Sweden

Tomoyuki Yamaya
Graduate School of Agricultural Science,Tohoku University,
1-1 Tsutsumidori-Amamiyamacha, Aoba-ku, Sendai 981-855, Japan

Preface

Nitrogen Acquisition and Assimilation in Higher Plants is the third volume in the *Plant Ecophysiology* series. The main aim of this volume is to integrate different research approaches for a better understanding of the regulation of nitrogen metabolism by environmental signals. The book is divided into three parts. Chapters 1-5 mainly deal with the processes involved, while chapters 6-8 are more focussed on regulatory aspects. The significance of nitrogen metabolism for food quality and security is the topic of chapters 9 and 10.

Nitrogen can be acquired by plant roots as nitrate (chapter 1) or ammonium (chapter 2). In both chapters the physiology and molecular biology of the uptake processes and regulation of its transporters are addressed. Chapter 1 also deals with the effect of nitrate in the environment on root development, and chapter 2 with the metabolic regulation of the assimilation of ammonium, including gene manipulation. Plants can acquire atmospheric nitrogen as well, from nitrogenous air pollutants (chapter 3) or by N_2 fixation (chapter 4). Whether atmospheric nitrogen is a pollutant or a fertiliser is the main topic of chapter 3. Chapter 4 stresses the importance of N_2 fixation, and deals with the latest developments in the cloning of genes involved in the legume-*Rhizobium* interaction. To what extent modelling can help in understanding the role of nitrogen metabolism in plant adaptation to the environment is the topic of chapter 5.

The role of light in transcriptional and post-transcriptional regulation of the genes involved in nitrate uptake, reduction and assimilation is reviewed in chapter 6. Chapter 7 reviews the modulation of nitrate reductase activity by environmental and internal signals, and its effect on the production of NO. Chapter 8 deals with the metabolic pathway of polyamines, molecules involved in abiotic stress signalling, and includes advanced genomic and proteomic approaches.

Chapter 9 is focussed on the significance of secondary nitrogen compounds for food quality. Chapter 10 highlights the implementation of conventional and recombinant-DNA breeding methods to increase both yield and agronomic N-use efficiency of rice, the most important cereal for human nutrition.

This volume is of interest for advanced students and junior researchers, and provides comprehensive information and the latest developments for scientists working in the field of nitrogen metabolism and readers interested in sustainable development.

Sara Amâncio
Ineke Stulen

Chapter 1

NITRATE UPTAKE BY ROOTS - TRANSPORTERS AND ROOT DEVELOPMENT

Bruno Touraine

INTRODUCTION

With the exception of legumes and a few other species (actinorhizal plants) that develop N_2-fixing symbioses, plants acquire N from the soil in the inorganic form, as NH_4^+ or NO_3^-. The subjects of NH_4^+ uptake and symbiotic fixation are addressed in chapters 2 and 4, respectively. Considering that in most soils NO_3^- is the predominant inorganic soluble N form, the uptake of this ion by higher plant roots is the main pathway for entry of N into the global food chain. NO_3^- uptake and its subsequent reduction in plant cells consume a significant portion of C and energy. For many plant species including crops, purely ammoniacal nutrition has unfavourable effects on plant growth, partly due to the fact that NH_4^+ is a cation while NO_3^- is an anion (Chaillou and Lamaze 2001). This chapter will focus on the determinants of NO_3^- acquisition by plant roots.

NITRATE UPTAKE, AN ESSENTIAL STEP FOR HIGHER PLANT N NUTRITION

Due to the radial root organisation and the apoplasm-symplasm dualism, NO_3^- ions that are acquired from the soil solution have to enter the root symplasm. "Uptake" or "absorption" refers to this specific step that consists of transport over the plasma membrane of a root cell. Once in a root cell, NO_3^- ions can diffuse within the symplasm, from cell to cell, through plasmodesmata. NO_3^- ions can be used in the root or be loaded into the xylem vessels for transport to the shoot. Since the vessels lack plasma membranes and protoplasts, the NO_3^- ions transported in the xylem arrive in the leaf apoplasm, and have to be absorbed through the leaf cell plasma membrane. Contrary to most other major ions, NO_3^- is not transported in the phloem sap (Allen and Raven 1987, Gessler *et al.* 1998). Therefore, the transport of N from source leaves to root and shoot sink organs totally depends on the assimilation of NO_3^- in roots or source leaves and subsequent transport of amino-N in the phloem.

In root as well as in a leaf cells, NO_3^- can serve both as osmotic compound and N source for the assimilation of organic molecules. NO_3^- is the main inorganic anion available for osmotic adjustment and its reduction leads to the synthesis of organic anions, while NH_4^+ assimilation is accompanied by decarboxylation reactions. Inadequate anion accumulation, and consequently cation accumulation, results in a higher susceptibility to water deficit of plants grown on NH_4^+ as the sole N source (Salsac *et al.* 1987). Most plants have a reduced growth rate in the absence of NO_3^- ("ammoniacal syndrome", Chaillou and Lamaze 2001). This dual role of NO_3^- depends on the operation of specific transport systems in the tonoplast, to store NO_3^- in the vacuole and, conversely, remobilise stored NO_3^- for cytosolic NO_3^- reduction and further assimilation. The assimilatory pathway and its regulation are described in chapters 2 and 6. NO_3^- nutrition involves a number of transport steps, including uptake by the root, transport over the tonoplast, loading into the xylem and absorption by leaf cells. Since NO_3^- uptake can be the bottleneck between the source in the soil solution and the assimilatory pathway in plant cells, it is likely to play a key role in whole plant N nutrition.

By definition NO_3^- uptake is a flux and is expressed as µmoles of NO_3^- taken up per unit root surface area, or unit root weight, and per unit of time. Therefore, the rate of NO_3^- acquisition by the plant also depends on the surface area of the root system and all environmental factors that can affect root growth will also affect NO_3^- uptake capacity. Furthermore, the architecture of the root system determines the volume of soil foraged by roots and thus the ability of roots to access soil NO_3^-. The root system is very plastic and NO_3^- availability itself strongly affects root development. This chapter mainly focuses on NO_3^- uptake and its regulation. In addition the N-related root development response will be presented. The interactions between N nutrition and a biotic factor of the rhizosphere, namely plant growth-promoting rhizobacteria (PGPR), will also be discussed because these microorganisms that colonise the root systems of all plant species are possibly interfering with both root development and NO_3^- uptake.

NITRATE UPTAKE - WHERE AND HOW?

Localisation of NO_3^- uptake in plant roots

Regardless of the root system, the taproot system found in dicotyledons or the fibrous root system of monocotyledons, all roots are essentially similar in their anatomy and functioning, and participate in nutrient uptake (Fitter 2002).

Within each root tissue heterogeneity is present along two planes: radially, *viz.* the different tissues that originate from specific stem cells in the root apical meristem, organised in concentric cell layers, and longitudinally, from the apex to the root-stem junction, *viz.* zones with cells that have reached different stages of development. Therefore, the localisation of nitrate absorption encompasses two questions: 1) which tissue(s) is (are) responsible for NO_3^- uptake? and 2) is NO_3^- uptake distributed all along the root or is it restricted to specific zones only?

The radial organisation of the root is dominated by the cortex-stele duality because of the formation of a hydrophobic barrier in the cell wall of the endodermis, the innermost cortical cell layer. This Casparian strip functionally separates the apoplasm of the outer cylinder (the cortex) from the apoplasm of the central cylinder (the stele). Therefore, water and nutrients, including N, have to be absorbed in the symplasm of the cortex before they can reach the symplasm within the stele (Steudle and Peterson 1998). As mentioned above, NO_3^- ions are transported from the root to the shoot in the xylem vessels while they are not translocated, or in a negligible amount, from the shoot to the root in the phloem. Therefore, the rate of NO_3^- transport to the leaves has to be controlled to prevent an excess of NO_3^- ions that would exceed the capacities of storage and assimilation under high NO_3^- availability and high transpiration rate. To enter the root symplasm, NO_3^- ions have to be transported through the plasma membrane of some cells in the root cortex. This transport step, to which "uptake" precisely refers, involves specific transport proteins, and it is under the control of specific regulatory mechanisms. Uptake is the checkpoint where ions are selectively acquired from the outer medium at a controlled rate.

Which specific cell layers within the cortex are involved in the uptake of mineral nutrients is still unclear. In the maturation zone of the root downstream the elongation zone, root hairs are formed in the epidermal cells. With regards to the large exchange surface area of root hairs, it has been thought that ion uptake should predominantly occur through these specific structures (Deane-Drummond and Gates 1987). In fact, to devote the role of nutrient absorption to root hairs looked so logical that it is still considered as a fact in many plant physiology textbooks, while there is no clear evidence to support this assumption. Anatomical observations have shown a low density of plasmodesmata between the epidermal cell layer and the adjacent cortical cells compared to a higher frequency of plasmodesmata connecting each cell of the epidermis to neighbouring epidermal cells or within the cortical parenchyma (Kragler *et al.* 1998). With fluorescent dyes that are unable to pass the plasma membrane it has been shown that a barrier exists between root epidermis and cortical parenchyma for dyes that move from cell to cell within each of these tissues (Erwee and Goodwin 1995). Anatomical evidence would thus indicate the existence of at least two different symplasmic domains in the root cortex, one that is restricted to the epidermis and one that spreads over the cortical parenchyma and the endodermis. The real uptake step would therefore occur at the site where ions enter this latter domain that extends to stele tissues and can deliver nutrients to the whole plant. This hypothesis would explain the discrepancy between the measurements of the root cytosolic NO_3^- pool made in barley roots using different methodologies (Siddiqi *et al.* 1991). This study has suggested that two distinct pools would exist, the smaller metabolic (nitrate reductase-containing) pool would possibly be located in the epidermal cells, while the larger pool, where nitrate absorbed from the external solution enters before following radial symplasmic transport in the root and xylem conduction to the shoot, would be located in the cortical parenchyma cells. Unfortunately, this thought-provoking hypothesis has not been checked by any direct experiment. Finally, only very scarce, circumstantial, and sometimes contradictory, physiological evidence is available and the question of which cell layer(s) is (are) responsible for NO_3^- uptake remains unresolved. Recently, functional promoter analysis in *Arabidopsis thaliana* revealed that the *Nrt2.1* gene, encoding a transporter involved in NO_3^- uptake (see below), is expressed in the entire cortex of mature regions of primary and lateral roots, with a higher expression level in the cortical parenchyma than either the epidermis or the endodermis (Nazoa *et al.* 2003). These results suggest that the three cortical tissues have the ability to absorb NO_3^-, but they do not allow determining whether they are involved in NO_3^- uptake *per se* or not. Furthermore, this investigation was performed in *Arabidopsis*, a plant species in which the root cortical parenchyma consists of a single cell layer, while this tissue consists of several cell layers in most other plant species. Whether the transporters involved in NO_3^- uptake in such plants are expressed in both the outer and inner cortical parenchyma cell layers or not is not known.

The second question about the localisation of NO_3^- uptake in the root - does it occur over the entire root surface or only in specific regions of the root? - has been sporadically debated. Essentially two methodologies, based upon the localised supply of $^{15}NO_3^-$ (Lazof *et al.* 1992, Nazoa *et al.* 2003) and the utilisation of NO_3^--selective microelectrodes (Henriksen *et al.* 1990, 1992, Colmer and Bloom 1998, Bertrand *et al.* 2000) respectively, have been used to measure the general pattern of NO_3^- uptake along roots. Overall, the experimental evidence obtained with both techniques shows that NO_3^- uptake occurs at all locations of the root surface, from the tip to the base, although a peak is usually measured in a sub-apical region. Due to the small size of this zone compared to the whole root surface in an actively growing plant, the older regions of the root should contribute to the total acquisition of NO_3^- for an important part. This conclusion is derived from data obtained with young seedlings, a few days to two weeks old. However, a similar pattern has been found for the primary and lateral roots of two-month-old hydroponically grown soybean plants with a forty-cm long taproot (Philippe Lobit and Bruno Touraine, unpublished results). Recent investigations revealed that the *Nrt2.1* promoter targets gene expression to the mature regions of the *Arabidopsis* root, but not to the apex or elongation zone (Nazoa *et al.* 2003). On the other hand, $^{15}NO_3^-$ influx in this latter zone was as fast as the influx measured in older root parts, suggesting that different NO_3^- transporters might be responsible for NO_3^- uptake in younger and older root parts, respectively.

Because both physiological and electrophysiological approaches require an easy access at the root surface and a free diffusion of the ion studied in an unstirred layer near the root surface, all the studies on NO_3^- uptake localisation have been performed with young hydroponically-grown plants. However, due to anatomical features, the NO_3^- uptake pattern along the root in soil-grown plants may differ from that in hydroponically-grown plants. During secondary growth, cork cells with suberised and ultimately lignified walls are formed between the cortex and the central cylinder of the root. In plants with secondary growth structures, only the young fine roots that have not yet developed a periderm are able to take up nutrients, including NO_3^- ions, from the soil solution. Furthermore, even in roots containing only tissues formed by the primary growth, the epidermis peels off and one can wonder to which extent the quasi-constant NO_3^- uptake rate observed along the roots of hydroponically-grown plants is representative of the pattern of NO_3^- uptake along the roots of field-grown plants.

The component fluxes of NO_3^- uptake

Using the short-lived radioisotope ^{13}N or the stable isotope ^{15}N, it has been shown that the rate of NO_3^- uptake is not a unidirectional flux, from the medium to the root symplasm, but the balance of two opposite fluxes: an influx from apoplasm to cytoplasm, and an efflux in the reverse direction. Because of the small size of the cytosolic pool of NO_3^-, a significant proportion of the NO_3^- absorbed is rapidly recycled out of the absorbing cell (Devienne *et al.* 1994, Muller *et al.* 1995). The cost of NO_3^- uptake is rather high (see below), so that the occurrence of such a futile cycle through the plasma membrane may be somewhat surprising. In principle, a possible physiological role for this leak of NO_3^- out of the root could be to regulate the net NO_3^- uptake rate. Indeed, a very efficient influx and a concomitant variable efflux might be, theoretically, the best way to precisely control the net flux and maintain plant N homeostasis despite large variations of the external NO_3^- concentration. However, experimental results dismiss this hypothesis. In most instances, the changes in net NO_3^- uptake, *e.g.* during the day/night cycle (Delhon *et al.* 1995) or in response to modifications of N availability (Muller *et al.* 1995), are reflected by changes in NO_3^- influx, while efflux responses are relatively minor. Overall, the patterns observed in earlier net uptake studies have been confirmed by NO_3^- influx measurements, thus supporting the idea that it is the regulation of influx that plays the major role in the control of NO_3^- uptake *in planta*. This led to efflux being ignored in most of the studies on NO_3^- uptake. On the other hand, no definitive evidence exists to conclude that the efflux has no role in this regulation, all the more since NO_3^- efflux is not invariable (Teyker et al. 1988, Van der Leij et al. 1998). Moreover, efflux is inherently relatively large in slow growing species, and this might explain the higher energy cost for NO_3^- uptake in these species compared to fast growing species (Scheurwater et al. 1998, 1999). Quantitative differences in the futile NO_3^- cycle across the root plasma membrane may thus be related to ecological differences. However, although the occurrence of protein-mediated NO_3^- efflux has been demonstrated using plasma membrane vesicles (Grouzis et al. 1997, Pouliquin et al. 1999, 2000), the transporters involved in NO_3^- efflux have not been isolated yet. The transport mechanisms involved in NO_3^- efflux, therefore, remains to be deciphered in detail. Below we will focus on the influx component of NO_3^- uptake.

The transport systems for NO_3^- influx

Physiological studies on NO_3^- influx in the plant roots provided evidence for the existence of at least three distinct transport systems for NO_3^-. At concentrations up to 0.5 mM, the curves representing NO_3^- influx vs. the external concentration of NO_3^- follow a saturable kinetic pattern with a Km value in the order of magnitude of 10 µM (Siddiqi et al. 1990). Comparing plants grown on a NO_3^--free medium prior the influx measurement with plants grown on a NO_3^--containing medium revealed that, in fact, two high-affinity transport systems (HATS) co-exist in the plasma membrane of root cells. Plants that have not been exposed to NO_3^- for several days prior to the uptake experiment absorb NO_3^- at a low rate within the first hour of exposure to NO_3^-. This rate subsequently increases to peak at 10- to 30-fold the initial level after a period that varies from 3 hours to a few days depending on plant species (Lee and Drew 1986, Siddiqi et al. 1990, Aslam et al. 1992, 1993, Kronzucker et al. 1995a). The fact that transcriptional and translational inhibitors blocked this increase of NO_3^- uptake rate (Tompkins et al. 1978, Lainé et al. 1995) suggested de novo synthesis of transporter protein. Moreover, precise determination of the kinetic parameters in the roots of NO_3^- induced and uninduced plants have shown that not only the Vmax value, but also the Km increased after exposure to NO_3^- (Lee and Drew 1986, Hole et al. 1990, Siddiqi et al. 1990, Aslam et al. 1992). Based on these observations, two different transport systems, namely a low capacity constitutive high affinity transport system (cHATS) and a high(er) capacity inducible high-affinity transport system (iHATS), are distinguished. The role of the cHATS is likely to enable the cytoplasmic concentration of NO_3^- to rise to a level sufficient for induction of the higher capacity transport system (Behl et al. 1988). It has been established that NO_3^- itself, but not NH_4^+ nor any product of its assimilation pathway is the inducer of the iHATS (King et al. 1993). NO_2^-, which is not found in significant amounts under natural conditions, is able to induce NO_3^-

uptake (Aslam *et al*. 1993). Induction of the NO_3^- transport system by its substrate is unique among the ion absorption systems recognised in plant roots. Furthermore, NO_3^- also induces several enzymes involved in N and C metabolism suggesting that this ion acts as a signal to initiate co-ordinated changes in N and C nutrition (Crawford 1995, Scheible *et al*. 1997a).

At external NO_3^- concentrations above those at which the HATS plateau is reached, another uptake system, referred to as the low affinity transport system (LATS), becomes apparent. According to most reports, this transport system shows a linear correlation with the external NO_3^- concentration, with no saturation up to 50 mM (Pace and McClure 1986, Siddiqi *et al*. 1990, Aslam *et al*. 1992, Kronzucker *et al*. 1995, Touraine and Glass 1997). Whatever the species considered, the significance of Km values cited for NO_3^- LATS therefore seems doubtful. The LATS is constitutive in the sense that it does not require induction by NO_3^-, as shown by both kinetic (Siddiqi *et al*. 1990) and electrophysiological (Glass *et al*. 1992) studies.

The three transport systems that have been identified in the root cell plasma membrane are likely to operate simultaneously. At any concentration, therefore, the NO_3^- uptake rate is the sum of the three components, cHATS, iHATS and LATS. However, for external NO_3^- concentrations below 200 to 500 µM, the LATS is negligible because of a relatively small slope compare with that of the HATS hyperbola, while at higher concentrations, the HATS rates have reached their plateau by 90% or more and the LATS rate become a significant component of NO_3^- influx. Although the threshold value for either HATS or LATS to be predominant varies with the plant species, the 200-500 µM range applies for most annual plants, monocots or dicots, studied. In woody species, identical transport systems have been characterised, though with significantly higher Km value for the iHATS (Kronzucker *et al*. 1995b).

The energisation of NO_3^- uptake

Thermodynamic calculations clearly show that the uptake of NO_3^- is uphill, indicating that NO_3^- influx is an active process. Different techniques, including ion-selective microelectrodes (Zhen *et al*. 1991, Van der Leij *et al*. 1998), pool size determination based on nitrate reductase activity (Robin *et al*. 1983, King *et al*. 1992), tracer influx profiles (Presland and MacNaughton 1984), and compartmental analysis by tracer efflux (Lee and Clarkson 1986, Siddiqi *et al*. 1991, Devienne *et al*. 1994, Muller *et al*. 1995, Kronzucker *et al*. 1999) have been used in order to estimate the NO_3^- concentration in the cytosol of root cells. The results reported varied, sometimes several-fold, from one study to another. Especially studies using ion-selective microelectrodes indicate that the cytosolic concentration of NO_3^- is low (3-5 mM) and invariable, whereas most other studies show that the cytosolic NO_3^- pool can vary several orders of magnitude. The reason for these discrepancies, and hence the actual NO_3^- concentration in the cytosol of root cells, is still a matter of debate (Siddiqi and Glass 2002, Britto and Kronzucker 2003). Whatever the precise values of the cytosolic NO_3^- concentration, we can assess that the ratio of this concentration to the external NO_3^- concentration is higher than 0.1, for any external NO_3^- concentration up to 50 mM. Therefore, given that the electrical potential difference through root cell plasma membrane is in the -150 to -300 mV range (Thibaud and Grignon 1981, Glass et al. 1992), NO_3^- ions must be actively transported into the cytoplasm with a minimum free energy requirement of 9 to 23 kJ mol^{-1}.

Classical studies on the effects of anoxia (Neyra and Hageman 1976), low temperature (Clarkson and Warner 1979) or metabolic inhibitors (Jackson et al. 1973, Rao and Rains 1976, Morgan et al. 1985) have confirmed that NO_3^- uptake is an energy-dependent process. Consistent with the chemiosmotic coupling model elucidated by Peter Mitchell in the 1960s, NO_3^- uptake (predominated by influx) is driven by the pH gradient (Thibaud and Grignon 1981). Conversely, NO_3^- efflux out of the root symplasm is down its electrochemical potential gradient, which was demonstrated by the finding that the loading of xylem vessels, which is a net efflux from xylem parenchyma cells, is passive and is driven by the electrical potential difference between the root symplasm and the xylem sap (Touraine and Grignon 1982). From root respiration analysis, it has been calculated that the absorption of 1 mol NO_3^- consumes 1 to 3 moles ATP, which represents more than 20% of the global cost of NO_3^- acquisition and assimilation in roots, or 5% of the energy supplied by respiration (Bloom et al. 1992). Such high values for the energy cost of NO_3^- uptake are likely to be due to both a high cost of NO_3^- influx, and the occurrence of relatively large efflux (Scheurwater et al. 1998, 1999).

The roots of plants fed with NO_3^- typically alkalinise the nutrient solution (Kirkby and Mengel 1967, Marschner and Röhmeld 1983, Touraine et al. 1988). Considering the quantitative importance of N intake compared to that of any other element, this is consistent with the operation of a NO_3^-/H^+ symport as hypothesised by the chemiosmotic coupling model. The transient depolarisation of the root cell plasma membrane observed upon NO_3^- supply (Ullrich and Novacky 1981) suggests that the stoichiometry of this system is $1NO_3^-/2H^+$ (McClure et al. 1990, Ruiz-Cristin and Briskin 1991, Glass et al. 1992). The membrane potential subsequently hyperpolarises (Thibaud and Grignon 1981), which has been attributed to repolarisation by enhanced H^+ pumping activity by the plasma membrane ATPase (Glass et al. 1992). Similar membrane potential patterns have been obtained for low and high external NO_3^- concentrations, indicating that both HATS and LATS are mediated by electrogenic $1NO_3^-/2H^+$ symports (Glass et al. 1992). Alternatively, NO_3^- influx could be mediated by a $1NO_3^-/2OH^-$ or a $1NO_3^-/2HCO_3^-$ antiport, since such transport systems would have identical effects on pH and electrical gradient across the plasma membrane. No direct evidence is available to favour one or other of these systems, although the positive relationships between the rates of organic acid decarboxylation in roots, alkalinisation of the solution, and NO_3^- uptake suggests that NO_3^- ions may be exchanged against HCO_3^- (Touraine *et al.* 1992). However, in the absence of definitive evidence to discriminate among the three transport systems, they are ascribed to NO_3^-/H^+ ($1NO_3^-/2H^+$ when stoichiometry specified) symport following the usual chemiosmotic terminology.

The nitrate transporters

Two families of genes coding for NO_3^- transporters in eukaryotes have been identified so far, namely the *Nrt1* and *Nrt2* families. The encoded proteins belong to the peptide transporter (PTR) and nitrate-nitrite porters (NNP) family respectively, both being assigned to the Major Facilitating Superfamily (MFS), a group of membrane proteins with two sets of six transmembrane domains linked by a cytosolic loop (Forde 2000). Both families are represented by multiple genes that may encode transporters with different properties and functions, different localisation, and which expression is differentially regulated. In *A. thaliana*, 4 members of the *Nrt1* family (Tsay *et al.* 1993, Hatzfeld and Saito 1999, Huang *et al.* 1999) and 7 members of the *Nrt2* family (Glass *et al.* 2001, Orsel *et al.* 2002) have been found, all of them presumably mediating an active transport of NO_3^- across the plasma membrane. However, until very recently the quasi-totality knowledge on plant NO_3^- transporters was restricted to one member of each group, namely the *Nrt1.1* and *Nrt2.1* genes from *A. thaliana*. Molecular data on these two genes, in relation to physiological knowledge on NO_3^- uptake, have been extensively reviewed in a number of papers (Crawford and Glass 1998, Daniel-Vedele *et al.* 1998, Forde and Clarkson 1999, Forde 2000, Galvan and Fernandez 2001, Glass *et al.* 2001, Touraine *et al.* 2001, Williams and Miller 2001), and the reader interested in a thorough description of *Nrt1.1* and *Nrt2.1* is referred to these excellent papers. Below, I will give a rather brief overview that aims at outlining the main features of these two transporters. In addition, recent data will be reported to update the knowledge on plant nitrate transporters. Since most of the results have been obtained in *A. thaliana*, the presentation is limited to this species.

The NRT1.1 and NRT2.2 transporters

The first NO_3^- transporter gene identified in plants, *Nrt1.1* (originally designated *CHL1*), was isolated from a T-DNA insertion mutant line of *A. thaliana* which was resistant to chlorate, a toxic analogue of NO_3^- (Tsay *et al.* 1993). Ironically, there is still controversy on the function of the only NO_3^- transporter gene that has been found from a functional test *in planta* (all the other *Nrt1* and *Nrt2* genes have been identified based upon sequence homologies, see below). Perhaps this might have been expected from the fact that chlorate is a poor analogue of NO_3^- for uptake processes (Guy *et al.* 1988, Kosola and Bloom 1996, Touraine and Glass 1997). In an initial screen for ClO_3^-

resistant mutants, only one class of mutants, the *chl1* mutants, was changed in ClO_3^- uptake (Oostindier-Braaksma and Feenstra 1973). At this time, characterisation of the *chl1* mutants indicated that the NO_3^- LATS was impaired (Doddema and Telkamp 1979, Scholten and Feenstra 1986). After isolation of the *Nrt1.1* gene, its product has first been characterized as a low-affinity NO_3^- transporter by expression in *Xenopus* oocytes (Huang *et al*. 1996). However, further characterisation of the *Arabidopsis chl1* mutants resulted in a more complex view of the CHL1 (NRT1.1) function. While it appeared to be responsible for most of the ClO_3^- uptake, its contribution to NO_3^- influx is modest (Touraine and Glass 1997). More precisely, the disruption of the *Nrt1.1* gene leads to reduced NO_3^- uptake in the LATS concentration range, but only when plants are grown on NH_4NO_3. In contrast, when plants were supplied with NO_3^- as the sole N source, no significant diminution of NO_3^- influx, either in the HATS or LATS concentration ranges, was detected (Touraine and Glass 1997). These observations indicated that NRT1.1 mediates NO_3^- influx via a LATS, but that it is not a major component of this transport system (even in plants grown on NH_4NO_3, a substantial NO_3^- low affinity influx remained) and that other low affinity transporters compensate for its deletion. Furthermore, *Nrt1.1* expression was induced by NO_3^- (Tsay *et al*. 1993) while LATS was thought to be constitutive (Siddiqi *et al*. 1990). Complexity was added by the finding that NRT1.1 is also involved in the NO_3^- HATS under certain growth conditions (Wang *et al*. 1998), leading to the proposal that it is a dual affinity NO_3^- transporter (Liu *et al*. 1999). In fact, the NRT1.1 function has become even less clear since the expression of the *Nrt1.1* gene has been shown not to be targeted to the epidermis and cortex, where NO_3^- uptake takes place (Guo *et al*. 2001). Furthermore, Guo *et al*. (2001) showed that the expression of *Nrt1.1* is also found in shoots. Finally, whether in roots or shoots, the expression of *Nrt1.1* was essentially localised in the nascent organs, suggesting a possible role in organogenesis rather than in NO_3^- uptake.

Contrary to *Nrt1.1*, the first *Nrt2* genes have been discovered by virtue of sequence homology with *Aspergillus nidulans* and *Chlamydomonas reinhardtii* genes (Zhuo *et al*. 1999) and by differential display (Filleur and Daniel-Vedele 1999). More recently, the isolation of a T-DNA mutant with a complete deletion of *Nrt2.1* and a partial deletion of *Nrt2.2* in its 3' end allowed functional characterisation of these genes in *A. thaliana* (Cerezo *et al*. 2001, Filleur *et al*. 2001). It has been shown that this mutant is severely impaired in the NO_3^- HATS but not in the LATS and the authors concluded that NRT2.1 or NRT2.2, or both of them, are the predominant transporters responsible for the iHATS involved in NO_3^- uptake. The *Nrt2.1* gene was expressed at a higher level than *Nrt2.2*, this latter being detectable by Northern-blot only in very young *Arabidopsis* seedlings (Zhuo *et al*. 1999). Furthermore, high correlations have been repeatedly found between the changes in *Nrt2.1* transcript abundance and NO_3^- influx by the iHATS in response to a variety of conditions (*e.g.* Lejay *et al*.

1999, Zhuo et al. 1999, Nazoa et al. 2003). Based on this evidence, it is proposed that NRT2.1 is the major transporter contributing to the iHATS component of NO_3^- influx in roots. The localisation of the *Nrt2.1* gene expression in the epidermal and cortical tissues of *Arabidopsis* roots is consistent with the involvement of NRT2.1 in NO_3^- uptake (Nazoa et al. 2003). However, a puzzling result was that *Nrt2.1* expression is limited to mature parts of the root, suggesting that other NO_3^- transporters must play its role in the younger root parts.

Table 1. Expression patterns of the *Nrt* genes in *Arabidopsis thaliana* (data from Okamoto et al. 2003). Regressions of *Nrt* transcript levels with HATS and LATS influx have been calculated during a 72 h induction-repression experiment (see text) on 100 μM and 5 mM NO_3^-, respectively. The correlation coefficient (r) had significant ($P<0.05$) positive values for HATS and LATS only for *Nrt1.1* and *Nrt2.1*, respectively (*a*). *Nrt2.3* and *Nrt2.5* gave a significant correlation with HATS and LATS, respectively, but r was negative (*b*). The root/shoot ratio of transcript abundance is expressed on a logarithmic scale.

Gene	Correlation coefficient (r)		Log (root/shoot) transcript abundance
	HATS	LATS	
Nrt1.1	+ 0.56	+ 0.88[a]	- 1.1
Nrt1.2	+ 0.20	+ 0.40	+ 0.1
Nrt1.3	- 0.10	- 0.52	- 0.6
Nrt1.4	- 0.25	+ 0.25	- 1.5
Nrt2.1	+ 0.74[a]	+ 0.78	+ 3.2
Nrt2.2	+ 0.46	- 0.20	+ 4.4
Nrt2.3	- 0.78[b]	- 0.41	+ 3.0
Nrt2.4	+ 0.10	+ 0.50	+ 2.0
Nrt2.5	- 0.33	- 0.85[b]	+ 3.0
Nrt2.6	- 0.47	+ 0.44	+ 0.5
Nrt2.7	- 0.17	+ 0.42	- 1.1

Which roles for the other NRTs?

Until recently, hardly any data was available on the other NRT1 and NRT2 members. The situation has changed with the publication of studies comparing the transcript abundances of the various *Nrt1* and *Nrt2* genes in root and shoot tissues under different conditions of N supply (Glass et al. 2001, Orsel et al. 2002, Okamoto et al. 2003). Especially, Okamoto et al. (2003) have exhaustively investigated the time-course changes of the transcript abundances of the 4 *Nrt1* and 7 *Nrt2* genes in roots and shoots after provision of NO_3^- to *Arabidopsis* plants previously grown on an N-free solution for 1 week. Classically, in such experiments, NO_3^- influx increases rapidly due to NO_3^- induction, peaks after a few hours and then decreases due to repression by the N status restored by the current uptake (Siddiqi et al. 1989, Zhuo et al. 1999). Based on the pattern of the transcript abundance time-course, Okamoto et al. (2003) attempted to classify the different genes as NO_3^--constitutive, NO_3^--inducible, or NO_3^--repressible. Furthermore, these authors established correlations between the transcript levels of the 11 *Nrt* genes, and the HATS and LATS NO_3^- influx estimated by the $^{13}NO_3^-$ influx from 100 μM and 5 mM external NO_3^- respectively. Finally, they concluded that the NO_3^- HATS and LATS involved in NO_3^- uptake are likely to be mediated primarily by NRT2.1 and NRT1.1, respectively, but they were unable to attribute specific functions to the other NRTs. Table 1 summarises some of the results obtained by Okamoto et al. (2003) on the expression patterns of the different *Nrt* genes.

The progress made since the first report of a putative NO_3^- transporter gene in plants (Tsay et al. 1993) is just beginning to provide new insights on the complexity of NO_3^- transport in plants. Indeed, most of the studies published up to now essentially correlated *Nrt* transcript abundance to known changes in NO_3^- influx under various conditions, most of which concerned either a change in N status or the day/night regulation. Assigning a role to the various members of the NRT1 and NRT2 families requires undertaking both additional functional studies (*e.g.* characterisation of mutants and transformants with altered expression of specific *Nrt* genes, localisation of gene expression using reporter gene fusion lines) and more mechanistic studies (*e.g.* promoter analysis, identification of specific transcription factors). In addition, the biochemistry of plant NO_3^- transporters is still a *terra incognita* to be explored.

REGULATION OF NITRATE UPTAKE

The regulation of nitrate uptake by plant nutritional demand

The physiological characterisation of NO_3^- transport systems that led to the HATS and LATS concept was based on a strict dependence of NO_3^- influx upon external NO_3^- concentration. This has been established with plants grown under identical conditions, including the same external NO_3^- concentration, that were transferred to a range of NO_3^- concentrations for the duration of the NO_3^- influx experiment (typically 1 to 5 min). Because these investigations aimed at evaluating the kinetic characteristics of the activity of transport proteins, it was essential that the transport protein equipment of root cell plasma membrane be identical in all plants. In contrast, to study the regulation of NO_3^- uptake requires either to analyse its changes in response to changes in an environmental factor over longer periods, or to compare plants grown under conditions that differ in one environmental factor. Such studies led to the conclusion that the rate of NO_3^- uptake is often independent of the current NO_3^- concentration at the time the uptake rate is measured. This is shown by two lines of evidence: 1) as long as

NO_3^- is not the limiting factor for plant growth, feeding plants with higher NO_3^- concentrations does not result in higher NO_3^- uptake rates (*e.g.* Clement *et al.* 1978), and 2) changing the plant's growth rate while external NO_3^- concentration remains invariable usually leads to concomitant changes in NO_3^- uptake rate and net photosynthesis rate or biomass accumulation (*e.g.* Gastal and Saugier 1989, Touraine unpublished results). These results indicate that NO_3^- uptake is under the control of regulatory mechanisms so that its rate may not change with external NO_3^- concentration (1), or may change while NO_3^- concentration is constant (2).

Most of the experiments on NO_3^- uptake regulation concern the net rate of NO_3^- uptake rather than the influx of NO_3^-. Although the changes observed in net NO_3^- uptake rate could theoretically be due to changes in either of its two components, it is generally thought that the target of NO_3^- uptake regulation is the NO_3^- influx. Consequently, experiments on NO_3^- uptake regulation where either the net uptake rate or the influx is measured have been interpreted in the same way, and any regulation demonstrated for NO_3^- uptake has been attributed explicitly or implicitly, to the influx.

The experiments usually referred as N deficiency experiments provide a well-documented example of how plants regulate NO_3^- uptake. These experiments consisted of growing plants on an N-free nutrient solution for periods of days, and resupplying them with NO_3^- for a few hours (typically 3 h) to ensure iHATS reinduction. The lack of any N source in the nutrient medium necessarily results in lowering some internal N pools since biomass production is maintained while no N can be taken up during the depletion period. However, although the terms starvation and deficiency are frequently used in these experiments, the time scale used is generally kept short enough so that no deficiency symptom is visible. Nevertheless, N depletion always resulted in a higher NO_3^- uptake rate than in control plants continuously fed with NO_3^-; moreover, the longer the depletion period, the higher the uptake rate (Doddema and Otten 1979, Lee and Rudge 1986, Bowman *et al.* 1989, Lee 1993). The diurnal regulation of NO_3^- uptake (Delhon *et al.* 1995) is another classical example that the demand of N by the shoot must control the rate of NO_3^- uptake in roots.

Since the activity of the NO_3^- transport systems depends on the external NO_3^- concentration, the regulation of the rate of NO_3^- uptake must involve changes in uptake capacity, *i.e.* the number of functional transport systems in the root cell plasma membrane. This is exemplified by experiments in which the relationships between the NO_3^- uptake rate and the external NO_3^- concentration were established for plants that were previously pre-treated with various NO_3^- concentrations for several days. Whatever the pre-treatment concentration, the plot obtained follows the HATS or LATS characteristic pattern, but the higher the pre-treatment concentration the lower the HATS Vmax and the LATS slope (Siddiqi *et al.* 1990). This indicates that the roots of plants grown with high NO_3^- concentrations must be equipped with a reduced number of NO_3^- transport systems per unit weight (or root length) in comparison with plants grown at low NO_3^- availability. Consistent with this hypothesis, both the NO_3^- influx and the expression of *Nrt2* gene were down-regulated in the roots of barley (Vidmar *et al.* 2000a) and *A. thaliana* (Nazoa *et al.* 2003) grown on high NO_3^- concentration.

The regulatory process responsible for the control of NO_3^- uptake is very powerful as shown by the fact that the rate of NO_3^- uptake by perennial ryegrass was unaffected by the NO_3^- concentration in a range from 14 µM to 14 mM (Clement *et al.* 1978). Of course, when NO_3^- availability is too low (below 14 µM in Clement's study), the synthesis of a larger number of NO_3^- transport systems may not be sufficient to compensate for their low activity. In this case, not only N homeostasis cannot be maintained but also N becomes limiting for growth when NO_3^- concentration decreases. Conversely when NO_3^- concentration is too high (above 14 mM in Clement's study), the NO_3^- uptake rate exceeds the requirement for growth so that the plant's N content increases while the growth rate doesn't when NO_3^- concentration increases.

Despite the wealth of physiological data showing the relative independence of NO_3^- uptake rate over NO_3^- concentration, the Michaelis-Menten formalism is still widely used for the NO_3^- acquisition term in ecophysiological plant growth models (Jeuffroy *et al.* 2002). Clearly this representation is unrealistic and does not represent the relationships between NO_3^- uptake and NO_3^- availability on the long term.

The mechanisms regulating NO_3^- uptake

NO_3^- uptake is under systemic control

As mentioned above, the restoration of NO_3^- supply to N-depleted plants results in a rapid increase of NO_3^- influx, which peaks within a few hours, and declines while the nutritional status is restored by current uptake (Siddiqi *et al.* 1989, Zhuo *et al.* 1999). This is indicative of two successive steps: the induction of transport systems by NO_3^- (the iHATS), then the repression of NO_3^- influx by increasing N status. Although NO_3^- uptake induction by its substrate is an interesting feature because of its uniqueness among root transports, its physiological meaning is unclear. Therefore, only the second mechanism, namely the nutritional status-dependent repression/derepression, will be considered below.

The changes in NO_3^- uptake rate in response to the changes in light intensity, or the diurnal regulation of NO_3^- uptake, clearly need interorgan signalling. In addition, split-root experiments revealed that the regulation of NO_3^- uptake by the plant's N demand is systemic: feeding one part of the root system with an N-free solution resulted in enhanced NO_3^- uptake rate by the other part of the root system, when compared with plants which both parts of the root system received NO_3^- (Edwards and Barber 1976, Burns 1991, Lainé *et al.* 1995). Different processes, involving different phloem-translocated signals, are likely to operate simultaneously. Some of these processes have been identified through physiological investigations in the 1970s and 1980s. In the last decade, molecular data have supported these physiological models, which depend upon metabolite control.

Organic acid signalling

Historically, the first model that envisaged the involvement of shoot-root translocation in the regulation of NO_3^- uptake was developed in the early 1970s to account for the impact of N metabolism on ionic and pH balance in plants (Ben Zioni *et al.* 1971). The starting point is that NO_3^- reduction results in the net release of 1 mol OH^- equivalent per mol NO_3^- reduced, and that the leaf cells cannot expel the basic ions (OH^- or HCO_3^-) because of the finite size of the leaf extracellular space (Raven and Smith 1976) and the reduced transport capacity of OH^- equivalents in the phloem sap (Raven 1977, Allen and Raven 1987). The pH homeostasis in the cytosol of leaf cells is thus achieved by biochemical pH-stats that synthesise strong organic acids (mainly malate) in response to small pH increase (Davies 1986). Ben Zioni *et al.* (1971) proposed that the organic acids that are produced stoichiometrically to amino acids in leaves are transported in the phloem to the roots, where their decarboxylation releases HCO_3^- ions that are excreted into the external medium. Numerous studies on the fate of the anion charge generated by NO_3^- assimilation have been reported, which demonstrated the validity of this interorgan circulation pattern in the species for which leaves are the major site of NO_3^- reduction (Touraine *et al.* 1988 and references therein). This model predicts the co-ordination of NO_3^- uptake to leaf nitrate reductase activity: the higher the NO_3^- reduction rate in leaves, the higher the synthesis and export rate of carboxylates to roots, and the higher the rate of HCO_3^- release in roots after decarboxylation. These ions could be exchanged with absorbed NO_3^- ions. Experimental evidence in favour of this model has been obtained in soybean (Touraine *et al.* 1992). Indeed, increasing the amounts of malate translocated into the phloem led to an increase in the NO_3^- uptake rate by roots, the alkalinisation rate of the medium and the release of inorganic C by roots. It is not known, however, whether the NO_3^--HCO_3^- ions exchange is mediated by the NO_3^- transporters or whether these proteins co-transport NO_3^- with H^+, this latter transport being possibly stimulated by the decarboxylation of malate in root cells. To investigate this would require detailed biochemistry studies on NO_3^- transporters that remain to be done. Whatever their precise mode of action on NO_3^- transporters, organic acids are likely to exert a regulation on the activity of these transporters, not on NO_3^- uptake capacity. Consistent with this assumption, organic acid treatments failed to increase the transcript abundances of the two NO_3^- transporters characterised to some extent in *A. thaliana*, *Nrt1.1* and *Nrt2.1* (unpublished data).

Sugar signalling

Experimentally, NO_3^- uptake can be increased in parallel with the growth rate by increasing light intensity (*e.g.* Gastal and Saugier 1989). The diurnal regulation of NO_3^- uptake (higher rate during the light period, lower rate during the dark period) has been considered as a specific example of its regulation by light (Delhon *et al.* 1995). The stimulation of NO_3^- uptake by light is attributed to photosynthesis, since it could be prevented by decreasing the atmospheric CO_2 concentration (Delhon *et al.* 1996). Moreover, the addition of carbohydrates to the nutrient solution is known to increase NO_3^- uptake (Hänisch Ten Cate and Breteler 1981). Blocking phloem translocation in soybean plants by means of stem girdling led to a decline in the rate of NO_3^- uptake, which is partially restored by adding glucose to the nutrient solution (Delhon *et al.* 1996). No consensus exists that an energetic shortage might occur in roots, even at night, and an increased energisation of the plasma membrane transporters by carbohydrate availability to roots is unlikely to be the explanation of these results. Alternatively, sucrose supplied to roots by the phloem would rather act as a signal that up-regulates NO_3^- uptake. Consistent with this hypothesis, the transcript abundance of both *Nrt1.1* and *Nrt2.1* genes of *A. thaliana* undergoes marked diurnal changes and can be rapidly increased in the dark period if sucrose is supplied (Delhon *et al.* 1996); these variations correlated with changes in NO_3^- influx. Contrary to the organic acid regulatory process, therefore, carbohydrates must be involved in the control of NO_3^- uptake capacity, *i.e.* the synthesis of new NO_3^- transporters, rather than in the activity of the existing NO_3^- transporters. Consistent with this hypothesis, the role of carbohydrates in signalling has already been demonstrated for both C and N metabolism. Sucrose and glucose especially are involved in C sensing and in the regulation of the expression of a series of genes involved in C and N assimilation (Coruzzi and Zhou 2001).

Amino acid signalling

The two models described above, involving the organic acids produced during the leaf reduction of NO_3^- and the carbohydrates produced by photosynthesis respectively, cannot explain the regulation of NO_3^- uptake by N demand itself. Several definitions of the term N demand can be proposed, but here I define N demand as the balance between the actual N amount in plant tissues and N requirement, the latter being defined as the product between optimal N content and growth rate. Photosynthates are a possible sensor of the growth rate, but not of specific N requirement, and organic acids are only related to NO_3^- reduction rate not N needs. For instance, in N-depleted plants NO_3^- uptake capacity is up-regulated whilst the current reduction rate is very low and the growth rate may be reduced compared to N-sufficient plants. The systemic regulation process by which N demand controls NO_3^- uptake must involve other molecules to inform the roots of the plant N status. It is generally agreed that these phloem-translocated signals must be amino acids, which would act as a sensor of the N status since their export rate from the source leaves must be determined by the difference between the assimilation rate and proteolysis on the one hand and the utilisation rate for proteogenesis on the other hand (Imsande and Touraine 1994). The increased NO_3^- uptake rate observed in N-depleted plants is considered to be the relief of the feedback control exerted by phloem-translocated amino acids on root NO_3^- transporters. The existence of an extensive pool of amino-N that continuously cycles between roots and shoots has been demonstrated (Touraine *et al.* 1988, Cooper and Clarkson 1989, Larsson 1992). This transport pool of amino N is relatively isolated, *i.e.* the exchange rate between this pool and the bulk tissue is slower than the exchange rate within this pool between xylem and phloem (Cooper and Clarkson 1989), which makes its composition representative of the whole plant N status. The inhibiting effect of amino acids on NO_3^- uptake has been demonstrated by using various procedures to increase the root amino acid content. Exogenous supply of amino acids in the nutrient solution led to a decrease of NO_3^- uptake rate (Doddema and Otten 1979, Breteler and Arnozis 1985, Muller and Touraine 1992, Muller *et al.* 1995) and *Nrt2* transcript abundance (Krapp *et al.* 1998, Zhuo *et al.* 1999, Nazoa *et al.* 2003). Providing the amino acids to soybean roots via the phloem also led to decreased NO_3^- uptake rate (Muller and Touraine 1992, Muller *et al.* 1995). Consistent with the effects of exogenously supplied amino acids, treatments with a glutamate synthase inhibitor to increase glutamine content in roots resulted in decreased NO_3^- uptake rate in dwarf bean (Breteler and Siegerist 1984), and led to a decline in both the NO_3^- influx and *Nrt2* transcript abundance in *Arabidopsis* (Zhuo *et al.* 1999) and barley (Vidmar *et al.* 2000b). Another approach to enrich the amino acid content of the phloem sap consisted of supplying N directly to the leaves using NO_2 or NH_3 fumigation (Muller *et al.* 1996, Gessler *et al.* 1998). Both studies showed a correlation between increased amino-N translocation from shoot to root via the phloem and decreased NO_3^- uptake rate. One of the main criticisms against exogenous amino acid application experiments is that, due to amino acid interconversions, it is not known what pool is actually modified in the root cells. A recent study showed that alanine, asparagine, aspartate, glutamine and glutamate treatments caused decreased NO_3^- influx and *Nrt2.1* transcript abundance, and inhibited the *Nrt2.1* promoter activity in the roots of *A. thaliana* (Nazoa *et al.* 2003). Analysing the amino acid content of these roots revealed that the *Nrt2.1* transcript abundance correlated (negatively) best with the glutamine concentration in the roots (Figure 1). These results strongly suggest that glutamine plays the main role in NO_3^- uptake regulation by the N demand and that this regulation is exerted at a transcriptional level. Consistent with this conclusion, the *Nrt2.1* gene, which expression is down-regulated by amino acids (see above), but not the *Nrt1.1* gene, which expression is unaffected by externally supplied amino acids (personal data), is derepressed in N-depleted *A. thaliana* (Lejay *et al.* 1999).

Despite the wealth of physiological and molecular data that support the amino acid signalling model for NO_3^- uptake regulation by N demand, a key assumption of the model remains to be verified. The model is based on the hypothesis that changes in leaf N status trigger changes in the export rate of all or some amino acids, but the difficulty in measuring the composition and translocation rate of phloem sap prevented accurate measurements of, for instance, the effects of N deprivation on the phloem transport of amino acids (Tillard et al. 1998). On the other hand, the experiments by Muller et al. (1996) and Gessler et al. (1998), already mentioned above, showed that the increase of leaf N status via NO_2 or NH_3 fumigation led to a reduction in NO_3^- uptake rate, indicating that a phloem-translocated signal conveys the information about shoot amino-N content to the roots.

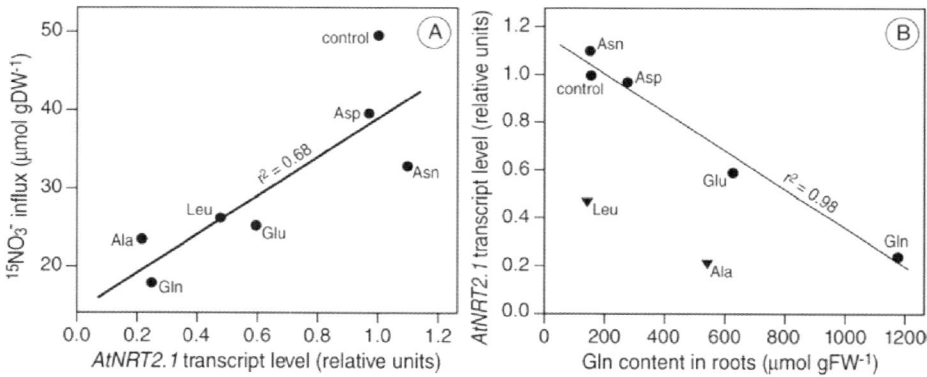

Figure 1. Correlations between the amount of *AtNRT2.1* mRNA and either $^{15}NO_3^-$ influx (A) or Gln content in roots (B). The plots represent data drawn from Nazoa et al. (2003). The *Arabidopsis* plants have been supplied with the amino acid indicated for 9 h, except for control that received no amino acid. The linear correlation in (B) has been calculated from Control, Asn, Asp, Glu and Gln, but not Ala and Leu, treatments.

Regulation of ion transport by the nutritional demand of the whole plant is not unique for NO_3^-, but has been shown to occur for most nutrients, *e.g.* K^+ (Drew and Saker 1984), $H_2PO_4^-$ (Clarkson and Scattergood 1982) and SO_4^{2-} (Clarkson et al. 1983, Lappartient and Touraine 1996). A few molecular evidence, the first of which for $H_2PO_4^-$ and SO_4^{2-} transporters, that supports the interorgan signalling model has been obtained. Tomato $H_2PO_4^-$ transporter genes *LePt1* and *LePt2* were up-regulated in split root experiments where half of the roots were fed with a P-free solution (Liu et al. 1998). Likewise, the expression of the *APS1* gene coding for an ATP sulfurylase and of the *AST68* gene coding for a SO_4^{2-} transporter increased in canola plants when a large part of the root system was deprived of S (Lappartient et al. 1999). More recently, developing the split root system for *Arabidopsis*, Gansel et al. (2001) showed that *Nrt2.1* transcript abundance increased in NO_3^--fed roots in response to N deprivation of the another part of the root system. This change correlated with $^{15}NO_3^-$ influx changes. Consistent with the a positive effect of high plant N demand, or low plant N status, on NO_3^- uptake, feeding *Arabidopsis* plants with a low NO_3^- concentration results in higher *Nrt2.1* transcript level and NO_3^- influx compared with plants fed with a high NO_3^- concentration (Gansel et al. 2001, Nazoa et al. 2003).

Beside its role in the adjustment of the NO_3^- uptake rate to the N status when the latter is modified by some environmental factor, the repression of NO_3^- influx in roots by phloem-translocated amino acids could also be involved in the control of NO_3^- uptake during plant development. This has been proposed by Imsande and Touraine (1994) for the transition between the vegetative and the pod-filling phase in legumes. Consistent with a developmental control of NO_3^- uptake capacity, *Nrt2.1* expression increased with age in young *Arabidopsis* seedlings, then decreased to become about undetectable at flowering stage (Nazoa et al. 2003). Considering that NO_3^- uptake rate is under the control of a systemic regulation, one may expect that some plant hormone would be involved, all the more since the main root NO_3^- influx transporter is developmentally regulated. Furthermore, both the auxin pathway and ABA are involved in the regulation of root development by N nutrition (see below), indicating that NO_3^- content in root and/or leaf does act as a signal that triggers hormonal signalling. However, up to now, no experimental evidence that hormone regulation could be involved in the regulation of NO_3^- uptake by N demand has been reported.

ROOT GROWTH AND DEVELOPMENT - REGULATION BY NITRATE AND N STATUS

Nitrate availability in soil is highly fluctuating with time, and plants adapt not only by changing the NO_3^- uptake capacity, as discussed above, but also by changing the architecture of the root system. This has been established for many plant species, which have been shown to promote root growth at the expense of shoot growth in response to nutrient limitation or deficiency, so that exploration of the soil for more nutrients is given priority, leading to increased root/shoot ratios (Ericsson 1995). Moreover, NO_3^-, like other nutrients, is frequently distributed unevenly within the soil, and many plant species can adapt to this situation by proliferating their roots preferentially within the nutrient-rich patches (Robinson 1994). In fact, the root plasticity responses to changes in NO_3^- availability and to heterogeneous distribution in the medium exploited by the root system are likely to be triggered by the same processes, as shown by split-root experiments (Burns 1991, Lainé et al. 1995).

Split-root experiments showed that plants attempt to maximise their ability to capture the available N by two successive processes. The first process, which begins within a few hours of exposing a root system to a localised NO_3^- supply, consists of a marked increase in NO_3^- uptake capacity in comparison with roots uniformly supplied with NO_3^-. A second response, which becomes apparent after a few days, consists of an increase in the number of lateral roots initiated in the NO_3^--containing zone and an increase in the growth rate of both new and existing lateral roots. This leads to a large increase in root density within the NO_3^--containing zone. The proliferation of roots in this zone is accompanied by a decrease in NO_3^- uptake rate (Burns 1991), suggesting that the rapid change in NO_3^- uptake capacity is an early response of plants to adjust NO_3^- acquisition to N demand but that the long-term adaptation to NO_3^- constraint is obtained through developmental changes that result in modifications of the root system architecture.

Occurrence of a dual, localised and systemic response

Split-root experiments showed that the lack of NO_3^- near some roots triggers the proliferation of roots in another zone where NO_3^- is present in the same concentration as for control plants that received a uniform supply of NO_3^- (Burns 1991, Lainé et al. 1995). In classical experiments, Drew and colleagues used flowing nutrient solutions to show that localised provision of NO_3^- stimulated root branching specifically within the NO_3^--rich zone (Drew et al. 1973, Drew 1975, Drew and Saker 1975). This could suggest that some signalling mechanism senses the lack of NO_3^- in a root region and stimulate the growth of root regions that develop in NO_3^--containing patches. In fact, a combination of physiological and molecular investigations in A. thaliana led Zhang and Forde (Zhang and Forde 1998, Zhang et al. 1999) to show that two different processes are involved in the root developmental control by localised NO_3^-. Growing A. thaliana seedlings on vertical agar Petri dishes with medium segmented in three zones that contained or lacked NO_3^- allowed them to reproduce the stimulation of lateral root elongation in the NO_3^--rich zone previously described in barley. Furthermore, the identification of the *ANR1* gene and its characterisation (see below) led Zhang and Forde to propose that two pathways co-exist, a localised pathway by which exogenous NO_3^- stimulates lateral root growth, and a systemic pathway by which shoot originating signals repress lateral root development in response to high N status. According to this model, which appears very robust, lateral root stimulation in NO_3^--rich patches is the combined result of the release of systemic inhibition due to low plant N status and the localised stimulation due to exogenous NO_3^-. While the localised response begins to be decrypted, the systemic response remains largely obscure.

The localised response of lateral roots to nitrate availability

In their work on *A. thaliana*, Zhang and Forde showed that a localised supply of NO_3^- could stimulate lateral root elongation without any significant increase in the size of mature root cells, indicating an increased meristematic activity in lateral root tips (Zhang *et al.* 1999). Since NO_3^-, but no other source of N (such as NH_4^+ or an amino acid) stimulates lateral root growth (Zhang *et al.* 1999, Tranbarger and Touraine, unpublished data), the signal that positively affects lateral root development must originate from NO_3^- itself, and not from an N metabolite or an effect of NO_3^- on the N status. The lateral root response to a localised NO_3^- supply was the same in the *nia1nia2* mutant, which is deficient in nitrate reductase, as in the wild-type, confirming that NO_3^- itself is a signal for stimulated lateral root proliferation, as it is a signal for NO_3^- transport system induction (see above).

The first component identified in the signal transduction pathway that links external NO_3^- to lateral root proliferation is the NO_3^--inducible root-specific *ANR1* gene, which encodes a member of the MADS-box family of transcription factors (Zhang and Forde, 1998). Transgenic *Arabidopsis* lines with down-regulated *ANR1* expression failed to display the lateral root growth response to localised NO_3^-, which showed that the *ANR1* gene product is involved in transducing the NO_3^- signal to elicit the developmental response leading to increased lateral root proliferation. Moreover, an analysis of the effect of localised NO_3^- treatments on lateral root proliferation in three auxin-resistant mutants (*aux1*, *axr2*, *axr4*) provided evidence for an involvement of the auxin signalling pathway in the NO_3^--related control of lateral root elongation (Zhang *et al.* 1999). While the *aux1* and *axr2* mutants showed wild-type responses to the localised NO_3^- treatment, no increase in lateral root elongation rate was seen in the *axr4* mutant.

The systemic response of lateral roots to the plant N status

Besides the positive effect of NO_3^- on lateral root proliferation, an inhibitory effect of high NO_3^- availability on lateral root development has also been reported (Scheible *et al.* 1997a, Zhang *et al.* 1999, Tranbarger *et al.* 2003a,b). Tobacco lines partially deficient in nitrate reductase grown on high NO_3^- concentrations had an increased shoot/root ratio that was at least partly due to a strong inhibition of root growth (Scheible *et al.* 1997a,b). Similarly, the lateral root development in the *nia1nia2* mutant *Arabidopsis* was more sensitive to NO_3^- inhibition than was the wild-type, showing that the assimilation of NO_3^- is not required for the inhibitory effect of NO_3^- (Zhang *et al.* 1999). The higher sensitivity of nitrate reductase deficient mutants to lateral root inhibition by NO_3^- indicates that this effect is triggered by NO_3^- accumulated in the plant rather than by external NO_3^-. Furthermore, when comparing a series of tobacco lines with a range of different nitrate reductase activities and grown at different external NO_3^- concentrations, a strong correlation between the leaf NO_3^- content and the shoot/root ratio was found (Scheible et al. 1997b). The results of a split root experiment confirmed that the signal for inhibition of root growth comes from the shoot: when one half of the root system was grown in low NO_3^- and the other half in high NO_3^-, the growth rate of the former root part was 8-20 times less than it was when the entire root system of the same line was grown in low NO_3^- (Scheible et al. 1997b).

The inhibitory effect appears not to affect the lateral root development at the same step as the stimulatory effect described above. While the latter acts on the elongation of mature lateral roots, the inhibitory effect blocks the lateral root primordia at a very early stage around lateral root emergence (Zhang et al. 1999, Tranbarger et al. 2003a,b). This stage coincides with a critical phase in lateral root development when the newly formed meristem becomes activated and elongation of the mature lateral root starts (Malamy and Benfey 1997). How NO_3^- levels are monitored in the shoot, which molecules are the phloem-translocated signals, and what are the underlying mechanisms responsible for the inhibitory effect on lateral root development is still not understood. However, recent studies using ABA insensitive mutants and ABA treatments have illustrated the central role played by this hormone in the regulation of root branching by the systemic nitrate inhibitory system (Signora et al. 2001, De Smet et al. 2003).

INTERACTION BETWEEN N NUTRITION AND BIOTIC FACTORS OF THE RHIZOSPHERE - PLANT GROWTH-PROMOTING RHIZOBACTERIA (PGPR)

Plant roots develop in a particular environment referred to as the rhizosphere, which can be defined as the portion of soil, the composition and structure of which is altered by root activity. Due to the high concentration of organic nutrients released as root exudates, the rhizosphere is characterised by the abundance of actively growing microbial populations. These microorganisms can be classified in three categories according to their effects on plants, the neutral microorganisms that have no known effect on plants, pathogens that have deleterious effects, and microorganisms that trigger beneficial effects. Of particular relevance for plant N nutrition are the symbiotic associations, especially the *Rhizobium*-legume nodules that supply the plant with N fixed by the bacteroids (see chapter 4). The present chapter will focus on a subset of bacteria that colonise the rhizosphere and have beneficial effects on plants, and are, therefore, called plant growth-promoting rhizobacteria (PGPR). In fact, this term covers two different types of interaction between soil-borne bacteria and plants. Pseudomonads and some other bacteria control pathogens or their effects on plants by various ways, including competition for colonisation sites or nutrients, antibiosis, degradation of pathogenic factors, and induction of plant resistance mechanisms (Induced Systemic Resistance, ISR, Bloemberg and Lugtenberg 2001, Whipps 2001, Persello-Cartieaux et al. 2003). Historically, the term PGPR has been proposed to designate these bacteria that indirectly increase plant growth via protection against pathogens action (Kloepper et al. 1980). However, *Azospirillum* spp. and other rhizobacteria (see chapters 4 and 10) stimulate plant growth more directly, by either developmental or nutritional effects, which have made them termed phytostimulators and biofertilisers, respectively (Bloemberg and Lugtenberg 2001). Bashan and Holguin (1998) proposed to use two different terms, namely biocontrol plant growth-promoting bacteria (biocontrol-PGPB) and plant growth-promoting bacteria (PGPB) in order to avoid confusion between two functionally different fields, even if in some cases the same bacterial strain may both suppress a pathogen and enhance plant growth more directly. Below I will focus on developmental and nutritional effects only and omit the biocontrol effects. Finally, because our scope is narrowed to rhizosphere bacteria, the term PGPR has been preferred to PGPB.

Considering the quantitative importance of N in plant mineral nutrition and given that both N availability and PGPR are known to trigger changes in both plant nutrition and root development, it is very likely that plant responses depend on interactions between these abiotic and biotic factors. Different aspects of this question have been discussed in a recent paper (Mantelin and Touraine 2003), and will be briefly presented below.

The effects of PGPR on root development

The PGPR elicit remarkable changes in root morphology, namely increased lateral root number and length, and increased root hair number and length (*e.g.* Tien et al. 1979, Pacovsky 1990, Okon and Vanderleyden 1997, Bertrand et al. 2000). It is generally assumed that these developmental responses are triggered by phytohormones released in the rhizosphere by the PGPR (Bloemberg and Lugtenberg 2001, Persello-Cartieaux et al. 2003). *Azospirillum* has been shown to be capable of producing three types of plant growth factors, auxins, cytokinins and gibberellins (Tien et al. 1979, Bottini et al. 1989). Some experimental results suggest that IAA would play the major role in the changes in root morphology elicited by PGPR inoculation. Firstly, *Azospirillum* mutants with decreased IAA production are not as efficient as the wild-type strain (Barbieri and Galli 1993, Dobbelaere et al. 1999). Secondly, a screen for *A. thaliana* mutants insensitive to *Pseudomonas* rhizobacteria resulted in the isolation of two mutants altered in the *Aux1* auxin influx transporter (Persello-Cartieaux et al. 2001). However, another possible effect of rhizosphere bacteria on the plant hormonal balance has been proposed based on the finding of an aminocyclopropane carboxylate deaminase in some PGPR strains (Shah et al.

1998). Such bacteria are likely to divert ACC, the precursor of ethylene, from the plant root, hence to reduce the inhibition of root growth by ethylene (Glick *et al.* 1998).

Whatever the phytohormones involved, it is assumed that the effects of PGPR on plant hormonal signalling pathways are responsible for the observed changes in root morphology, which in turn result in enhanced mineral uptake, and that the improved nutrition is responsible for the stimulated plant growth (Okon and Kapulnik 1986). Consistent with this idea, plants inoculated with *Azospirillum* take up nutrients, including N, more efficiently from the soil (Okon and Vanderleyden 1997). However, such observations do not allow determining whether the enhancement of mineral acquisition by PGPR is a consequence of the increase in root surface area or whether the PGPR can more directly stimulate the ion uptake systems (see below).

When considering the relationships between PGPR and NO_3^- acquisition, it is worth reminding that NO_3^- availability affects root branching (see above). Since NO_3^- availability and PGPR both affect the same developmental processes, the question of the interactions between the responses to NO_3^- and those to PGPR is posed. However, a more detailed observation of NO_3^- and PGPR effects on root morphology makes the possibility that these responses share common pathways very unlikely. Indeed, as mentioned above, the systemic effect of high NO_3^- consists of blocking lateral root emergence rather than decreasing the growth rate of emerged lateral root, while PGPR have been reported to alter (positively) the lateral root elongation. For instance, the inoculation of *A. thaliana* with a very efficient PGPR strain isolated from the rhizosphere of *Brassica napus* (Bertrand *et al.* 2001) led to a slight increase in lateral root number but mainly a marked increase in lateral root length (unpublished results). In contrast, external NO_3^- concentration affects lateral root development at a stage after the primordium initiation and before emergence (Tranbarger *et al.* 2003a,b). Recently, combining NO_3^- treatments with *Phyllobacterium* inoculums at various doses, we confirmed that high N status and PGPR affect lateral root development independently from each other: the rhizobacteria promoted lateral root growth whatever the NO_3^- concentration in the nutrient medium; conversely, increasing the NO_3^- concentration repressed lateral root growth whether the plants were inoculated or not (S. Mantelin and B. Touraine, unpublished results). Morphologically, the effects of PGPR on lateral root development is more like the localised response to NO_3^- since the increased root branching is mainly a consequence of an enhanced elongation of lateral roots in both cases. However, it is very unlikely that PGPR could elicit a pathway overlapping with the first steps of the signalling pathway involved in the localised response to NO_3^-. Indeed, under most experimental conditions used in PGPR experiments, whether on soil or in vitro, NO_3^- was present at sufficient concentrations to trigger *ANR1* induction (Zhang and Forde 1998). However, given that the effect of PGPR on root morphogenesis has been attributed to the release of auxin by the PGPR and that a link has been found between *ANR1* and the auxin pathway (Zhang *et al.* 1999), the PGPR and NO_3^- could elicit different sensing mechanisms that converge on the same auxin-related pathway. Considering the rapid progresses in our knowledge on the complex networks triggered by auxin signalling, new biological tools (*e.g.* well-characterised mutants altered in specific steps of the IAA-related pathways) and realistic regulatory models should help to decipher the interconnections between such abiotic and biotic developmental responses.

Plant growth-promoting rhizobacteria and nitrate uptake

It is widely accepted that PGPR enhance "nutrient uptake", but this effect has usually been attributed to the observed increase in size of the root system of inoculated plants. However, this rationale confuses nutrient acquisition rate with uptake rate: while the former can be estimated as the amount of mineral nutrient acquired from the soil by a plant per unit of time, the actual uptake is a flux that has to be expressed per unit of root surface or weight. This distinction is very important because the controls of uptake and root surface area appeal to totally distinct processes. With regards to the effects of PGPR, if ever a link has to be found between NO_3^- uptake and root development, the question to be asked is to determine whether PGPR increase the acquisition rate of NO_3^- (hence N delivered to shoot) as a consequence of the increased root surface area or whether they have also a positive effect on NO_3^- uptake itself. Only sketchy data are available to try to answer this question. Inoculation with PGPR has been reported to result in either unchanged (Dobbelaere *et al.* 2002) or higher (Saudibet *et al.* 2002) plant N content, leading the authors to conclude at an unaffected or enhanced NO_3^- uptake, respectively. However, one must remember that NO_3^- uptake regulation should normally reduce NO_3^- uptake rate when root growth is enhanced, so that N homeostasis is maintained (Burns 1991, and see above). Therefore, two possibilities can be evoked: 1) NO_3^- is limiting and increasing the root surface would result in increased NO_3^- acquisition rate because no feedback control would operate, and 2) the feedback mechanisms, or NO_3^- uptake itself, are targets of PGPR. Considering the variety of culture conditions used in the different studies published, plant N nutrition is unlikely to be always limited when plant growth promotion by rhizobacteria was observed. Therefore, the second possibility, *i.e.* that PGPR do have a positive effect on NO_3^- uptake has to be considered. The effect of PGPR on NO_3^- uptake has been investigated in *Brassica napus* inoculated with an *Achromobacter* strain (Bertrand *et al.* 2000). As frequently observed with PGPR, this strain led to a higher increase in shoot biomass than root biomass, implying that the total amount of NO_3^- acquired by the plant has increased more than the root surface did. Using ion-specific microelectrodes, it has been shown that the uptake rate of NO_3^- did increase in response to the inoculation with *Achromobacter* (Bertrand *et al.* 2000). However, it is remarkable that the uptake rate of the other ion tested, K^+, and the efflux of H^+ increased similarly. This suggests that the rhizobacteria affected mineral ion uptake globally due to a stimulation of the proton pump ATPase activity. Alternatively the *Achromobacter* strain could simultaneously enhance the activities of several ion transporters as a consequence of changes in plant's nutritional demand. In this case, the increase in shoot/root ratio would be an effect of the inoculation with the PGPR and the increased nutrient uptake would just be a consequence of this effect. If this is correct, the mechanisms through which shoot development is enhanced by PGPR have still to be discovered.

Finally, root hairs are another factor to be considered when looking at nutrient uptake by PGPR-inoculated roots. Several PGPR strains have been shown to increase root hair size and number (De Freitas and Germida 1990, Okon and Vanderleyden 1997, Bertrand *et al.* 2000), which is likely to increase the absorbing surface much more than root weight. The difficulty here is that the specific role of root hair cells in NO_3^- uptake is not known (see above). The progress made in the identification of NO_3^- transporters should provide tools to determine whether the PGPR can or cannot affect activity and/or regulation of NO_3^- transporters involved in NO_3^- uptake.

CONCLUSIONS

Physiological investigations have provided a wealth of data that gave the framework to build models that describe how the transport systems responsible for NO_3^- uptake must work and how they are regulated. Since the identification of the first gene encoding a NO_3^- transporter in plants, ten years ago, the results obtained in a series of molecular studies, most of which concerned the *Nrt1.1* and *Nrt2.1* genes of *A. thaliana*, supported the physiological models established before. To get a more detailed insight on the proteins mediating the various NO_3^- transport steps in a higher plant, their specific roles and their regulation, it is now crucial to undertake functional investigations on the various members of the NRT1 and NRT2 families in *A. thaliana*. Reverse genetics using knockout mutants or transgenic plants with altered expression of the *Nrt* genes should help reaching this objective. Concerning the regulation of NO_3^- uptake, many reports show that it is largely exerted at a transcriptional level, but there is very limited information on the mechanisms involved in this regulation. Promoter studies should be developed in order to identify the *cis*-element and the transcription factors for the regulation of NO_3^- transporters. The transduction pathways involved in NO_3^- uptake regulation may then be deciphered, which should lead to the identification of the sensors of plant's N demand and the signals involved in the systemic control of NO_3^- uptake. Although focusing on the NRT transporters certainly may be very productive, many other important questions about NO_3^- uptake and transport in plants remain unresolved, *e.g.* the identification of NO_3^- efflux transporters.

In comparison with NO_3^- transport, the control of root development by NO_3^- sensing has progressed much recently, leading to new models that integrate N sensing and hormone signalling. This should help understanding both N signalling in plants and the basic mechanisms for root development plasticity. The study on the effects of PGPR on NO_3^- nutrition is a widely opened research field that should provide interesting systems to integrate functional and developmental plasticity of plants in response to environmental changes.

REFERENCES

Allen S., Raven J.A. 1987. Intracellular pH regulation in *Ricinus communis* grown with ammonium or nitrate as N source: the role of long-distance transport. - J. Exp. Bot. 38: 580-596.
Aslam M., Travis R.L., Huffaker R.C. 1992. Comparative kinetics and reciprocal inhibition of nitrate and nitrite uptake in roots of uninduced and induced barley (*Hordeum vulgare* L.) seedlings. - Plant Physiol. 99: 1124-1133.
Aslam M., Travis R.L., Huffaker R.C. 1993. Comparative induction of nitrate and nitrite uptake and reduction systems by ambient nitrate and nitrite in intact roots of barley (*Hordeum vulgare* L.) seedlings. - Plant Physiol. 102: 811-819.
Barbieri P., Galli E. 1993. Effect on wheat root development of inoculation with an *Azospirillum brasilense* mutant with altered indole-3-acetic acid production. - Res. Microbiol. 144: 69-75.
Behl R., Tischner R., Raschke K. 1988. Induction of a high capacity nitrate uptake mechanism in barley roots prompted by nitrate uptake through a constitutive low-capacity mechanism. - Planta 176: 235-240.
Ben Zioni A., Vaadia Y., Lips S.H. 1971. Nitrate uptake by roots as regulated by nitrate reduction products of the shoot. - Physiol. Plant. 24: 288-290.
Bertrand H., Nalin R., Bally R., Cleyet-Marel J.-C. 2001. Isolation and identification of the most efficient plant growth-promoting bacteria associated with canola (*Brassica napus*). - Biol. Fertil. Soils 33: 152-156.
Bertrand H., Plassard C., Pinochet X., Touraine B., Normand P., Cleyet-Marel J.-C. 2000. Stimulation of the ionic transport system in *Brassica napus* by a plant growth-promoting rhizobacterium (*Achromobacter* sp.). - Can. J. Microbiol. 46: 229-236.
Bloemberg G.V., Lugtenberg B.J.J. 2001. Molecular basis of plant growth promotion and biocontrol by rhizobia. - Curr. Opin. Plant Biol. 4: 343-350.
Bloom A.J., Sukrapanna S.S., Warner R.L. 1992. Root respiration associated with ammonium and nitrate absorption and assimilation by barley. - Plant Physiol. 99: 1294-1301.
Bottini R., Fulchieri M., Pearce D., Pharis R.P. 1989. Identification of gibberellins A_1, A_3 and iso-A_3 in cultures of *Azospirillum lipoferum*. - Plant Physiol. 90: 45-47.
Bowman D.C., Paul J.L., Davis W.B. 1989. Nitrate and ammonium uptake by nitrogen-deficient perennial ryegrass and Kentucky bluegrass turf. - J. Am. Soc. Hortic. Sci. 114: 421-426.
Breteler H., Arnozis P.A. 1985. Effect of amino compounds on nitrate utilization by roots of dwarf bean. - Phytochemistry 24: 653-658.
Breteler H., Siegerist M. 1984. Effect of ammonium on nitrate utilization by roots of dwarf bean. - Plant Physiol. 75: 1099-1103.
Britto D.T., Kronzucker H.J. 2003. The case for cytosolic heterostasis: a critique of a recently proposed model. - Plant Cell Environ. 26: 83-188.

Burns I. 1991. Short and long-term effects of a change in the spatial distribution of nitrate in the root zone on N uptake, growth and root development of young lettuce plants. - Plant Cell Environ. 14: 21-33.

Cerezo M., Tillard P., Filleur S., Munos S., Daniel-Vedele F., Gojon A. 2001. Major alterations of the regulation of root NO_3^- uptake was associated with the mutation of *Nrt2.1* and *Nrt2.2* genes in *Arabidopsis*. - Plant Physiol. 127: 262-271.

Chaillou S., Lamaze T. 2001. Ammoniacal nutrition of plants. - In: Morot-Gaudry J.-F. (Ed.) Nitrogen assimilation by plants - Physiological, biochemical and molecular aspects. - Enfield (NH, USA), Plymouth (UK), Science Publishers Inc., pp. 53-69.

Clarkson D.T., Scattergood C.B. 1982. Growth and phosphate transport in barley and tomato plants during the development of, and recovery from phosphate stress. - J. Exp. Bot. 33: 865-875.

Clarkson D.T., Smith F.W., Van den Berg P.J. 1983. Regulation of sulphate transport in a tropical legume, *Macroptilium atropurpureum*, cv. Siratro. - J. Exp. Bot. 34: 1463-1483.

Clarkson D.T., Warner A. 1979. Relationships between root temperature and transport of ammonium and nitrate ions by Italian and perennial ryegrass *Lolium multiflorum* and *Lolium perenne*. - Plant Physiol. 64: 557-561.

Clement C., Hopper M.J., Jones L.H.P. 1978. The uptake of nitrate by *Lolium perenne* from flowing nutrient solution. I. Effect of NO_3^- concentration. - J. Exp. Bot. 29: 453-464.

Colmer T.D., Bloom A.J. 1998. A comparison of NH_4^+ and NO_3^- net fluxes along roots of rice and maize. - Plant Cell Environ. 21: 240-246.

Cooper H.D., Clarkson D.T. 1989. Cycling of amino-nitrogen and other nutrients between shoots and roots in cereals - A possible mechanism integrating shoot and root in the regulation of nutrient uptake. - J. Exp. Bot. 40: 753-762.

Coruzzi G.M., Zhou L. 2001. Carbon and nitrogen sensing and signaling in plants: emerging "matrix effects". - Curr. Opin. Plant Biol. 4: 247-253.

Crawford N.M. 1995. Nitrate: nutrient and signal for plant growth. - Plant Cell 7: 859-868.

Crawford N.M., Glass A.D.M. 1998. Molecular and physiological aspects of nitrate uptake in plants. - Trends Plant Sci. 3: 389-395.

Daniel-Vedele F., Filleur S., Caboche M. 1998. Nitrate transport: a key step in nitrate assimilation. - Curr. Opin. Plant Biol. 1: 235-239.

Davies D.D. 1986. The fine control of cytosolic pH. - Physiol. Plant. 67: 702-706.

De Freitas J.R., Germida J.J. 1990. Plant growth-promoting rhizobacteria for winter wheat. - Can. J. Microbiol. 36: 265-272.

De Smet I., Signora L., Beeckman T., Inzé D., Foyer C.H., Zhang H. 2003. An abscisic acid-sensitive checkpoint in lateral root development of *Arabidopsis*. - Plant J. 33: 543-555.

Deane-Drummond C.E., Gates P. 1987. A novel technique for identification of sites of anion transport in intact cells and tissues using a fluorescent probe. - Plant Cell Environ. 10: 221-227.

Delhon P., Gojon A., Tillard P., Passama L. 1995. Diurnal regulation of NO_3^- uptake in soybean plants. I. Changes in NO_3^- influx, efflux, and N utilization in the plant during the day/night cycle. - J. Exp. Bot. 46: 1585-1594.

Delhon P., Gojon A., Tillard P., Passama L. 1996. Diurnal regulation of NO_3^- uptake in soybean plants. IV. Dependence on current photosynthesis and sugar availability to the roots. - J. Exp. Bot. 47: 893-900.

Devienne F., Mary B., Lamaze T. 1994. Nitrate transport in intact wheat roots. I. Estimation of cellular fluxes and NO_3^- distribution using compartmental analysis from data of $^{15}NO_3^-$ efflux. - J. Exp. Bot. 45: 667-676.

Dobbelaere S., Croonenborghs A., Thys A., Ptacek D., Okon Y., Vanderleyden J. 2002. Effect of inoculation with wild type *Azospirillum brasilense* and *A. irakense* strains on development and nitrogen uptake of spring wheat and grain maize. - Biol. Fertil. Soils 36: 284-297.

Dobbelaere S., Croonenborghs A., Thys A., Vande Broek A., Vanderleyden J. 1999. Phytostimulatory effect of *Azospirillum brasilense* wild type and mutant strains altered in IAA production on wheat. - Plant Soil 212: 155-164.

Doddema H., Otten H. 1979. Uptake of nitrate by mutants of *Arabidopsis thaliana*, disturbed in uptake or reduction of nitrate. III. Regulation. - Physiol. Plant. 45: 339-346.

Doddema H., Telkamp G.P. 1979. Uptake of nitrate by mutants of *Arabidopsis thaliana*, disturbed in uptake or reduction of nitrate. II. Kinetics. - Physiol. Plant. 45: 332-338.

Drew M.C. 1975. Comparison of the effects of a localized supply of phosphate, nitrate, ammonium and potassium on the growth of the seminal root system, and the shoot, in barley. - New Phytol. 75: 479-490.

Drew M.C., Saker L.R. 1975. Nutrient supply and the growth of the seminal root system of barley. II. Localized, compensatory increases in lateral root growth and rates of nitrate uptake when nitrate supply is restricted to only part of the root system. - J. Exp. Bot. 26: 79-90.

Drew M.C., Saker L.R. 1984. Uptake and long-distance transport of phosphate, potassium and chloride in relation to internal ion concentrations in barley: evidence for a non-allosteric regulation. - Planta 160: 500-507.

Drew M.C., Saker L.R., Ashley T.W. 1973. Nutrient supply and the growth of the seminal root system in barley. I. The effect of nitrate concentration on the growth of axes and laterals. - J. Exp. Bot. 24: 1189-1202.

Edwards J.H., Barber S.A. 1976. Nitrogen flux into corn roots as influenced by shoot requirement. - Agron. J. 68: 471-473.

Ericsson T. 1995. Growth and shoot: root ratio of seedlings in relation to nutrient availability. - Plant Soil 168-169: 205-214.

Erwee M.G., Goodwin P.B. 1995. Symplast domains in extrastelar tissues of *Egeria densa* Planch. - Planta 163: 9-19.

Filleur S., Daniel-Vedele F. 1999. Expression analysis of a high-affinity nitrate transporter isolated from *Arabidopsis thaliana* by differential display. - Planta 207: 461-469.

Filleur S., Dorbe M.F., Cerezo M., Orsel M., Granier F., Gojon A., Daniel-Vedele F. 2001. An *Arabidopsis* T-DNA mutant affected in *Nrt2* genes is impaired in nitrate uptake. - FEBS Lett. 489: 220-224.

Fitter A. 2002. Characteristics and functions of root systems. In: Waisel Y., Eshel Y., Kafkaki U. (Eds.) Plant roots: the hidden half, 3rd edition. - New York, USA, Marcel Dekker, pp. 15-32.

Forde B.G. 2000. Nitrate transporters in plants: structure, function and regulation. - Biochim. Biophys. Acta 1465: 219-235.

Forde B.G., Clarkson D.T. 1999. Nitrate and ammonium nutrition of plants: Physiological and molecular perspectives. - Adv. Bot. Res. 30: 1-90.

Galvan A., Fernandez E. 2001. Eukaryotic nitrate and nitrite transporters. - Cell. Mol. Life Sci. 58: 225-233.

Gansel X., Muños S., Tillard P., Gojon A. 2001. Differential regulation of the NO_3^- and NH_4^+ transporter genes *AtNrt2.1* and *AtAmt1.1* in *Arabidopsis*: relation with long-distance and local controls by N status of the plant. - Plant J. 26: 143-155.

Gastal F., Saugier B. 1989. Relationships between nitrogen uptake and carbon assimilation in whole plants of tall fescue. - Plant Cell Environ. 12: 407-418.

Gessler A., Schneider S., Weber P., Hanemann U., Rennenberg H. 1998. Soluble N compounds in trees exposed to high loads of N: a comparison between the roots of Norway spruce (*Picea abies*) and beech (*Fagus sylvatica*) trees grown under field conditions. - New Phytol. 138: 385-399.

Gessler A., Schultze M., Schrempp S., Rennenberg H. 1998. Interaction of phloem-translocated amino compounds with nitrate net uptake by the roots of beech (*Fagus sylvatica*) seedlings. - J. Exp. Bot. 49: 1529-1537.

Glass A.D.M., Britto D.T., Kaiser B.N., Kronzucker H.J., Kumar A., Okamoto M., Rawat S.R., Siddiqi M.Y., Silim S.M., Vidmar J.J., Zhuo D. 2001. Nitrogen transport in plants, with an emphasis on the regulation of fluxes to match plant demand. - J. Plant Nutr. Soil Sci. 164: 199-207.

Glass A.D.M., Shaff J.E., Kochian L.V. 1992. Studies of the uptake of nitrate in barley. IV. Electrophysiology. - Plant Physiol. 99: 456-463.

Glick B.R., Penrose D.M., Li J. 1998. A model for the lowering of plant ethylene concentrations by plant growth-promoting bacteria. - J. Theor. Biol. 190: 63-8.

Grouzis J.-P., Pouliquin P., Rigaud J., Grignon C., Gibrat R. 1997. In vitro study of passive nitrate transport by native and reconstituted plasma membrane vesicles from corn root cells. - Biochim. Biophys. Acta 1325: 329-342.

Guo F.Q., Wang R., Chen M., Crawford N.M. 2001. The *Arabidopsis* dual-affinity nitrate transporter gene *AtNRT1.1* (*CHL1*) is activated and functions in nascent organ development during vegetative and reproductive growth. - Plant Cell 13: 1761-1777.

Guy M., Zabala G., Filner P. 1988. The kinetics of chlorate uptake by XD tobacco cells. - Plant Physiol. 86: 817-821.

Hänisch Ten Cate C.H., Breteler H. 1981. Role of sugars in nitrate utilization by roots of dwarf bean. - Plant Physiol. 52: 129-135.

Hatzfeld Y., Saito K. 1999. Identification of two putative nitrate transporters highly homologous to CHL1 from *Arabidopsis thaliana* (Accession Nos. AJ011604 and AJ131464) (PGR 99-018). - Plant Physiol. 119: 805.

Henriksen G.H., Bloom A.J., Spanswick R.M. 1990. Measurement of net fluxes of ammonium and nitrate at the surface of barley roots using ion-selective microelectrodes. - Plant Physiol. 93: 271-280.

Henriksen G.H., Raman D.R., Walker L.P., Spanswick R.M. 1992. Measurement of net fluxes of ammonium and nitrate at the surface of barley roots using ion-selective microelectrodes. II. Patterns of uptake along the root axis and evaluation of the microelectrode flux estimation technique. - Plant Physiol. 99: 734-747.

Hole D.J., Emran A.M., Fares Y., Drew M. 1990. Induction of nitrate transport in maize roots, and kinetics of influx, measured with nitrogen-13. - Plant Physiol. 93: 642-647.

Huang N.C., Chiang C.S., Crawford N.M., Tsay Y.F. 1996. CHL1 encodes a component of the low-affinity nitrate uptake system in Arabidopsis and shows cell type-specific expression in roots. - Plant Cell 8: 2183-91.

Huang N.-C., Liu K.-H., Lo H.-J., Tsay Y.-F. 1999. Cloning and functional characterization of an *Arabidopsis* nitrate transporter gene that encodes a constitutive component of low-affinity uptake. - Plant Cell 11: 1381-1392.

Imsande J., Touraine B. 1994. N demand and the regulation of nitrate uptake. - Plant Physiol. 105: 3-7.

Jackson W.A., Flesher D., Hageman R.H. 1973. Nitrate uptake by dark-grown corn seedlings. Some characteristics of apparent induction. - Plant Physiol. 51: 120-127.

Jeuffroy M.H., Ney B., Ourry A. 2002. Integrated physiological and agronomic modelling of N capture and use within the plant. - J. Exp. Bot. 53: 809-823.

King B.J., Siddiqi M.Y., Glass A.D.M. 1992. Studies of the uptake of nitrate in barley. V. Estimation of root cytoplasmic nitrate concentration using nitrate reductase activity - implications for nitrate influx. - Plant Physiol. 99: 1582-1589.

King B.J., Siddiqi M.Y., Ruth T.J., Warner R.L., Glass A.D.M. 1993. Feedback regulation of nitrate influx in barley roots by nitrate, nitrite, and ammonium. - Plant Physiol. 102: 1279-1286.

Kirkby E.A., Mengel K. 1967. Ionic balance in different tissues of the tomato plant in relation to nitrate, urea or ammonium nutrition. - Plant Physiol. 42: 6-14.

Kloepper J.W., Leong J., Teintze M., Schroth M.N. 1980. Enhanced plant growth by siderophores produced by plant growth- promoting rhizobacteria. - Nature 286: 885-886.

Kosola K.R., Bloom A.J. 1996. Chlorate as a transport analog for nitrate absorption by roots of tomato. - Plant Physiol. 110: 1293-1299.

Kragler F., Lucas W.J., Monzer J. 1998. Plasmodesmata: dynamics, domains and patterning. - Ann. Bot. 81: 1-10.
Krapp A., Fraisier V., Scheible W.-R., Quesada A., Gojon A., Stitt M., Caboche M., Daniel-Vedele F. 1998. Expression studies of *Nrt2:1Np*, a putative high-affinity nitrate transporter: evidence for its role in nitrate uptake. - Plant J. 14: 723-731.
Kronzucker H.J., Glass A.D.M., Siddiqi M.Y. 1995. Nitrate induction in spruce: an approach using compartmental analysis. - Planta 196: 683-690.
Kronzucker H.J., Glass A.D.M., Siddiqi M.Y. 1999. Inhibition of nitrate uptake by ammonium in barley. Analysis of component fluxes. - Plant Physiol. 120: 283-291.
Kronzucker H.J., Siddiqi M.Y., Glass A.D.M. 1995. Kinetics of NO_3^- influx in spruce. - Plant Physiol. 109: 319-326.
Lainé P., Ourry A., Boucaud J. 1995. Shoot control of nitrate uptake rates by roots of *Brassica napus* L.: effects of localized nitrate supply. - Planta 196: 77-83.
Lappartient A.G., Touraine B. 1996. Demand-driven control of root ATP sulfurylase activity and SO_4^{2-} uptake in intact canola. The role of phloem-translocated glutathione. - Plant Physiol. 111: 147-157.
Lappartient A.G., Vidmar J.J., Leustek T., Glass A.D.M., Touraine B. 1999. Inter-organ signaling in plants: regulation of ATP sulfurylase and sulfate transporter genes expression in roots mediated by phloem-translocated compound. - Plant J. 18: 89-95.
Larsson M. 1992. Translocation of nitrogen in osmotically stressed wheat seedlings. - Plant Cell Environ. 15: 447-453.
Lazof D.B., Rufty T.W., Redinbaugh M.G. 1992. Localization of nitrate absorption and translocation within morphological regions of the corn root. - Plant Physiol. 100: 1251-1258.
Lee R.B. 1993. Control of net uptake of nutrients by regulation of influx in barley plants recovering from nutrient deficiency. - Ann. Bot. 72: 223-230.
Lee R.B., Clarkson D.T. 1986. Nitrogen-13 studies of nitrate fluxes in barley roots. I. Compartmental analysis from measurements of ^{13}N efflux. - J. Exp. Bot. 37: 1753-1767.
Lee R.B., Drew M.C. 1986. Nitrogen-13 studies of nitrate fluxes in barley roots. II. Effect of plant N-status on the kinetic parameters of nitrate influx. - J. Exp. Bot. 185: 1768-1779.
Lee R.B., Rudge K. 1986. Effects of nitrogen deficiency on the absorptions of nitrate and ammonium by barley plants. - Ann. Bot. 57: 471-486.
Lejay L., Tillard P., Lepetit M., Olive F.D., Filleur S., Daniel-Vedele F., Gojon A. 1999. Molecular and functional regulation of two NO_3^- uptake systems by N- and C-status of *Arabidopsis* plants. - Plant J. 18: 509-519.
Liu C., Muchhal U.S., Uthappa M., Kononowicz A.K., Raghothama K.G. 1998. Tomato phosphate transporter genes are differentially regulated in plant tissues by phosphorus. - Plant Physiol. 116: 91-99.
Liu K.-H., Huang C.-Y., Tsay Y.-F. 1999. CHL1 is a dual-affinity nitrate transporter of *Arabisopsis* involved in multiple phases of nitrate uptake. - Plant Cell 11: 865-874.
Malamy J.E., Benfey P.N. 1997. Down and out in Arabidopsis: the formation of lateral roots. - Trends Plant Sci 2: 390-396.
Mantelin S., Touraine B. 2004. Plant growth-promoting bacteria and nitrate availability: Impacts on root development and nitrate uptake. - J. Exp. Bot. 55: 27-34.
Marschner H., Röhmeld V. 1983. *In vivo* measurement of root-induced pH changes at the soil-root interface: effect of plant species and nitrogen source. - Z. Pflanzenphysiol. 111: 241-251.
McClure P.R., Kochian L.V., Spanswick R.M., Shaff J.E. 1990. Evidence for cotransport of nitrate and protons in maize roots. 1. Effects of nitrate on the membrane potential. - Plant Physiol. 93: 281-289.
Morgan M.A., Volk R.J., Jackson W.A. 1985. p-Fluorophenylalanine-induced restriction of ion uptake and assimilation by maize roots. - Plant Physiol. 77: 718-722.
Muller B., Tillard P., Touraine B. 1995. Nitrate fluxes in soybean seedling roots and their response to amino acids: an approach using ^{15}N. - Plant Cell Environ. 18: 1267-1279.

Muller B., Touraine B. 1992. Inhibition of NO_3^- uptake by various phloem-translocated amino acids in soybean seedlings. - J. Exp. Bot. 43: 617-623.
Muller B., Touraine B., Rennenberg H. 1996. Interaction between atmospheric and pedospheric nitrogen nutrition in spruce (*Picea abies* L. Karst) seedlings. - Plant Cell Environ. 19: 345-355.
Nazoa P., Vidmar J.J., Tranbarger T.J., Mouline K., Damiani I., Tillard P., Zhuo D., Glass A.D.M., Touraine B. 2003. Regulation of the nitrate transporter gene *AtNRT2.1* in *Arabidopsis thaliana*: responses to nitrate, amino acids and developmental stage. - Plant Mol. Biol. 52: 689-703.
Neyra C.A., Hageman R.H. 1976. Relationships between carbon dioxide, malate and nitrate accumulation and reduction in corn (*Zea mays* L.) seedlings. - Plant Physiol. 58: 726-730.
Okamoto M., Vidmar J.J., Glass A.D.M. 2003. Regulation of *NRT1* and *NRT2* gene families of *Arabidopsis thaliana*: responses to nitrate provision. - Plant Cell Physiol. 44: 304-317.
Okon Y., Kapulnik Y. 1986. Development and function of *Azospirillum*-inoculated roots. - Plant Soil 90: 3-16.
Okon Y., Vanderleyden J. 1997. Root-associated *Azospirillum* species can stimulate plants. - ASM News 63: 366-370.
Oostindier-Braaksma F., Feenstra F. 1973. Isolation and characterisation of chlorate-resistant mutants of *Arabidopsis thaliana*. - Mutation Research 19: 175-185.
Orsel M., Krapp A., Daniel-Vedele F. 2002. Analysis of the *NRT2* nitrate transporter family in *Arabidopsis*. Structure and gene expression. - Plant Physiol. 129: 886-896.
Pace G.M., McClure P.R. 1986. Comparison of nitrate uptake kinetic parameters across maize inbred lines. - J. Plant Nutr. 9: 1095-1111.
Pacovsky R.S. 1990. Development and growth effects in the *Sorghum-Azospirillum* association. - J. Appl. Bacteriol. 68: 555-563.
Persello-Cartieaux F., David P., Sarrobert C., Thibaud M.C., Achouak W., Robaglia C., Nussaume L. 2001. Utilization of mutants to analyze the interaction between *Arabidopsis thaliana* and its naturally root-associated *Pseudomonas*. - Planta 212: 190-8.
Persello-Cartieaux F., Nussaume L., Robaglia C. 2003. Tales from the underground: molecular plant-rhizobia interactions. - Plant Cell Environ. 26: 189-199.
Pouliquin P., Boyer J.-C., Grouzis J.-P., Gibrat R. 2000. Passive nitrate transport by root plasma membrane vesicles exhibits an acidic optimal pH like the H^+-ATPase. - Plant Physiol. 122: 265-273.
Pouliquin P., Grouzis J.-P., Gibrat R. 1999. Electrophysiological study with oxonol VI of passive NO_3^- transport by isolated plant root plasma membrane. - Biophys. J. 76: 360-373.
Presland M.R., MacNaughton G.S. 1984. Whole plant studies using radioactive 13-nitrogen. II. A compartmental model for the uptake and transport of nitrate ions by *Zea mays*. - J. Exp. Bot. 35: 1277-1288.
Rao K.P., Rains D.W. 1976. Nitrate absorption by barley. 1. Kinetics and energetics. - Plant Physiol. 57: 55-58.
Raven J.A. 1977. H^+ and Ca^{2+} in phloem and symplast: relation of relative immobility of the ions to the cytoplasmic nature of the transport paths. - New Phytol. 79: 465-480.
Raven J.A., Smith F.A. 1976. Nitrogen assimilation and transport in vascular land plants in relation to intracellular pH regulation. - New Phytol. 76: 415-431.
Robin P., Conejero G., Passama L., Salsac L. 1983. Evaluation de la fraction métabolisable du nitrate par la mesure *in situ* de sa réduction. - Physiol. Vég. 21: 115-122.
Robinson D. 1994. The responses of plants to non-uniform supplies of nutrients. - New Phytol. 127: 635-674.
Ruiz-Cristin J., Briskin D.P. 1991. Characterization of a H^+/NO_3^- symport associated with plasma membrane vesicles of maize roots using $^{36}ClO_3^-$ as a radiotracer analog. - Arch. Biochem. Biophys. 285: 74-82.
Salsac L., Chaillou S., Morot-Gaudry J.-F., Lesaint C., Jolivet E. 1987. Nitrate and ammonium nutrition in plants. - Plant Physiol. Biochem. 25: 805-812.

Saudibet M.I., Fatta N., Barneix A.J. 2002. The effect of inoculation with *Azospirillum brasilense* on growth and nitrogen utilization by wheat plants. - Plant Soil 245: 215-222.

Scheible W.R., Gonzalez Fontes A., Lauerer M., Mueller Roeber B., Caboche M., Stitt M. 1997a. Nitrate acts as a signal to induce organic acid metabolism and repress starch metabolism in tobacco. - Plant Cell 9: 783-798.

Scheible W.R., Lauerer M., Schulze E.D., Caboche M., Stitt M. 1997b. Accumulation of nitrate in the shoot acts as a signal to regulate shoot-root allocation in tobacco. - Plant J. 11: 671-691.

Scheurwater I., Clarkson D.T., Purves J.V., Van Rijt G., Saker L.R., Welschen R., Lambers H. 1999. Relatively large nitrate efflux can account for the high respiratory costs for nitrate transport in slow-growing grass species. - Plant Soil 215: 123-134.

Scheurwater I., Cornelissen C., Dictus F., Welschen R., Lambers H. 1998. Why do fast- and slow-growing grass species differ so little in their rate of root respiration, considering the large differences in rate of growth and ion uptake? - Plant Cell Environ. 21: 995-1005.

Scholten H.J., Feenstra W.J. 1986. Uptake of chlorate and other ions in seedlings of the nitrate uptake mutant B1 of *Arabidopsis*. - Physiol. Plant. 66: 265-269.

Shah S., Li J., Moffatt B.A., Glick B.R. 1998. Isolation and characterization of ACC deaminase genes from two different plant growth-promoting rhizobacteria. - Can. J. Microbiol. 44: 833-843.

Siddiqi M.Y., Glass A.D.M. 2002. An evaluation of the evidence for, and implications of, cytoplasmic nitrate homeostasis. - Plant Cell Environ. 25: 1211-1217.

Siddiqi M.Y., Glass A.D.M., Ruth T.J. 1991. Studies of the uptake of nitrate in barley. III. Compartmentation of NO_3^-. - J. Exp. Bot. 42: 1455-1463.

Siddiqi M.Y., Glass A.D.M., Ruth T.J., Fernando M. 1989. Studies of the regulation of nitrate influx by barley seedlings using $^{13}NO_3^-$. - Plant Physiol. 90: 806-813.

Siddiqi M.Y., Glass A.D.M., Ruth T.J., Rufty T. 1990. Studies of the uptake of nitrate in barley. I. Kinetics of $^{13}NO_3^-$ influx. - Plant Physiol. 93: 1426-1432.

Steudle E., Peterson C.A. 1998. How does water get through roots? J. Exp. Bot. 49: 775-788.

Signora L., De Smet I., Foyer C.H., Zhang H. 2001. ABA plays a central role in mediating the regulatory effects of nitrate on root branching in *Arabidopsis*. - Plant J. 28: 655-662.

Teyker R.H., Jackson W.A., Volk R.J., Moll R. 1988. Exogenous $^{15}NO_3^-$ influx and endogenous $^{14}NO_3^-$ efflux by two maize (*Zea mays* L.) inbreds during nitrogen deprivation. - Plant Physiol. 86: 778-781.

Thibaud J.-B., Grignon C. 1981. Mechanism of nitrate uptake in corn roots. - Plant Sci. Lett. 22: 279-289.

Tien T.M., Gaskins M.H., Hubbell D.H. 1979. Plant growth substances produced by *Azospirillum brasilense* and their effect on the growth of pearl millet (*Pennisetum americanum*). - Appl. Environ. Microbiol. 37: 1016-1024.

Tillard P., Passama L., Gojon A. 1998. Are phloem amino acids involved in the shoot to root control of NO_3^- uptake in *Ricinus communis* plants? - J. Exp. Bot. 49: 1371-1379.

Tompkins G.A., Jackson W.A., Volk R.J. 1978. Accelerated nitrate uptake in wheat seedlings. Effects of ammonium and nitrite pretreatments and of 6-methylpurine and puromycin. - Physiol. Plant. 43: 166-171.

Touraine B., Daniel-Vedele F., Forde B.G. 2001. Nitrate uptake and its regulation. - In: Lea P.J., Morot-Gaudry J.-F. (Eds.) Plant nitrogen. - Berlin-Heidelberg, INRA-Editions and Springer-Verlag, pp. 1-36.

Touraine B., Glass A.D.M. 1997. NO_3^- and ClO_3^- fluxes in the *chl1-5* mutant of *Arabidopsis thaliana*. Does the *CHL1-5* gene encode a low-affinity NO_3^- transporter? - Plant Physiol. 114: 137-144.

Touraine B., Grignon C. 1982. Potassium effect on nitrate secretion into the xylem of corn roots. - Physiol. Vég. 20: 23-31.

Touraine B., Grignon N., Grignon C. 1988. Charge balance in NO_3^--fed soybean. Estimation of K^+ and carboxylate recirculation. - Plant Physiol. 88: 605-612.

Touraine B., Muller B., Grignon C. 1992. Effect of phloem-translocated malate on NO_3^- uptake by roots of intact soybean plants. - Plant Physiol. 93: 1118-1123.
Tranbarger T.J., Al-Ghazi Y., Muller B., Teyssendier de la Serve B., Doumas P., Touraine B. 2003a. A macro-array-based screening approach to identify transcriptional factors involved in the nitrogen-related root plasticity response of *Arabidopsis thaliana*. - Agronomie 23: 519-528.
Tranbarger T.J., Al-Ghazi Y., Muller B., Teyssendier de la Serve B., Doumas P., Touraine B. 2003b. Transcription factor genes with expression correlated to nitrate-related root plasticity of *Arabidopsis thaliana*. - Plant Cell Environ. 26: 459-469.
Tsay Y.F., Schroeder J.I., Feldmann K.A., Crawford N.M. 1993. The herbicide sensitivity gene *CHL1* of *Arabidopsis* encodes a nitrate-inducible nitrate transporter. - Cell 72: 705-13.
Ullrich W.R., Novacky A. 1981. Nitrate-dependent membrane potential changes and their induction in *Lemna gibba*. - Plant Sci. Lett. 22: 211-217.
Van der Leij M., Smith S.J., Miller A.J. 1998. Remobilization of vacuolar stored nitrate in barley root cells. - Planta 205: 64-72.
Vidmar J.J., Zhuo D., Siddiqi M.Y., Glass A.D.M. 2000a. Isolation and characterization of *HvNRT2.3* and *HvNRT2.4*, cDNAs encoding high-affinity nitrate transporters from roots of barley. - Plant Physiol. 122: 783-792.
Vidmar J.J., Zhuo D., Siddiqi M.Y., Schoerring J.K., Touraine B., Glass A.D.M. 2000b. Regulation of high-affinity nitrate transporter genes and high-affinity nitrate influx by nitrogen pools in roots of barley. - Pant Physiol. 123: 307-318.
Wang R.C., Liu D., Crawford N.M. 1998. The *Arabidopsis* CHL1 protein plays a major role in high-affinity nitrate uptake. - Proc. Natl. Acad. Sci. USA 95: 15134-15139.
Whipps J.W. 2001. Microbial interactions and biocontrol in the rhizosphere. - J. Exp. Bot. 52: 487-511.
Williams L.E., Miller A.J. 2001. Transporters responsible for the uptake and partitioning of nitrogenous solutes. - Annu. Rev. Plant Physiol. Plant Mol. Biol. 52: 659-688.
Zhang H., Forde B.G. 1998. An Arabidopsis MADS box gene that controls nutrient-induced changes in root architecture. - Science 279: 407-9.
Zhang H., Jennings A., Barlow P.W., Forde B.G. 1999. Dual pathways for regulation of root branching by nitrate. - Proc. Natl. Acad. Sci. USA 96: 6529-6534.
Zhen R.G., Kyoro H.W., Leigh R.A., Tomos A.D., Miller A.J. 1991. Compartmental nitrate concentrations in barley root cells measured with nitrate-selective microelectrodes and by single-cell sap sampling. - Planta 185: 356-361.
Zhuo D., Okamoto M., Vidmar J.J., Glass A.D.M. 1999. Regulation of a putative high-affinity nitrate transporter (*Nrt2;1At*) in roots of *Arabidopsis thaliana*. - Plant J. 17: 563-558.

Chapter 2

METABOLIC REGULATION OF AMMONIUM UPTAKE AND ASSIMILATION

Tomoyuki Yamaya and Ann Oaks

INTRODUCTION

Plants take up inorganic nitrogen as NO_3^- or NH_4^+ ions, or as N_2 fixed by symbiotic bacteria. The subjects of nitrate uptake as well as symbiotic N_2 fixation will be discussed in chapter 1 and 4, respectively. N-containing molecules as amino acids, proteins, chlorophyll and nucleic acids, are all products of NH_4^+ assimilation. Aspects of the synthesis and regulation of amino acids and amides (Lam *et al.* 1996), the interaction of sucrose, amides and NO_3^- with the assimilation of NH_4^+ (Sivasankar and Oaks 1996) and the role of NO_3^- as a signal regulating metabolism and growth (Crawford and Glass 1998, Stitt *et al.* 2002) have already been discussed in detail, but will be considered in the light of current developments in our understanding of NH_4^+ assimilation. This chapter will focus on NH_4^+ uptake, its assimilation into glutamine and glutamate and its metabolic regulation. Recent progress in molecular biology, gene manipulation, and post-genomic research on NH_4^+ metabolism will be described.

SOURCE OF AMMONIUM IONS

Ammonium ions are the major form of inorganic N in certain soils, such as flooded paddy rice fields and mature conifer forest soils (Kronzucker *et al.* 1997). NH_3 is a weak base that protonates rapidly to form NH_4^+ with a dissociation constant of $10^{-9.25}$ (Kleiner 1981). According to the equation proposed by Freney *et al.* (1985), 99.4% of NH_3 is in the protonated form in water at pH 7.0, and its proportion increases with decreasing pH. NH_4^+, therefore, is the major species in paddy fields and acid soils. In oxidative upland soils, on the other hand, NH_4^+ is quickly converted to NO_2^- and then to NO_3^- by nitrifying bacteria so that NO_3^- is the major form of N for most upland crops.

S. Amâncio and I. Stulen (eds.),
Nitrogen Acquisition and Assimilation in Higher Plants, 35-63.

In the plant, NH_4^+ is liberated in a number of metabolic processes, as the reduction of NO_3^-, symbiotic N_2 fixation, the conversion of glycine to serine in photorespiratory N metabolism, the deamination of phenylalanine catalysed by phenylalanine ammonia lyase in the phenylpropanoid pathway or the biosynthesis and catabolism of amino acids (Joy 1988, Sechley et al. 1992, Lea and Ireland 1999). The rate of NH_4^+ production through photorespiratory N metabolism in C_3 leaves is calculated to be 80 µmol h^{-1} g fresh weight^{-1}, which is 10 times that of normal NO_3^- reduction (Sechley et al. 1992, Lea and Ireland 1999). Recent measurements showed that NH_4^+ concentrations in leaf tissue water of oilseed rape and tomato plants ranged from 0.6-1.2 mM (Schjoerring et al. 2002). Therefore, NH_4^+ liberated in photorespiration and other metabolic processes is apparently efficiently reassimilated. The reassimilation of NH_4^+ is important for optimal functioning of photosynthesis in leaves (Sechley et al. 1992, Good et al. 1966). Re-assimilation efficiency appears to be a significant factor in the enhanced nitrogen use efficiency (NUE) in C_4 relative to C_3 plants (Oaks 1994).

AMMONIUM UPTAKE

When NH_4^+ is taken up by plant roots, a system for its transport across the cell membrane is required. When NH_4^+ is generated by any of the metabolic processes in cells, mechanisms are required for movement of ions between organelles, viz. between cytosol and vacuole for temporary storage, between mitochondria and chloroplasts during photorespiration, and between apoplast and phloem companion cells or parenchyma cells for long distance transport. Physiological studies on NH_4^+ uptake in plant roots provided evidence for the existence of two transport systems for NH_4^+, a high-affinity transport system (HATS) and a low-affinity transport system (LATS) (Wang et al. 1993, Karasawa et al. 1994, Mäck and Tischner 1994, Kronzucker et al. 1996).

Transporters

In *Saccharomyces cerevisiae*, ammonium transporters (MEP1, MEP2 and MEP3) have been cloned and characterised (Marini et al. 1997). MEP1 and MEP2 encode HATS with Km values of 5-10 µM and 1-2 µM, respectively, while MEP3 encodes LATS with Km values of 1.4-2.1 mM. Ninnemann et al. (1994) first identified the gene encoding a high-affinity NH_4^+ transporter, AtAMT1;1, from *Arabidopsis thaliana* using functional complementation of a yeast mutant defective in NH_4^+ uptake. Since then, the isolation of further AMT1 homologues from *A. thaliana* (*AtAMT1;1*, *AtAMT1;2*, *AtAMT1;3*) (Gazzarrini et al. 1999, Rawat et al. 1999), tomato (Lauter et al. 1996, von Wirén et al. 2000) and rice (von Wirén et al. 1997, Sonoda et al. 2003) have shown that the AMT1 gene family in plants consists of at least 3-5 members. Most AMT1 proteins act as functional NH_4^+ transporters in yeast mutants, having 10-11 membrane spanning domains estimated from their sequences. The Km value for AtAMT1;1 was estimated to be < 0.5 μM and its gene was expressed in both roots and shoots of *A. thaliana*. Other HATSs, *AtAMT1;2* and *AtAMT1;3*, were preferentially expressed in root tissues and Km values for NH_4^+ were approximately 40 μM. In tomato, *LeAMT;1* and *LeAMT;2* were expressed in roots, while *LeAMT;3* is preferentially expressed in shoots (von Wirén *et al.* 2000). The AMT1 family in rice has been characterised (*OsAMT1;1*, *OsAMT1;2* and *OsAMT1;3*) where *OsAMT1;2* is specifically expressed in roots (Sonoda *et al.* 2003). Another type of *AMT*, which is more similar to yeast *MEP* than plant *AMT1*, has been cloned from *A. thaliana* and designated as *AtAMT2* (Sohlenkamp *et al.* 2000). Two genes/cDNA for the AMT2 family were also cloned from rice (*OsAMT2;1* and *OsAMT3;1*) (Suenaga *et al.* 2003). Transcripts for *AMT2* were more abundant in shoots than in roots. When a database search was carried out against the amino-acid sequence of OsAMT2;1 in rice, 10 *OsAMT* homologues were found. Kinetic properties of the AMT2 family and other AMT1 species should be examined in further detail. At present, no gene encoding LATS has been detected.

Regulation

Under N-deprivation, transcripts of *AtAMT1;1, AtAMT1;3, and LeAMT1;1* increased within a few days, while abundance of transcripts for *LeAMT1;2* and *OsAMT1;2* increased after N was supplied. To understand the physiological function and regulation mechanisms of individual AMT families, a more detailed examination of the cellular/subcellular localisation and gene manipulation/knockout mutants is required. Recent studies with *in situ* mRNA detection and functional promoter analysis in rice revealed that within 30 min of supplying NH_4^+, the *OsAMT1;2* is expressed in the surface cells, as well as in the central cylinder of the root (Sonoda *et al.* 2003). Cytosolic glutamine synthetase was detected in all cell types of rice roots including the surface cells (Ishiyama *et al.* 1998). Because NADH-glutamate synthase protein, a key enzyme for NH_4^+ assimilation in rice (see below) accumulated in epidermis and exodermis of rice root tips within 3 hr (Ishiyama *et al.* 1998), these compartments could be responsible for the assimilation of ions taken up by roots. The *OsAMT1;2* species is probably important in NH_4^+ uptake in rice. A T-DNA inserted knockout mutant has recently been isolated from *A. thaliana* that fails to express *AtAMT1;1* (Glass *et al.* 2001). Disruption of this gene caused only 20-30% reduction in NH_4^+ influx, although *AtAMT1;1* showed the strongest response to N-deprivation and had the highest affinity for NH_4^+. This suggests that there is a compensation for the disruption of *AtAMT1;1*.

AN OVERVIEW OF AMMONIUM ASSIMILATION

Glutamine was identified as an early product of NH_4^+ assimilation by Yoneyama and Kumazawa (1974) and as a major metabolite in supplying reduced N to other reactions. Glutamine synthetase (GS, EC 6.3.1.2) was also identified as the dominant catalyst mediating this reaction (O' Neal and Joy 1973). Nevertheless, the identification of a reaction for the synthesis of glutamate, a substrate in the GS reaction, was still missing. When Sims and Folkes (1964) examined the assimilation of NH_4^+ in the fungus *Candida utilis*, they identified glutamate dehydrogenase (GDH, EC 1.4.1.2) as being involved in the incorporation of NH_4^+ into glutamate by amination of α-ketoglutarate, and they postulated a similar reaction in higher plants. The NADH-dependent GDH catalyses the following reversible reaction:

$$\alpha\text{-ketoglutarate} + NH_4^+ + NADH + H^+ \leftrightarrow \text{glutamate} + H_2O + NAD^+$$

If GDH were to supply the glutamate required in the synthesis of glutamine a logistical problem had to be resolved: the level of GDH is higher in plant roots than in leaves, yet in most plants rates of glutamine synthesis are usually highest in the leaves. The identification of an NAD(P)H dependent glutamate synthase in bacteria presented an alternative route for the synthesis of glutamate (Tempest *et al.* 1970), the glutamate synthase reaction (GOGAT). Subsequently two seminal reports from Miflin's group (Lea and Miflin, 1974, Keys *et al.* 1978) established that this alternative reaction for the synthesis of glutamate was present in plants, that in green leaves ferredoxin (Fd) reduced by the transfer of electrons from the non-cyclic photosynthetic electron transfer chain replaced NAD(P)H as reductant, and that via a series of reactions NH_4^+ released in photorespiration could be re-assimilated via reactions involving GS and Fd-GOGAT (EC 1.4.7.1), the GS/GOGAT cycle (Sechley *et al.* 1992). GS catalyses an ATP-dependent conversion of glutamate to glutamine with NH_4^+ as substrate:

$$\text{glutamate} + NH_4^+ + ATP \xrightarrow{Mg^{2+}} \text{glutamine} + H_2O + ADP + Pi$$

GOGAT catalyses the reductant-dependent conversion of glutamine and α-ketoglutarate to 2 molecules of glutamate:

$$\text{glutamine} + \alpha\text{-ketoglutarate} + 2\ Fd_{red}\ (NADH + H^+) \rightarrow 2\ \text{glutamate} + 2\ Fd_{ox}\ (NAD^+)$$

According to this cycle one of the two resulting glutamates is shunted back to the chloroplast to serve as substrate for GS and the other glutamate serves as N donor for serine production in the peroxisome.

Experiments with metabolic inhibitors (Keys *et al.* 1978) and mutants lacking GOGAT (Somerville and Ogren, 1980) and GS (Blackwell *et al.* 1988, Wallsgrove *et al.* 1987) established unequivocally that GDH could not replace GS in the assimilation of NH_4^+ produced in photorespiration. Further research established that other amino acids, most notably alanine, could supply the N required for serine biosynthesis in the peroxisome (Joy 1988, Ireland and Lea 1999). These results explained the drainage of C required to support enhanced protein synthesis in leaves exposed to light or for export of amino acids to other parts of the plant. Our current understanding of the GS/GOGAT cycle has been discussed by Somerville and Ogren (1980), Miflin and Lea (1977), and by Ireland and Lea (1999). Bowsher *et al.* (1992) also identified a Fd-GOGAT in plastids in pea roots. Their experiments showed that GS/GOGAT functions in root plastids and that the reductant for Fd is supplied by the pentose phosphate shunt. In maize roots most of the GS is in the cytosol (Sechley *et al.* 1992). There is a Fd-GOGAT in maize roots (Sakakibara *et al.* 1992b) and probably also an active GS/GOGAT cycle. Unlike leaves NH_4^+ levels can accumulate in roots to fairly high levels without adversely affecting viability of the plants.

Strategies for root metabolism differ from leaves in significant ways and this has not been considered in detail in research papers or reviews (Oaks and Hirel 1985). In legumes for instance, asparagine is accumulated and is the dominant form of reduced N in the xylem sap (Pate, 1980). In cereals, on the other hand, nitrate or glutamine is prominent in the xylem sap (Oaks 1992) and in paddy rice where NH_4^- is the dominant source of N, only glutamine is found in the xylem sap (Fukumorita and Chino 1982).

GLUTAMINE SYNTHETASE

Characteristics and localisation

Biochemical studies clearly show that GS in higher plants is an octameric protein with a native molecular mass of 350-400 kD (Ireland and Lea 1999). Leaves of most plants contain two isoenzymes of GS, which are readily separated by ion-exchange chromatography (McNally *et al.* 1983). The two forms were termed GS1 and GS2 from their order of elution. Tissue fractionation studies have shown that GS1 is localised in the cytosol while GS2 is in the chloroplasts/plastids (Mann *et al.* 1979, Hirel and Gadal 1981). Direct evidence to support the sub-cellular localisation of GS1 and GS2 was obtained by immunogold labelling techniques in several species (Botella *et al.* 1988, Branjeon *et al.* 1989, Carvalho *et al.* 1992, Pereira *et al.* 1992, Peat and Tobin 1996). There is only one gene encoding GS2. The GS2 protein is first translated into the precursor protein that possesses chloroplast/plastid-targeting sequences at the N-terminus region. On the other hand, GS1 is encoded by a small multigene family, which varies from 2-6 genes, in monocotyledons and dicotyledons. Each gene synthesises a different subunit, which may be assembled into either a homooctametric and heterooctameric form (Walker and Coruzzi 1989, Sechley *et al*. 1992, Ireland and Lea 1999). The fact that regulation of GS gene families differs with cell type and environmental conditions indicates that each GS could have a non-overlapping function (Tingey *et al*. 1987, Edwards *et al*. 1990, Kamachi *et al*. 1992b, Lam *et al*. 1996).

GS2, the major isoenzyme in green leaves, is located in chloroplasts of mesophyll cells. Because mutants lacking GS2 were able to grow normally under non-photorespiratory conditions (Blackwell *et al*. 1988, Wallsgrove *et al*. 1987), GS1 in leaves could be important in the synthesis of glutamine for normal growth and development. Promoter-deletion analyses revealed that the 323-bp region in pea *GS2* was able to express a reporter gene as a mesophyll cell-specific and light-responsive protein in transgenic tobacco and *A. thaliana* (Tjaden *et al*. 1995). Cellular localisation studies showed that GS1 is abundant in vascular tissues. For example, immunolocalisation studies indicate that GS1 protein is localised in phloem companion cells and related vascular cells of senescing rice leaf blades (Kamachi *et al*. 1992b, Sakurai *et al*. 1996) and sclerenchyma, xylem parenchyma, and guard cells of developing rice leaves (Sakurai *et al*. 2001). In tobacco, one of two *GS1* genes is expressed in the vascular tissues of the stem and leaf midrib (Dubois *et al*. 1996). Promoter analyses also showed that one of three *GS1* genes in pea (Edward *et al*. 1990) and kidney bean (Watson and Cullimore 1996), respectively, directed vascular-tissue specific expression in the heterologous transformation system. The localisation studies suggest that GS1 is important in the synthesis of glutamine, which is the major form of N in phloem sap in rice (Hayashi and Chino 1990), and for export from mature and senescing leaves (Tobin and Yamaya 2001). In developing leaves, GS1 is probably involved in assimilating NH_4^+ released from the reaction of phenylalanine ammonia lyase during lignin synthesis (Sakurai *et al*. 2001). Transgenic tobacco expressing antisense RNA for a GS1 (*Gln1-5*) showed a pronounced stress phenotype associated with a decrease in proline content, suggesting that GS1 in phloem plays a major role in regulating proline production (Brugière *et al*. 1999). On the other hand, GS1 protein was detected both in vascular tissues and mesophyll cells of barley primary leaves (Tobin and Yamaya 2001) and in mesophyll cells in tobacco leaves during senescence (Brugière *et al*. 2000). There are two *GS1* genes in rice, *i.e. GS1* preferentially expressed in leaves and *GSr*, which is mainly expressed in roots (Sakamoto *et al*. 1989). Promoter analysis of the *GS1* gene of rice showed a preferential expression in vascular tissues of leaf blade, leaf sheath, roots and ear, but also expression in the wall-region of anther loculus and embryo in transgenic rice (Hanzawa *et al*. 2002). One of the five *GS1* maize genes is preferentially expressed in the pedicels of developing kernels (Rastogi *et al*. 1998).

Non-legumes using NH_4^+ as the major N source, would be expected to have an efficient NH_4^+ assimilation system. Conversely there is an additional Casparian strip between exodermis and sclerenchyma in rice roots and therefore apoplastic transport of NH_4^+ is unlikely (Ishiyama et al. 1998). NH_4^+ taken up by rice roots is transported through the xylem to the shoots, mainly as glutamine (Fukumorita and Chino 1982). Immunolocalisation using GS1-specific IgG, which cross-reacts with both GS1 and GSr, showed that these cytosolic GS forms are distributed throughout the rice roots with apparent homogeneity within the epidermis, exodermis, cortex, and central cylinder (Ishiyama et al. 1998). To distinguish between GS1 and GSr in rice roots, GS1-promoter was fused with a GUS reporter gene and the chimeric gene was introduced into rice. Preliminary data showed that the GS1 promoter is active in the central cylinder and cortex of the main roots as well as in the central cylinder of lateral roots (Hanzawa et al. 2002). Thus, GSr in epidermis and exodermis is probably responsible for the assimilation of NH_4^+ taken up by rice roots.

In contrast to rice, most plants growing on well-aerated soils use NO_3^- as their main N source. In the root NO_3^- is partly reduced to NH_4^+ before being assimilated via GS/GOGAT, but the proportion varies with plant species, plant age, and N availability (Tobin and Yamaya 2001). In roots of hydroponically grown barley seedlings, a GS1 (42 kD in subunit molecular mass) was constitutively detected in all root sections. Two additional polypeptides were detected in mature roots of NH_4^+-grown seedlings and one additional GS1 in those grown on NO_3^- (Peat and Tobin 1996). Changes in GS1 polypeptides in response to NH_4^+ were also observed in barley roots (Mäck 1995), although these differ slightly in molecular mass from the observation of Peat and Tobin (1996). Immunogold localisation studies indicated a higher concentration of GS1 in cortical parenchyma than in vascular stele cells (Peat and Tobin 1996). In these cell types the apparent concentration of GS1 was highest in N-deficient plants. In addition, significant labelling of GS protein was detected in plastids of cortical and vascular parenchyma cells of barley root apical cells (Peat and Tobin 1996). When barley plants were grown on 15 mM NO_3^-, there was an increase in immunogold labelling in the "tubular" and "flat" plastids in the roots (Tobin and Yamaya 2001).

In maize roots, a constitutive GS1 polypeptide (40 kD) and an N inducible form (GSr, 39 kD) were detected (Sakakibara et al. 1992a). Two genes, *GS1a* and *GS1b*, encoded the GS1, while GSr was the product from *GS1c* and *GS1d*. Rapid accumulation of mRNA for *GS1c* and *GS1d* was observed when NH_4^+ was supplied, whereas the mRNA for *GS1a* and *GS1b* decreased under these conditions (Sakakibara et al. 1996). The GSr enzyme had a higher ratio of synthetase/transferase activities than the GS1 enzyme. Sakakibara et al. (1996) proposed that the GSr isoenzyme was more important in the assimilation of external NH_4^+. GS2 mRNA was also detected in maize roots exposed to low NO_3^- concentrations (Redinbaugh and Campbell 1993). Three putative genes for

GS1 and one for GS2 were expressed in roots of *A. thaliana* (Peterman and Goodman 1991). *Gln1-5* from tobacco was preferentially expressed in the root vascular system, whereas *Gln1-3* was detected in all root tissues (Dubois *et al.* 1996). More conclusive evidence, such as the use of knockout mutants, is required to establish the function of individual GS1s and GS2 isoenzymes in roots.

Regulation

Regulation of GS activity within a cell is determined by a number of events, such as sensing of signals, signal transduction, transcription factors and *cis*-elements on the promoter, rate and abundance of transcripts, translation, and post-translational modification of GS proteins. Genes for GS1 and GS2 and corresponding cDNAs have been isolated from a number of plant species (Ireland and Lea 1999). As mentioned above, GS2 is encoded by only one gene and is expressed in the chloroplast or root plastid. GS1, on the other hand, is encoded by a small subfamily of genes that varies in number from 2 in rice (Sakamoto *et al.* 1989) to 4 (Sakakibara *et al.* 1992b) or 5 (Li *et al.* 1993) in maize.

Exogenous N often results in changes in steady-state abundance of mRNA and protein for some isoenzymes, but underlying molecular mechanisms are not clearly understood. Glutamine (Glass *et al.* 2002) or other amino acids (Sivasankar and Oaks 1996, Crawford and Glass 1998, Stitt *et al.* 2002) may act as signals of plant-N status, but neither sensor protein nor any member related to signal transduction has been identified. Homologues of PII protein, which involve a signal transduction system after the sensing of the glutamine/α-ketoglutarate status in *E. coli*, were identified in *A. thaliana* and *Ricinus communis* (Hsieh *et al.* 1998). A similar system may be involved in the signal transduction system of GS gene expression.

Steady-state abundance of GS transcripts may not be determining actual protein contents and activity of GS. For example, the protein content of the GS-δ isoenzyme in kidney bean did not correlate with the abundance of transcripts from *GS-δ* during leaf senescence (Cock *et al.* 1991). No correlation between mRNA and protein for either GS1 or GS2 was found in rice leaves during senescence (Kamachi *et al.* 1992a). Transgenic alfalfa transformed with 35S-promoter:GS1 accumulated transcripts without an increase in the corresponding enzyme (Ortega *et al.* 2001). These results suggest that post-translational regulation is also important in controlling GS activity *in vivo*. One mechanism could be the phosphorylation of GS protein followed by the interaction of phosphorylated GS protein with 14-3-3 proteins (Moorehead *et al.* 1999). A reversible control mechanism of GS1 by phosphorylation/dephosphorylation and 14-3-3 binding was proposed by Finnemann and Schjoerring (2000). When ATP/AMP is high, such as in the dark, GS1 is phosphorylated and binds to 14-3-

3 proteins, which protect it from proteolysis. Phosphorylation of GS2 protein has also been observed in tobacco (Riedel *et al*. 2001). For more details on post-translational regulation by light see chapter 6.

GLUTAMATE SYNTHASE

Characteristics and localisation of Fd-GOGAT

There are two molecular species of GOGAT: Fd-GOGAT (EC 1.4.7.1) and NADH-GOGAT (EC 1.4.1.14) (Ireland and Lea 1999). Fd-GOGAT was first isolated from pea leaves (Lea and Miflin 1974). Fd-GOGAT is monomeric; the molecular mass has been determined in a number of species and ranges from 145 kD in pea to 168 kD in pine (Ireland and Lea 1999). The spinach enzyme contains one FMN, one FAD, and one [3Fe-4S] cluster per molecule (Knaff *et al*. 1991). In maize a single cDNA for Fd-GOGAT was first isolated and sequenced by Sakakibara *et al*. (1991). The maize Fd-GOGAT is synthesised as a 1616 amino acid protein, including the 97 amino-acid presequence required for targeting chloroplasts and only a single copy of the gene was detected. Full-length cDNA for Fd-GOGAT in *A. thaliana* has been sequenced by Suzuki and Rothstein (1997). The enzyme consists of 1648 amino acids including the 131 amino acids of a transit peptide. Recently Hayakawa *et al*. (2003) sequenced a 5.8 kb cDNA for Fd-GOGAT in rice encoding 1616 amino acids including 96 amino acids as a transit peptide and described the structure of a gene for Fd-GOGAT. The transcribed region for the rice Fd-GOGAT gene consisted of 33 exons separated by 32 introns. A putative glutamate-binding site based on similarities of the sequence with *pur F*-type amidotransferases, as well as flavin and Fe-S cluster binding domains were identified in all of these Fd-GOGAT cDNAs. Although only one copy of the Fd-GOGAT gene has been detected in most plant species (Ireland and Lea 1999), two Fd-GOGAT genes have recently been identified in *A. thaliana* (Coschigano *et al*. 1998).

In leaves Fd-GOGAT is localised in the chloroplasts (Lea and Miflin 1974, Wallsgrove *et al*. 1979, Suzuki and Gadal 1984), and in roots in plastids (Bowsher *et al*. 1992). Immunogold labelling showed that Fd-GOGAT protein was detected in the chloroplast stroma of mesophyll, xylem parenchyma, and upper epidermal cells of tomato leaves (Botella *et al*. 1988). In barley leaves, Fd-GOGAT protein was detected specifically in chloroplasts of mesophyll and vascular cells (Tobin and Yamaya 2001). Using light microscopy at lower resolution, immunolabelling showed that Fd-GOGAT protein was predominantly localised in mesophyll cells in rice (Hayakawa *et al*. 1994) and barley (Tobin and Yamaya 2001) leaves and in the bundle sheath cells of maize leaves (Becker *et al*. 1993). In rice leaf blades, Fd-GOGAT was also detected in parenchyma sheath cells and parenchyma cells surrounding protoxylem lacunae

(Hayakawa et al. 1994). In young rice grains, Fd-GOGAT was detected mainly in pericarp cross-cells where chloroplasts are localised (Hayakawa et al. 1994). Protein content and activity of Fd-GOGAT were highest in fully expanded mature leaf blades of rice and were greatly reduced in leaf sheaths and developing non-green leaf blades (Yamaya et al. 1992). Research with mutants lacking Fd-GOGAT in *A. thaliana* (Somerville and Ogren 1980) and barley (Kendall et al. 1986) established that the major function of Fd-GOGAT in leaves is the re-assimilation of NH_4^+ released during photorespiration. The localisation of Fd-GOGAT in chloroplasts is consistent with these results with mutants.

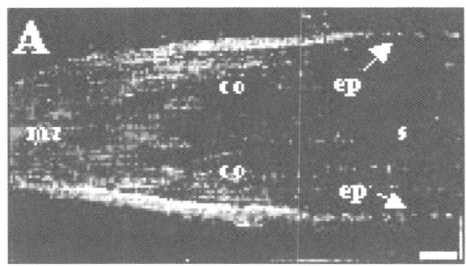 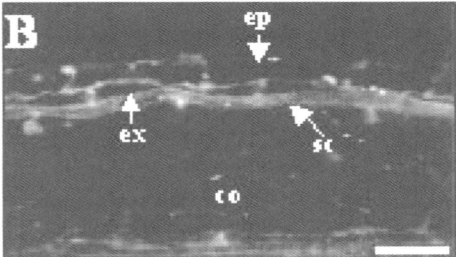

Figure 1. Distinct cellular localisation of two GOGAT proteins in rice root tips. (A, B) Longitudinal sections of roots treated with 1 mM NH₄Cl were double-stained with affinity purified anti-rabbit NADH-GOGAT IgG (Texas Red) and anti-mouse Fd-GOGAT IgG (FITC). co, cortex; ep, epidermis; ex, exodermis; mz, meristem zone; s, stele; sc, sclerenchyma. Bar = 50 μm in A and 12.5 μm in B. (Adapted from Ishiyama et al. 2003).

Fd-GOGAT was also found in non-photosynthetic tissues (Morigasaki et al. 1990, Matsumura et al. 1997). In such tissues Fd can be reduced in plastids through the oxidative pentose phosphate pathway (Bowsher et al. 1992). In *A. thaliana*, one of the two Fd-GOGAT genes was constitutively expressed, relatively high in roots and low in leaves (Coschigano et al. 1998). In rice roots where the Fd-GOGAT protein content changes with root development, it was highest in tip sections, and found mainly in the apical meristem and central cylinder (Ishiyama et al. 1998). Although NADH-GOGAT (see below) protein in rice roots was detected in the same areas, double immunostaining of both GOGAT proteins clearly showed that these two proteins were localised in distinct cell types of root-tips (Figure 1), where NADH-GOGAT protein was in the root surface while Fd-GOGAT was detected inside (Ishiyama et al. 2003). In nitrate-grown barley roots, there is an increase in Fd-GOGAT (Tobin and Yamaya 2001). The importance of Fd-GOGAT in root nitrate assimilation is indicated by analysis of barley mutants which resulted in increased concentrations of glutamine in roots and a three-fold increase in the export of glutamine to the shoot (Joy *et al.* 1992). In rice roots, activity and protein contents of Fd-GOGAT are not influenced by the availability of N (Yamaya *et al.* 1995). Although some information on Fd-GOGAT in non-photosynthetic tissues is available as described, more work is required to argue its precise function in these tissues.

Characteristics and localisation of NADH-GOGAT

Only a few species have been used to purify NADH-GOGAT protein and to isolate its cDNA. In alfalfa nodules Anderson *et al.* (1989) and Gregerson *et al.* (1993) purified the protein and isolated the cDNA, respectively; in rice, the protein was purified by Hayakawa *et al.* (1992) and cDNA was isolated by Goto *et al.* (1998). Since N metabolism in symbiotic root nodules is described in chapter 4, characteristics of NADH-GOGAT will be limited to our observations in rice. The protein was purified to homogeneity from rice cells suspension (Hayakawa *et al.* 1992). This enzyme is a monomer with a molecular mass of 196 kD and the pure form uses only NADH as a reductant. In rice plants the antibody for the rice NADH-GOGAT was used for immunological studies. A 7047 bp long clone consisting of full-length cDNA for NADH-GOGAT and genomic clones for the enzyme were sequenced (Goto *et al.* 1998). The presumed transcribed region (11.7 kb) consisted of 23 exons. Rice NADH-GOGAT is synthesised as a 2166 amino acid protein with a molecular mass of 236.7 kD that includes a 99 amino acid presequence. Only one gene was detected in rice. The organisation and structure is very similar to the alfalfa NADH-GOGAT gene (Gregerson *et al.* 1993, Vance *et al.* 1995). Several conserved sequences that are likely to be involved in the binding of glutamine, [3Fe-4S] cluster, FMN, and NADH were identified in both genes. The presequence could be involved in targeting of the enzyme to plastids, since we have shown with immunogold labelling that NADH-GOGAT is localised in plastids in rice roots (Hayakawa *et al.* 1999).

In green leaves, NADH-GOGAT activity is very low compared to Fd-GOGAT (Matoh and Takahashi 1981, Yamaya *et al.* 1992). During development of pea seedlings, NADH-GOGAT activity was replaced by Fd-GOGAT (Matoh and Takahashi 1981). In rice leaves, high activities of NADH-GOGAT and high levels of its protein are present in non green and developing leaf blades (Yamaya *et al.* 1992). In rice spikelets at the early stage of ripening, a large increase in NADH-GOGAT activity and protein content was found (Hayakawa *et al.* 1993). A similar increase in enzyme activity was also found in maize endosperm during 20 days after pollination (Muhitch 1991). High NADH-GOGAT activity has also been detected in roots of a range of plants (Suzuki *et al.* 1981, Matoh and Takahashi 1982, Yamaya *et al.* 1995, Lancien *et al.* 2002). N supply enhanced the expression of NADH-GOGAT in roots. For example, when N-starved rice seedlings were transferred to 1 mM NH_4Cl, both activity and protein content of

NADH-GOGAT increased more than 10-fold within 24 hours (Yamaya et al. 1995). Prior to this, NADH-GOGAT mRNA increased dramatically within 3 hr (Hirose et al. 1997). Microarray analysis also showed that NADH-GOGAT gene expression was stimulated by nitrate in A. thaliana (Wang et al. 2000). In developing rice leaf blades, NADH-GOGAT protein was localised in phloem and xylem parenchyma cells and mestome sheath cells of vascular bundles (Hayakawa et al. 1994). Immunolocalisation also showed a cell-type specific expression of the *NADH-GOGAT* gene which was confirmed by using transgenic rice. The expressing fusion gene consisted of *NADH-GOGAT*-promoter:*GUS* reporter gene (Kojima et al. 2000). Because these specific cell types where NADH-GOGAT protein was localised could be a route for the transport of solutes, this enzyme is considered important in the reutilisation of glutamine transported through vascular systems of developing organs of rice plants. In rice roots e.g., the content of NADH-GOGAT protein in all sequential segments along the crown roots was markedly increased after supplying 1 mM NH_4Cl (Ishiyama et al. 1998). The NH_4^+-induced NADH-GOGAT protein accumulation was observed specifically in root epidermis and exodermis (Ishiyama et al. 1998). Since solute movement from exodermis to the cortex is symplastic, because of the presence of a Casparian strip between these cell types, NH_4^+ ions would be taken up by both cell layers and then readily assimilated by the cytosolic GSr/NADH-GOGAT pathway. A detailed comparison of rice with other plants is not possible at present, since cellular localisation has not been studied extensively (Tobin and Yamaya 2001).

Regulation

Although light and N availability are major environmental factors causing an increase in Fd-GOGAT activity in leaves (Lam et al. 1996, Ireland and Lea 1999, chapter 6), molecular mechanisms that regulate its gene expression have not yet been examined.

The expression of the *NADH-GOGAT* gene appears to be, at least in rice plants, regulated in a cell-type specific, age-specific and N responsive manner. Deletion experiments showed that a 149 bp region of the promoter sequence was essential for the specific expression in vascular tissue in rice (Kojima et al. 2000). In this region, there were three putative *cis*-elements that were able to interact with nuclear proteins from rice cells (Kojima et al. 2002). In rice roots, *NADH-GOGAT* gene expression is induced by exogenous NH_4^+ (Yamaya et al. 1995). Induction by NH_4^+ was probably indirect, since glutamine appears to be the signal for induction of the *NADH-GOGAT* gene (Hirose et al. 1997). However, the use of exogenous glutamine may mask the effect of other amino acids or metabolites which could act as signals of the N status in plants (Stitt et al. 2002). Pharmacological approaches suggested that okadaic acid-sensitive protein phosphatases could have a role in mediating the metabolite regulation of

NADH-GOGAT gene expression in rice (Hirose and Yamaya 1999). As described above, cellular localisation studies indicate that NADH-GOGAT is important in the re-utilisation of glutamine transported via vascular tissues in developing leaves and grains. The enzyme appears to be important in primary NH_4^+ assimilation in rice roots, and this observation is supported by results with *NADH-GOGAT* gene knockout mutants in *A. thaliana* (Lancien *et al.* 2002).

GLUTAMATE DEHYDROGENASE

Characteristics and localisation

When *Chlorella sorokiniana* cells cultured on NO_3^- are transferred to NH_4^+, there is a rapid loss of GS and GOGAT activities and a concomitant increase in chloroplast-localised NADPH-GDH activity (Tischner and Lorenzen 1980). This suggests that the major route of NH_4^+ assimilation shifts from the GS/GOGAT pathway to NADPH-GDH when these cells are transferred from NO_3^- to NH_4^+. Bascomb and Schmidt (1987) successfully purified the two isoforms of NADPH-GDH from *C. sorokiniana*. The α-isoform, which has a Km value between 0.02 and 3.5 mM for NH_4^+, is induced at low NH_4^+ concentrations (< 2 mM), whereas the β-isoform with a Km of 75 mM is induced at high NH_4^+ concentrations (29 mM). The molecular mass of the α- and β-subunits was 55.5 kD and 53.0 kD, respectively. The mature enzyme protein is a hexamer of randomly associated subunit polypeptides. Recent molecular studies indicated that these two subunits are the products of alternative splicing of a precursor-mRNA encoded by the single NADPH-GDH gene (Miller *et al.* 1998).

In higher plants, physiological and biochemical characteristics of the NADH-GDH have been reviewed by Stewart *et al.* (1980), Srivastava and Singh (1987), Yamaya and Oaks (1987), and more recently by Loulakakis and Roubelakis-Angelakis (2001), Miflin and Habash (2002) and Dubois *et al.* (2003). The enzyme is located in mitochondria where the amination reaction is enhanced by Ca^{2+} (Yamaya *et al.* 1984) and calmodulin (Das *et al.* 1989). NADH-GDH in maize shoots is loosely associated with the mitochondrial membrane. It has an apparent Km of 4.8 mM for NH_4^+ when the assay levels of NADH are similar to those normally detected in mitochondria (Yamaya *et al.* 1984). Mitochondria prepared from maize shoots contained 3.6-5.0 mM NH_4^+ (Yamaya *et al.* 1984). Using *in vivo* concentrations of NH_4^+, NADH, NAD, and Ca^{2+} in the presence of 5 mM each of glutamate and α-ketoglutarate, the amination reaction was observed in isolated mitochondria at a pH range of 6.0-9.0 (Yamaya *et al.* 1984). When respiration was tested, malate and succinate supported a good rate of O_2 uptake, but citrate, α-ketoglutarate, and glutamate were also oxidised (Yamaya and Matsumoto 1985). With the transamination inhibitor, aminooxyacetate, glutamate was no longer a suitable substrate (Yamaya and Matsumoto 1985,

Yamaya and Oaks 1987). These results suggest that a transamination is important in the synthesis of α-ketoglutarate. This conclusion has recently been corroborated by Shultz *et al.* (1998) using transamination mutants of *A. thaliana*. In mature maize leaves, NADH-GDH activity is detected in bundle sheath strands (Yamaya and Oaks 1988). Recent immunolocalisation studies showed that the GDH protein in rice leaves (Maki *et al.* 2002), in grapevine leaves and flowers (Paczek *et al.* 2002) and in maize and tobacco leaves (Dubois *et al.* 2003) was mainly found in companion cells of vascular bundles. Since these cell types contain little or no chloroplasts, one would expect little direct interaction between GDH and either photorespiration or nitrate reduction.

There are two alleles for NADH-GDH in maize (Pryor 1974) and in *A. thaliana* (Melo-Oliviera *et al.* 1996). In each species one allele is mutated suggesting that the second allele has some essential function in plant metabolism (Melo-Oliviera *et al.* 1996). Native gels show seven activity bands in wild type maize (Ju *et al.* 1997) and in grapevine callus (Loulakakis and Roubelakis-Angelakis 1991). The fast moving a-band is enhanced when NH_4^+ levels are high as a result of external application or endogenous shifts in metabolism such as senescence (Loulakakis *et al.* 2002). Cd treatment results in an accumulation of NH_4^+ and increased GDH activity, together with an enhanced synthesis of γ-glutamyl peptides (Ju *et al.* 1997, Boussama *et al.* 1999). In the GDH1-null mutant of maize only the fast moving activity band was detected and the level of its activity increased with Cd treatment. Cd had no effect on GDH protein bands, but pyruvate carboxylase (PEPC) protein levels increased (Ju *et al.* 1997). These results suggest that the effect of Cd on GDH activity appears to be mediated by the activation of the GDH protein and not by its synthesis. When maize seedlings are supplied with NO_3^- the levels of synthetic reactions involving the TCA cycle are activated and there is *de novo* synthesis of PEPC (Takei *et al.* 2002). The results of Ju *et al.* (1997) point in the same direction. Since γ-glutamyl peptides are being made, NH_4^+ levels are high, and the fast form of GDH is present, it appears that GDH is mediating the synthesis of glutamate destined for γ-glutamyl peptide formation, which takes place in root tissue (Ju *et al.* 1997). This point will be considered in the next section. Melo-Oliviera *et al.* (1996) also suggest that the amination reaction was favoured in their mutants under conditions of ammonia and carbohydrate excess.

Function of GDH *in vivo*

The GS/GOGAT cycle in plant tissues is well documented (Ireland and Lea 1999). *In vitro*, GDH can catalyse the amination of α-ketoglutarate to form glutamate and the deamination of glutamate to form α-ketoglutarate and NH_4^+. Since NADH- or Fd-GOGAT is located in chloroplasts in leaf tissue or plastids in roots, it is unlikely that GDH could replace the GOGAT function as it does in bacteria. Mitochondria in photorespiring leaves generate large amounts of NH_4^+.

If there were an adequate supply of α-ketoglutarate, it might be expected this NH_4^+ could be re-assimilated (Hodges 2002, Mooney et al. 2002). Experiments with ^{15}N-glycine or ^{15}N-NH_4^+ in isolated mitochondria, showed only minor synthesis of ^{15}N-glutamate. Since the mitochondria were actively respiring and had an adequate supply of both citric and malic acids, it appears that the amination reaction of GDH was not active (Yamaya et al. 1986).

Evidence to support the deamination reaction of GDH in vivo has been presented and discussed in two short articles (Fox et al. 1995, Oaks 1995). Pahlich (1996) also contributed to the discussion and emphasised that the localisation of GDH in mitochondrial membranes could have a profound effect on the amination/deamination reaction. Results obtained by Robinson et al. (1991, 1992) and recently by Aubert et al. (2001) in their NMR and labelling studies indicated that GDH catalyses the deamination reaction. ^{15}N-NH_4^+ feeding studies coupled with the use of inhibitor studies in GDH1-null mutants of maize suggest that GDH does not catalyse the amination reaction (Magalhães et al. 1990). Experiments with GDH1-null mutants of A. thaliana (Melo-Olivera et al. 1996) suggested that GDH1 functioned in the assimilation of NH_4^+ under conditions of excess of inorganic N and C, and in the direction of glutamate catabolism under C-limiting conditions. There are several results that support the operation of the amination reaction of GDH under stress conditions. The deduced amino acid sequence for grapevine GDH is homologous to archeabacterial GDH, indicating the evolutionary relationship in plant GDH to the stress-related function of GDH (Syntichaki et al. 1996). For example, in Cd-stressed maize (Boussama et al. 1999, Ju et al. 1997) and herbicide-treated peanut (Osuji 1997), GDH activity was increased and a glutamate derivative was synthesised. These experiments would have been clearer had the researchers been able to perform ^{15}N feeding experiments as performed by Magalhães et al. (1990). To clarify the situation, knock-out mutants, particularly the GDH1-null and GDH2-null double mutant, would determine whether GDH has an essential role in plant metabolism or serves only as a redundancy shunt. In addition ^{15}N-feeding experiments under stress conditions might supply a definitive answer to the role of GDH in roots.

NITROGEN USE EFFICIENCY (NUE)

Interactions between photosynthesis and nitrogen

Since the eighties of the last century it is well known that growth conditions leading to higher levels of endogenous NH_4^+ and/or relatively low levels of carbohydrates result in relatively high levels of GDH. At that time the regulation of nitrate reductase appeared to be fairly straight forward, but regulation of NH_4^+ assimilation appeared to be complex (Oaks 1986). That photosynthesis supplies the C, energy and reductant to drive this assimilation and that light is required for the synthesis of glutamate but not glutamine is also a well-established fact. The easy explanation for this is that cytosolic GS derives its ATP from mitochondrial energy production and that Fd-GOGAT is dependent on photosynthesis to supply the reductant to reduce Fd. The nagging question was, and perhaps still is, that GDH could be a light activated enzyme in leaves. From the previous section it is clear that root GDH is active in the amination reaction under certain growth conditions. Reactions with leaf mitochondria or ^{15}N feeding experiments, on the other hand, do not support this idea. Thus it appears that the GS/GOGAT system is operational in the assimilation of NH_4^+ in leaves.

Assimilation of NH_4^+ is complicated by photorespiration which leads to a reduced C flow from pyruvate to citric acid and α-ketoglutarate and to the accumulation of NO_3^- in C_3 plants (Oaks 1995). When photorespiration is eliminated by the use of specific metabolic inhibitors, by high levels of ambient CO_2 or by low levels of O_2, C flow through citrate resumes, NO_3^- levels decline and flow of N into leaf proteins is more efficient. Under these conditions, NUE in C_3 plants is similar to that in C_4 plants. One advantage of C_4 plants is that reactions mediating photorespiration are localised in the bundle sheath cells, whereas the reactions involved in NH_4^+ production via nitrate reduction are in the mesophyll cells. It appears therefore that levels of NH_4^+ influence NUE, 1) by altering C flow through citrate, an effect reduced in C_4 plants by virtue of cell structure, and 2) possibly by an effect on N flow from NO_3^- to protein. This point needs to be explored further.

Genetic manipulation for the improvement of NUE

Recent progress on genetic manipulation allows us to attempt the improvement of NH_4^+ assimilation by the overexpression of GS, GOGAT, or GDH genes. Transgenic experiments are, however, sometimes complicated, because the positions in genome as well as the number of transgenes incorporated are not yet controlled by the current transformation procedures. In addition, differences in promoters used and in host plants can reform in a very complex situation. Furthermore analysis at T0 generation (regenerated plants just after the transformation) can be influenced by "somaclonal variation", from which various phenotype and characteristics can result, in particular lines, while segregation occurs in the T1 or T2 generation. Therefore, it is recommended to focus on a progeny of transformants, which contain a minimum number of homozygously fixed transgenes.

Nevertheless, in the literature mixed results have been reported for the overexpression of GS isoenzymes. For example, transgenic *Lotus corniculatus* overexpressing soybean *GS1* under the control of 35S promoter accelerated their growth rate and leaf senescence (Vincent *et al*. 1997). Growth improvements have been reported for poplar trees expressing a conifer *35S:GS1* (Gallardo *et al*.

1999), for tobacco expressing *RbcS:GS2* (Migge *et al.* 2000), and for tobacco expressing *35S:GS1* (Fuentes *et al.* 2001, Oliveira *et al.* 2002). In contrast, transgenic alfalfa expressing *35S:GS1* (Ortega *et al.* 2001) had no difference in growth and transgenic rice expressing *35S:GS2* enhanced salt tolerance without changing growth properties (Hoshisa *et al.* 2000). Localisation and biochemical studies led to the hypothesis that NADH-GOGAT might be important in the reutilisation of glutamine, transported via vascular tissues in rice plants. This hypothesis has been tested by gene manipulation techniques. To maintain the specificity of *NADH-GOGAT* gene expression in the transformants, the cDNA was fused with the promoter for *NADH-GOGAT* gene from Sasanishiki (a *japonica* cultivar of rice) and transformed into Kasalath (*indica* cultivar), which has less NADH-GOGAT protein in developing leaf blades (Obara *et al.* 2000). Several T0 transgenic lines overproducing NADH-GOGAT showed an increase in grain weight of maximal 80% (Yamaya *et al.* 2002). The results indicate that NADH-GOGAT is a key enzyme for N utilisation and grain filling in rice. Overexpression of alfalfa nodule *NADH-GOGAT* under the control of 35S promoter in transgenic tobacco also showed an increase in shoot dry weight associated with increased total C and N contents (Chichkova *et al.* 2001).

Amenziane *et al.* (2000) introduced the *gdhA* gene from *Escherichia coli* encoding a NADPH-GDH into tobacco plants and expressed under the control of 35S promoter. The introduced NADPH-GDH was apparently able to assimilate some of the excess NH_4^+, resulting from the increase in free amino acids, carbohydrates, and biomass production in the transgenic plants grown under controlled conditions or in the field. Transgenic wheat plants expressing the α-isoform of the *Chlorella NADPH-GDH* gene (aminating GDH) showed an increase in grain yield (Schmidt R., personal communication). Since the gene product contains a chloroplast-targeting transit peptide, the translated α-isoform GDH is localised in chloroplasts in the transgenic wheat. It has been shown that NH_3 emission from barley leaves (Mattsson and Schjoerring 1996) and oil seed rape (Mattsson *et al.* 1997) occurs in plants grown hydroponically with 2 mM NH_4^+ in the light, but not with NO_3^-. An inhibitor of GS increased the emission, while an inhibitor of photorespiration reduced it. These results suggest that GS2 in chloroplasts is just enough to reassimilate photorespiratory NH_4^+ in the NO_3^--grown plants. These plants are, however, not able to (re)assimilate excess NH_4^+ that is probably caused by low GS/GOGAT or C limitation to generate glutamate for the assimilation of NH_4^+. Better growth or biomass production found in transgenic plants overexpressing GS1 or GS2 as described above or in transgenic tobacco or wheat expressing aminating NADPH-GDH could result from an increase in the capacity for the assimilation of NH_4^+. Overexpression of GS or the NADPH-GDH in C_4 plants may not result in higher yields, because C_4 plants have greater NUE than C_3 plants, which is probably caused by the presence of two compartments, *i.e.* mesophyll cells and bundle sheath cells (Oaks 1994).

Quantitative-trait loci mapping for NUE

The connection between N metabolism and yield is poorly quantified (Lawlor 2002). This is mainly because changing N assimilation requires modifications in too many processes to effect improvements in the whole system. Genetic engineering alterations to single steps are unlikely to achieve this, because of the complexity of the interactions between processes and environment at different developmental stages of plants. A new genetic approach assisted with DNA marker technologies is now available to analyse crop traits and link them to regions in the genome (Tanksley and McCouch 1997). Agronomic traits, such as the size of seeds and time for heading, and physiological traits, such as N content and activity of GS1, could be determined by multiple gene functions. In maize, Hirel *et al.* (2001) have analysed recombinant inbred lines for physiological traits. Comparison of the variation in physiological traits and yield components showed a positive correlation between nitrate content, GS activity and yield. Quantitative trait loci (QTL) were determined on the map of the maize chromosomes. Coincidences of the QTLs for yield components, *GS1* gene and leaf GS activity were detected. In addition, six QTLs for leaf GS activity were detected on chromosomes 1, 5, and 9. QTLs associated with GS1 protein contents in senescing leaf blades have also been detected using backcross-inbred lines developed between *japonica* and *indica* cultivars of rice (Obara *et al.* 2001). Immunoblotting analysis showed a continuous distribution of GS1 contents with one peak showing transgressive segregations toward lower or greater contents in these lines. Seven QTLs for GS1 protein contents were mapped on chromosomes 2, 4, 8, and 11. Some of these QTLs, especially on chromosome 2, were co-located in QTL regions for various agronomic traits, such as a spikelet weight and panicle weight. These results suggest that GS1 could be a key component for NUE and yield in maize and rice. QTL regions, which do not coincide with structural *GS1* gene, could contain genes important in the regulation of GS1. If these genes are identified they could provide novel information on GS1 regulation.

Genetic approaches were also employed to detect putative QTLs associated with NADH-GOGAT protein content (Obara *et al.* 2001) in rice. *Oryza sativa* is widespread and there are three types or subspecies identified for cultivated plants, *i.e. japonica*, *indica* and *javanica*. In general, total biomass production of *indica* cultivars is greater than that of *japonica*, whereas grain yield is lower (Yamaya *et al.* 2002). Based on total leaf area N, Obara *et al.* (2000) showed that most *indica* cultivars contained less NADH-GOGAT protein content in developing young leaf blades than *japonica*. Transgressive segregations of NADH-GOGAT protein content in developing leaf blades were detected in backcross-inbred lines developed between *japonica* and *indica* cultivars (Obara *et al.* 2001). Six chromosomal QTL regions for NADH-GOGAT protein content were detected. One QTL on chromosome 1 was located close to a marker encoding the structural gene of NADH-GOGAT. Therefore, other 5 QTLs regulate NADH-GOGAT protein content in developing rice leaf blade. One QTL on chromosome 2 was co-located in the QTL region for agronomic traits, such as spikelet number on the main stem and spikelet weight. Thus, as in QTLs for GS1, this region could contain genes important in regulating both NADH-GOGAT contents and agronomic traits. Identification of such genes and survey of genetic resources containing more NADH-GOGAT protein will probably contribute to the improvement of NUE in rice.

Using a series of maize recombinant inbred lines, Dubois *et al.* (2003) successfully mapped three QTLs (two for the aminating and one for the deaminating reaction) for GDH activity in source leaves on maize chromosomes. The three QTLs for GDH activity co-localised with QTLs for grain yield and its components. The results suggest that GDH is involved in controlling the translocation of assimilates during the remobilisation phase. This possible role of GDH is also supported by localisation studies showing that GDH protein is mostly concentrated in the vascular tissues of a number of species (Maki *et al.* 2002, Paczek *et al.* 2002, Dubois *et al.* 2003).

NITROGEN METABOLISM COMES OF AGE - DO BIOCHEMISTRY AND MOLECULAR BIOLOGY MAKE A DIFFERENCE?

Since the proposal of the GS/GOGAT cycle 30 years ago, studies related to the pathway, compartmentation and regulation of NH_4^+ assimilation, have been done by many groups using different plant species. Most of these studies clearly show that GS/GOGAT is the major pathway in the primary assimilation of NH_4^+ in both leaves and roots. However, when the demand for glutamate increases as under stress conditions, GDH particularly at least appears to act in the amination reaction in roots. Distinct metabolic functions of isoenzymes of GS, GOGAT and GDH have been suggested. Characteristics of GS, GOGAT and GDH in different organs of rice are summarised in Table 1. There are still many "?" in the column of putative functions, which must be solved in order to understand the whole system of N metabolism. Recent progress in transgenic techniques, either decreasing or overexpressing the target gene, should help in understanding the complex metabolism. However, transgenic experiments can be of low reproducibility due to the lack of control of the insertion site and copy number of transgene. Genetic approaches will be important in the complete understanding of NH_4^+ assimilation. In addition, there is only limited information on transcriptional factors and *cis*-elements for the regulation of NH_4^+-assimilating genes. Recent development of exhaustive analyses, *i.e.* transcriptome, proteome, and metabolome, will certainly provide useful information. To obtain conclusive evidence, knockout mutants of a target gene, by insertion of either T-DNA or (retro)transposon into the gene, will be important tools to use. Finally, as a post-genomic approach, QTL analysis should indicate which reactions to manipulate in order to archieve an enhanced NUE and an enhanced crop inprovement.

Table 1. Localisation and regulation of GS, GOGAT, and GDH in rice.

Enzyme	Localisation	Putative function
Developing leaves		
GS1	cells active in secondary metabolism - xylem parenchyma - sclerenchyma	coupled with PAL
NADH-GOGAT	cells active in solute transport - vascular parenchyma - mestome sheath	reutilisation of Gln from senescing organs
GS2/Fd-GOGAT	cells active in photosynthesis - mesophyll	reassimilation of photorespiratory NH_4^+
NADH-GDH	companion cells	?
Mature/senescing		
GS1	cells active in N export - companion cells - phloem parenchyma	remobilisation of Gln through phloem
NADH-GOGAT	not detected	-
GS2/Fd-GOGAT	cells active in photosynthesis - mesophyll	reassimilation of photorespiratory NH_4^+
NADH-GDH	companion cells	?
Developing spikelets		
GS1	promoter active in embryo	?
NADH-GOGAT	cells active in solute transport - dorsal vascular bundles - nucellar projection/ epidermis cells	reutilisation of Gln from senescing organs
Fd-GOGAT	cells containing chloroplasts - cross cells in pericarp	?
GS2	not examined	-
NADH-GOGAT	companion cells	?
Roots		
GSr + GS1	all cell types	primary assimilation
GS1	promoter active in central cylinder and cortex cells	-
NADH-GOGAT	cells active in N uptake - epidermis/exodermis	primary assimilation induced by Gln or metabolite
Fd-GOGAT	central cylinder and cortex cells	?

ACKNOWLEDGEMENTS

The authors wish to express their gratitude to Dr. Bertrand Hirel for supplying the pre-print of Paczek *et al.* (2002) on cellular localisation of NADH-glutamate dehydrogenase. Financial supports to T.Y. from the CREST of JST and from the Ministry of Education, Culture, Sport, Science and Technology of Japan (Grant-in-Aid for Scientific Research: Nos. 14360035 and 14654160) are greatly appreciated.

REFERENCES

Ameziane R., Bernhard K., Lightfoot D. 2000. Expression of the bacterial *gdhA* gene encoding a NADPH glutamate dehydrogenase in tobacco affects plant growth and development. - Plant Soil 221: 47-57.
Anderson M.P., Vance C.P., Heichel G.H., Miller S.S. 1989. Purification and characterization of NADH-glutamate synthase from alfalfa root nodules. - Plant Physiol. 90: 351-358.
Aubert S., Bligny R., Douce R., Gout E., Ratcliffe R.G., Roberts J.K.M. 2001. Contribution of glutamate dehydrogenase to mitochondrial glutamate metabolism studied by C-13 and P-31 nuclear magnetic resonance. - J. Exp. Bot. 52: 37-45.
Bascomb N.F., Schmidt R.R. 1987. Purification and partial kinetic and physical characterization of two chloroplast-localized NADP-specific glutamate dehydrogenase isoenzymes and their preferential accumulation in *Chlorella sorokiniana* cells cultured at low or high ammonium levels. - Plant Physiol. 83: 75-84.
Becker T.W., Perrot-Rechemann C., Suzuki A., Hirel B. 1993. Subcellular and immunocytochemical localization of the enzymes involved in ammonia assimilation in mesophyll and bundle-sheath cells of maize leaves. - Planta 191: 129-136.
Blackwell R.D., Murray A.J.S., Lea P.J., Joy K.W. 1988. Photorespiratory amino donors, sucrose synthesis and the induction of CO_2 fixation in barley deficient in glutamine syntherase and/or glutamate synthase. - J. Exp. Bot. 38: 1799-1809.
Botella J.R., Verbelen J.P., Valpuesta V. 1988. Immunocytolocalization of ferredoxin-GOGAT in the cells of green leaves and cotyledons of *Lycopersicon esculentum*. - Plant Physiol. 87: 255-257.
Boussama N., Ouariti O., Suzuki A., Ghorbal M.H. 1999. Cd-stress on nitrogen assimilation. - J. Plant Physiol. 155: 310-317.
Bowsher C.G., Boulton E.L., Roses J., Nayagam S., Emes M.J. 1992. Reductant for glutamate synthase is generated by the oxidative pentose phosphate pathway in nonphotosynthetic root plastids. - Plant J. 2: 893-898.
Branjeon J., Hirel B., Forchioni A. 1989. Immunogold localisation of glutamine synthetase in soybean leaves, roots, and nodules. - Protoplasma 151: 88-97.
Brugière N., Dubois F., Limami A.M., Lelandais M., Roux Y., Sangwan R.S., Hirel B. 1999. Glutamine synthetase in the phloem plays a major role in controlling proline production. - Plant Cell 11: 1995-2011.
Carvalho H., Pereira S., Sunkel C., Salema R. 1992. Detection of a cytosolic glutamine synthetase in leaves of *Nicotiana tabacum* L. by immunocytochemical methods. - Plant Physiol. 100: 1591-1594.
Chichkova S., Arellano J., Vance C.P., Hernandez G. 2001. Transgenic tobacco plants that overexpress alfalfa NADH-glutamate synthase have higher carbon and nitrogen content. - J. Exp. Bot. 52: 2079-2087.

Cock J.M., Brock I.W., Watson A.T., Swarup R., Morby A.P., Cullimore J.V. 1991. Regulation of glutamine synthetase genes in leaves of *Phaseolus vulgaris*. - Plant. Mol. Biol. 17: 761-771.
Coschigano K.T., Melo-Oliveira R., Lim J., Coruzzi G.M. 1998. Arabidopsis *gls* mutants and distinct Fd-GOGAT genes: Implications for photorespiration and primary nitrogen assimilation. - Plant Cell 10: 741-752.
Crawford N.M., Glass A.D.M. 1998. Molecular and physiological aspects of nitrate uptake in plants. - Trends Plant Sci. 3: 389-395.
Das R., Sharma A.K., Sopory S.K. 1989. Regulation of NADH-glutamate dehydrogenase activity by phytochrome, calcium and calmodulin in *Zea mays*. - Plant Cell Physiol. 30: 317-323.
Dubois F., Brugière N., Sangwan R.S., Hirel B. 1996. Localization of tobacco cytosolic glutamine synthetase enzymes and the corresponding transcripts shows organ- and cell-specific patterns of protein synthesis and gene expression. - Plant. Mol. Biol. 31: 803-817.
Dubois F., Terecé-Laforgue T., Gonzalez-Moro M.-B., Estavillo J.-M., Sangwan R., Gallais, A., Hirel B. 2003. Glutamate dehydrogenase in plants: is there a new story for an old enzyme? - Plant Physiol. Biochem. 41: 565-576.
Edwards J.M., Walker E.L., Coruzzi G.M. 1990. Cell-specific expression in transgenic plants reveals non-overlapping roles for chloroplast and cytosolic glutamine synthetase. - Proc. Natl. Acad. Sci. U.S.A. 87: 3459-3463.
Finnemann J., Schjoerring J.K. 2000. Post-translational regulation of cytosolic glutamine synthetase by reversible phosphorylation and 14-3-3 protein interaction. - Plant J. 24: 171-181.
Fox G.G., Ratcliffe R.G., Robinson S.A., Stewart G.R. 1995. Evidence for deamination by glutamate dehydrogenase in higher plants: Commentary. - Can. J. Bot. 72: 739-750.
Freney J.R., Leuning R., Simpton J.R., Denmead O.T., Muirhead W.A. 1985. Estimating ammonia volatilization from flooded rice fields by simplified techniques. - Soil Sci. Soc. Am. J. 49: 1049-1054.
Fuentes S.I., Allen D.J., Ortiz-Lopez A., Hernández G. 2001. Over-expression of cytosolic glutamine synthetase increases photosynthesis and growth at low nitrogen concentration. - J. Exp. Bot. 52: 1071-1081.
Fukumorita T., Chino M. 1982. Sugar, amino acid, and inorganic contents in rice phloem sap. - Plant Cell Physiol. 23: 273-283.
Gallardo F., Fu J., Cantón F.R., García-Gutiérrez A. Cánovas F.M., Kirby E.G. 1999. Expression of a conifer glutamine synthetase gene in transgenic poplar. - Planta 210: 19-26.
Gazzarrini S., Lejay L., Gojon A., Ninnemann O., Frommer W.B., von Wirén, N. 1999. Three functional transporters for constitutive, diurnally regulated, and starvation-induced uptake of ammonium into *Arabidopsis* roots. - Plant Cell 11: 937-947.
Glass A.D.M., Britto D.T., Kaiser B.N., Kronzucker, H.J., Kumar A., Okamoto M., Rawat S.R., Siddiqi M.Y., Silim S.M., Vidmar J.J., Zhuo D. 2001. Nitrogen transport in plants, with emphasis on the regulation of fluxes to match plant demand. - Zeit. Pflanz. Bod. 1604: 199-207.
Glass A.D.M., Britto D.T., Kaiser B.N., Kinghorn J.R., Kronzucker H.J., Kumar A., Okamoto M., Rawat S., Siddiqi M.Y., Unkles S.E., Vidmar J.J. 2002. The regulation of nitrate and ammonium transport systems in plants. - J. Exp. Bot. 53: 855-864.
Good N., Izawa S., Hind G. 1966. Uncoupling and energy transfer inhibition in photophosphorylation. - Curr. Top. Bioenerg. 1: 75-112.
Goto S., Akagawa T., Kojima S., Hayakawa T., Yamaya T. 1998. Organization and structure of NADH-dependent glutamate synthase gene from rice plants. - Biochim. Biophys. Acta 1387: 298-308.
Gregerson R.G., Miller S.S., Twary S.N., Gantt J.S., Vance C.P. 1993. Molecular characterization of NADH-dependent glutamate synthase from alfalfa nodules. - Plant Cell 5: 215-226.
Hanzawa S., Matsumura S., Hayakawa T., Yamaya T. 2002. Transgenic rice expressing either *GUS* or cDNA for the cytosolic glutamine synthetase (GS1) under the control of GS1 promoter from rice. - Plant Cell Physiol. 43 (Suppl.): 71.

Hayakawa T., Yamaya T., Mae T., Ojima K. 1992. Purification, characterization, and immunological properties of NADH-dependent glutamate synthase from rice cell cultures. - Plant Physiol. 98: 1317-1322.

Hayakawa T., Yamaya T., Mae T., Ojima K. 1993. Changes in the content of two glutamate synthase proteins in spikelets of rice (*Oryza sativa*) plants during ripening. - Plant Physiol. 101: 1257-1262.

Hayakawa T., Nakamura T., Hattori F., Mae T., Ojima K., Yamaya T. 1994. Cellular localization of NADH-dependent glutamate-synthase protein in vascular bundles of unexpanded leaf blades and young grains of rice plants. - Planta 193: 455-460.

Hayakawa T., Hopkins L., Peat L.J., Yamaya T., Tobin A.K. 1999. Quantitative intercellular localization of NADH-dependent glutamate synthase protein in different types of root cells in rice plants. - Plant Physiol. 119: 409-416.

Hayakawa T., Sakai T., Ishiyama K., Hirose N., Nakajima H., Takezawa M., Naito K., Hino-Nakayama M., Akagawa T., Goto S., Yamaya T. 2003. Organization and structure of ferredoxin-dependent glutamate synthase gene from rice plants. - Plant Biotechnology 20: 43-55.

Hayashi H., Chino M. 1990. Chemical composition of phloem sap from the upper most internode of the rice plant. - Plant Cell Physiol. 31: 247-251.

Hirel B., Gadal P. 1981. Glutamine synthetase isoforms in pea leaves. Intracellular localization. - Z. Pflanzenphysiol. 102: 315-319.

Hirel B., Bertin P., Quillere I., Bourdoncle W., Attagnant C., Dellay C., Gouy A., Cadiou S., Retailliau C., Falque M., Gallais A. 2001. Towards a better understanding of the genetic and physiological basis for nitrogen use efficiency in maize. - Plant Physiol. 125: 1258-1270.

Hirose N., Yamaya T. 1999. Okadaic acid mimics nitrogen-stimulated transcription of NADH-glutamate synthase gene in rice cell cultures. - Plant Physiol. 121: 805-812.

Hirose N., Hayakawa T., Yamaya T. 1997. Inducible accumulation of mRNA for NADH-dependent glutamate synthase in rice roots in response to ammonium ions. - Plant Cell Physiol. 38: 1295-1297.

Hodges M. 2002. Enzyme redundancy and the importance of 2-oxoglutarate in plant ammonium assimilation. - J. Exp. Bot. 53: 905-916.

Hoshida H., Tanaka Y., Hibino T., Hayashi Y., Tanaka A., Takabe T., Takabe T. 2000. Enhanced tolerance to salt stress in transgenic rice that overexpresses chloroplast glutamine synthetase. - Plant Mol. Biol. 43: 103-111.

Hsieh B., Lam H.M., van de Loo F.J., Coruzzi G. 1998. A PII-like protein in *Arabidopsis*: putative role in nitrogen sensing. - Proc. Nat. Acad. Sci. U.S.A. 95: 13965-13970.

Ireland, R.J., Lea P.J. 1999. The enzymes of glutamine, glutamate, asparagine, and aspartate metabolism. - Singh B.K. (Ed.) Plant amino acids; biochemistry and biotechnology - New York, U.S.A., Marcel Dekker, pp. 49-109.

Ishiyama K., Hayakawa T., Yamaya T. 1998. Expression of NADH-dependent glutamate synthase protein in the epidermis and exodermis of rice roots in response to the supply of nitrogen. - Planta 204: 288-294.

Ishiyama K., Kojima S., Takahashi H., Hayakawa T., Yamaya, T. 2003. Cell-type distinct accumulation of mRNA and protein for NADH-glutamate synthase in rice roots in response to the supply of NH_4^+. - Plant Physiol. Biochem. 41: 643-647.

Joy K.W. 1988. Ammonia, glutamine and asparagine: A carbon-nitrogen interface. - Can. J. Bot. 66: 2103-2109.

Joy K.W., Blackwell R.D., Lea P.J. 1992. Assimilation of nitrogen in mutants lacking enzymes of the glutamate synthase cycle. - J. Exp. Bot. 43: 139-145.

Ju G.C., Li X.-Z., Rauser W. E., Oaks A. 1997. Influence of cadmium on the production of gamma-glutamylcysteine peptides and enzymes of nitrogen assimilation in *Zea mays* seedlings. - Physiol. Plant. 101: 793-799.

Kamachi K., Yamaya T., Hayakawa T., Mae T., Ojima K. 1992a. Changes in cytosolic glutamine synthetase polypeptide and its mRNA in a leaf blade of rice plants during natural senescence. - Plant Physiol. 98: 1323-1329.

Kamachi K., Yamaya T., Hayakawa T., Mae T., Ojima K. 1992b. Vascular bundle-specific localization of cytosolic glutamine synthetase in rice leaves. - Plant Physiol. 99: 1481-1486.

Karasawa T., Hayakawa T., Mae T., Ojima K., Yamaya, T. 1994. Characteristics of ammonium uptake by rice cells in suspension culture. - Soil Sci. Plant Nutr. 40: 333-338.

Kendall A.C., Wallsgrove R.M., Hall N.P., Turner J.C., Lea P.J. 1986. Carbon and nitrogen metabolism in barley (*Hordeum vulgare* L.) mutants lacking ferredoxin-dependent glutamate synthase. - Planta 168: 316-323.

Keys A.J., Bird I.F., Cornelius M.J., Lea P.J., Wallsgrove R.M., Miflin B.J. 1978. The photorespiratory nitrogen cycle. - Nature 275: 741-743.

Kleiner D. 1981. The transport of NH_3 and NH_4^+ across biological membranes. - Biochim. Biophys. Acta 639: 41-52.

Knaff D.B., Hirasawa M., Ameyibor E., Fu W., Johnson M.K. 1991. Spectroscopic evidence for a (3Fe-4S) cluster in spinach glutamate synthase. - J. Biol. Chem. 266: 15080-15084.

Kojima S., Kimura M., Nozaki Y., Yamaya T. 2000. Analysis of a promoter for NADH-glutamate synthase gene in rice (*Oryza sativa*): Cell-type specific expression in developing organs of transgenic rice plants. - Aust. J. Plant Physiol. 27: 787-793.

Kojima S., Hayamawa T., Yamaya T. 2002. *cis*-Elements that regulate vascular-bundle specific expression of NADH-GOGAT gene in rice. - Plant Cell Physiol. 43 (Suppl.): 127.

Kronzucker H.J., Siddiqi M.Y., Glass A.D.M. 1996. Kinetics of NH_4^+ influx in spruce. - Plant Physiol. 110: 773-779.

Kronzucker H.J., Siddiqi M.Y., Glass A.D.M. 1997. Root discrimination against soil nitrate and the ecology of forest succession. - Nature 385: 59-61.

Lam H.-M., Coschigano K.T., Oliviera I.C., Melo-Oliveira R., Coruzzi G. M. 1996. The molecular-genetics of nitrogen assimilation into amino acids in higher plants. - Annu. Rev. Plant Physiol. Plant Mol. Biol. 47: 569-593.

Lancien M., Martin M., Hsieh M.-H., Leustek T., Goodman H., Coruzzi G.M. 2002. Arabidopsis glt1-T mutant defines a role for NADH-GOGAT in the non-photorespiratory ammonium assimilatory pathway. - Plant J. 29: 347-358.

Lauter F.R., Ninnemann O., Bucher M., Riesmeier J., Frommer W.B. 1996. Preferential expression of an ammonium transporter and two putative nitrate transporters in root hairs of tomato. - Proc. Natl. Acad. Sci. USA 93: 8139-8144.

Lea P.J., Ireland, R.J. 1999. Nitrogen metabolism in higher plants. – In: Singh B.K. (Ed.) Plant amino acids; biochemistry and biotechnology - New York, U.S.A., Marcel Dekker, pp. 1-47.

Lea P.J., Miflin B.J. 1974. Alternative route for nitrogen assimilation in higher plants. - Nature 251: 614-616.

Li M.G., Villemur R., Hussey P.J., Silflow C.D., Gantt J.S., Snustad D.P. 1993. Differential expression of six glutamine synthetase genes in *Zea mays*. - Plant Mol. Biol. 23: 401-407.

Lawlor D.W. 2002. Carbon and nitrogen assimilation in relation to yield: mechanisms are the key to understanding production systems. - J. Exp. Bot. 53: 773-787.

Loulakakis K.A., Roubelakis-Angelakis K.A. 1991. Plant NAD(H) glutamate dehydrogenase consists of two subunit polypeptides and their participation in the seven isoenzymes occurs in an ordered ratio. - Plant Physiol. 97: 104-111.

Loulakakis, K.A., Roubelakis-Angelakis. K.A. 2001. Ammonia assimilating genes in *Vitis vinifera* L. - In: Roubelakis-Angelakis KA. (Ed.) Molecular biology and biotechnology of the grapevine - Dordrecht, The Netherlands, Kluwer Academic Publishers, pp. 59-108.

Loulakakis K.A., Primikirios N.I., Nikolantonakis M.A., Roubelakis-Angelakis K.A. 2002. Immunocharacterization of *Vitis vinifera* L. ferredoxin-dependent glutamate synthase and its spatial and temporal changes during leaf development. - Planta 215:630-638.

Mäck G. 1995. Organ-specific changes in the activity and subunit composition of glutamine-synthetase isoforms of barley (*Hordeum vulgare* L.) after growth on different levels of NH_4^+. - Planta 196: 231-238.

Mäck G., Tischner R. 1994. Constitutive and inducible net NH_4^+ uptake of barley (*Hordeum vulgare* L.) seedlings. - J. Plant Physiol. 144: 351-357.

Magalhães J.R., Ju G.C., Rich P.J., Rhodes D. 1990. Kinetics of $^{15}NH_4^+$ assimilation in *Zea mays*. - Plant Physiol. 94: 647-656.

Maki H., Ushioda A., Abiko T., Hayakawa T., Yamaya T. 2002. Cloning and expression analysis of NADH-glutamate dehydrogenase in rice plants. - Plant Cell Physiol. 43 (Suppl.): 127.

Mann A.F., Fentem P.A., Stewart G.R. 1979. Identification of two forms of glutamine synthetase in barley (*Hordeum vulgare* L.). - Biochem. Biophys. Res. Commun. 88: 515-521.

Marini A.-M., Springael J.-Y., Frommer W.B., André, B. 1997. A family of ammonium transporters in *Saccharomyces cerevisiae*. - Mol. Cell. Biol. 17: 4248-4293.

Matoh T., Takahashi E. 1981. Glutamate synthase in greening pea shoots. - Plant Cell Physiol. 22: 727-731.

Matoh T., Takahashi E. 1982. Changes in the activities of ferredoxin-glutamate synthase and NADH-glutamate synthase during seedling development of peas. - Planta 154: 289-294.

Mattsson M., Husted S., Schjoerring J.K. 1997. Influence of nitrogen nutrition and metabolism on ammonia volatilization in plants. - Nutr. Cycling Agroecosyst. 51: 35-40.

Mattsson M., Schjoerring J.K. 1996. Ammonia emmision from young barley plants: Influence of N source, light/dark cycles and inhibition of glutamine synthetase. - J. Exp. Bot. 47: 477-484.

Matsumura T., Sakakibara H., Nakano R., Kimata Y., Sugiyama T., Hase T. 1997. A nitrate-inducible ferredoxin in maize roots: Genomic organization and differential expression of two nonphotosynthetic ferredoxin isoproteins. - Plant Physiol. 114: 653-660.

McNally S.F., Hirel B., Gadal P., Mann A.F., Stewart G.R. 1983. Glutamine synthetase of higher plants. Evidence for a specific isoform content related to their possible physiological role and their compartmentation within the leaf. - Plant Physiol. 72: 22-25.

Melo-Oliveira R., Oliveira I.C., Coruzzi G.M. 1996. Arabidopsis mutant analysis and gene regulation define a nonredundant role for glutamate dehydrogenase in nitrogen assimilation. - Proc. Natl. Acad. Sci. U.S.A. 93: 4718-4723.

Miflin B.J., Lea P.J. 1976. The pathway of nitrogen assimilation in plants. - Phytochemistry 15: 873-885.

Miflin B.J., Habash D.Z. 2002. The role of glutamine synthetase and glutamate dehydrogenase in nitrogen assimilation and possibilities for improvement in the nitrogen utilization of crops. - J. Exp. Bot. 53: 979-987.

Migge A., Carrayol E., Hirel B., Becker T. 2000. Leaf-specific overexpression of plastidic glutamine synthetase stimulates the growth of transgenic tobacco seedlings. - Planta 210: 252-260.

Miller P.W., Dunn W.I., Schmidt R. 1998. Alternative splicing of a precursor-mRNA encoded by the *Chlorella sorokiniana* NADP-specific glutamate dehydrogenase gene yields mRNAs for precursor proteins of isoenzyme subunits with different ammonium affinities. - Plant Mol. Biol. 37: 243-263.

Moorhead G., Douglas P., Cotelle V., Harthill J., Morrice N., Meek S., Deiting U., Stitt M., Scarabel M., Aitken A., MacKintosh C. 1999. Phosphorylation-dependent interactions between enzymes of plant metabolism and 14-3-3 proteins. - Plant J. 18: 1-12.

Mooney B.P., Miernyk J.A., Randall D.D. 2002. The complex fate of α-ketoacids. - Annu. Rev. Plant Biol. 53: 357-375.

Morigasaki S., Takada K., Sanada Y., Wada K., Yee B.C., Shin S., Buchanan B.B. 1990. Novel forms of ferredoxin and ferredoxin-$NADP^+$ reductase from spinach roots. - Arch. Biochem. Biophys. 283: 75-80.

Muhitch M.J. 1991. Tissue distribution and developmental patterns of NADH-dependent and ferredoxin-dependent glutamate synthase activities in maize (*Zea mays*) kernels. - Physiol. Plant. 81: 481-488.

Ninnemann O., Jauniaux J.-C., Frommer W.B. 1994. Identification of a high affinity NH_4^+ transporter from plants. - EMBO J. 13: 3464-3471.
Oaks A. 1992. A Re-evaluation of nitrogen assimilation in roots. - Bioscience 42: 103-111.
Oaks A. 1994. Efficiency of nitrogen utilization in C3 and C4 cereals. - Plant Physiol. 106: 407-414.
Oaks A. 1995. Evidence for deamination by glutamate dehydrogenase in higher plants: Reply. - Can. J. Bot. 73: 1116-1117.
Oaks A. 1986. Biochemical aspects of nitrogen in a whole plant context. In: Lambers H., Neeteson J.J., Stulen I. (Eds.) Fundamental, ecological and agricultural aspects of nitrogen metabolism in higher plants. - Dordrecht, The Netherlands, Martinus Nijhoff Publishers, pp. 133-151.
Oaks A., Hirel B. 1985. Nitrogen metabolism in roots. - Ann. Rev. Plant Physiol. 36: 345-365.
Obara M., Sato T., Yamaya T. 2000. High content of cytosolic glutamine synthetase does not accompany with a high activity of the enzyme in rice (*Oryza sativa* L) leaves of indica cultivars during the life span. - Physiol. Plant. 108: 11-18.
Obara M., Kajiura M., Fukuta Y., Yano M., Hayashi M., Yamaya T., Sato T. 2001. Mapping of QTLs associated with cytosolic glutamine synthetase and NADH-glutamate synthase in rice (*Oryza sativa* L.). - J. Exp. Bot. 52: 1209-1217.
Oliveira I.C., Brears T., Knight T.J., Clark A., Coruzzi G.M. 2002. Overexpression of cytosolic glutamine synthetase. Relation to nitrogen, light, and photorespiration. - Plant Physiol. 129: 1170-1180.
O'Neal D., Joy K.W. 1973. Glutamine synthetase of pea leaves. Purification, stabilization, and pH optima. - Arch. Biochem. Biohpys. 159: 113-122.
Ortega J.L., Temple S.J., Sengupta-Gopalan C. 2001. Constitutive overexpression of cytosolic glutamine synthetase (GS1) gene in transgenic alfalfa demonstrates that GS1 may be regulated at the level of RNA stability and protein turnover. - Plant Physiol. 126: 109-121.
Osuji G.O. 1997. Peanut glutamate dehydrogenase: A target site of herbicide action. - Soil Sci. Plant Nutr. 43: 1159-1164.
Paczek V., Dubois F., Sangwan R., Morot-Gaudry J.F., Roubelakis-Angelakis K.A., Hirel B. 2002. Cellular and subcellular localization of glutamine synthetase and glutamate dehydrogenase in grapes gives new insights on the regulation of carbon and nitrogen metabolism. - Planta 216: 245-254.
Pahlich E. 1996. Remarks concerning the dispute related to the function of glutamate dehydrogenase: Commentary. - Can. J. Bot. 74: 512-515.
Pate J.S. 1980. Transport and partitioning of nitrogenous solutes. Annu. Rev. Plant Physiol. 31: 313-340.
Peat L.J., Tobin A.K. 1996. The effect of nitrogen nutrition on the cellular localization of glutamine synthetase isoforms in barley roots. - Plant Physiol. 111: 1109-1117.
Pereira S., Cavalho H., Sunkel C., Salema R. 1992. Immunocytolocalization of glutamine synthetase in mesophyll and phloem of leaves of *Solanum tuberosum* L. - Protoplasma 167: 66-73.
Peterman T.K., Goodman H.M. 1991. The glutamine synthetase gene family of *Arabidopsis thaliana*: light-regulation and differential expression in leaves, roots and seeds. - Mol. Gen. Genet. 230: 145-154.
Pryor A.J. 1974. Allelic glutamic dehydrogenase isoenzymes in maize: A single hybrid isoenzyme in heterozygotes? - Heredity 32: 397-419.
Rastogi R., Chourey P.S., Muhitch M.J. 1998. The maize glutamine synthetase GS1-2 gene is preferentially expressed in kernel pedicels and is developmentally-regulated. - Plant Cell Physiol. 39: 443-446.
Rawat S.R., Silim S.N., Kronzucker H.J., Siddiqi M.Y., Glass A.D.M. 1999. *AtAMT1* gene expression and NH_4^+ uptake in roots of *Arabidopsis thaliana*: evidence for regulation by root glutamine levels. - Plant J. 19: 143-152.

Redinbaugh M.G., Campbell W.H. 1993. Glutamine synthetase and ferredoxin-dependent glutamate synthase expression in the maize (*Zea mays*) root primary response to nitrate. Evidence for an organ-specific response. - Plant Physiol. 101: 1249-1255.

Riedel J., Tischner R., Mäck G. 2001. The chloroplastic glutamine synthetase (GS-2) of tobacco is phosphorylated and associated with 14-3-3 proteins inside the chloroplast. - Planta 213: 396-401.

Robinson S.A., Slade A.P., Fox G.G., Philips R., Ratcliffe R.G., Stewart G.R. 1991. The role of glutamate dehydrogenase in plant nitrogen metabolism. - Plant Physiol. 95: 509-516.

Robinson S.A., Stewart G.R., Philips R. 1992. Regulation of glutamate dehydrogenase activity in relation to carbon limitation and protein catabolism in carrot cell suspension cultures. - Plant Physiol. 98: 1190-1195.

Sakakibara H., Watanabe M., Hase T., Sugiyama T. 1991. Molecular cloning and characterization of complementary DNA encoding for ferredoxin-dependent glutamate synthase in maize leaf. - J. Biol. Chem. 266: 2028-2035.

Sakakibara H., Kawabata S., Hase T., Sugiyama T. 1992a. Differential effects of nitrate and light on the expression of glutamine synthetases and ferredoxin-dependent glutamate synthase in maize. - Plant Cell Physiol. 33: 1193-1198.

Sakakibara H., Kawabata S., Takahashi H., Hase T., Sugiyama T. 1992b. Molecular cloning of the family of glutamine synthetase genes from maize: Expression of genes for glutamine synthetase and ferredoxin-dependent glutamate synthase in photosynthetic and non-photosynthetic tissues. - Plant Cell Physiol. 33: 49-58.

Sakakibara H., Shimizu H., Hase T., Yamazaki Y., Takao T., Shimonishi Y., Sugiyama T. 1996. Molecular identification and characterization of cytosolic isoforms of glutamine synthetase in maize roots. - J. Biol. Chem. 271: 29561-29568.

Sakamoto A., Ogawa M., Masumura T., Shibata D., Takeba G., Tanaka K., Fujii S. 1989. Three cDNA sequences coding for glutamine synthetase polypeptides in *Oryza sativa* L. - Plant. Mol. Biol. 13: 611-614.

Sakurai N., Hayakawa T., Nakamura T., Yamaya T. 1996. Changes in the cellular localization of cytosolic glutamine synthetase protein in vascular bundles of rice leaves at various stages of development. - Planta 200: 306-311.

Sakurai N., Katayama Y., Yamaya T. 2001. Overlapping expression of cytosolic glutamine synthetase and phenylalanine ammonia lyase in immature leaf blades of rice. - Physiol. Plant. 113: 400-408.

Schjoerring J.K., Husted S., Mäck G., Mattsson M. 2002. The regulation of ammonium translocator in plants. - J. Exp. Bot. 53: 883-890.

Schultz C.J., Hsu M., Miesak B., Coruzzi G.M. 1998. *Arabidopsis* mutants define an in vivo role for isoenzymes of aspartate aminotransferase in plant nitrogen assimilation. - Genetics 149: 491-499.

Sechley K.A., Yamaya T., Oaks A. 1992. Compartmentation of nitrogen assimilation in higher plants. - Int. Rev. Cytol. 134: 85-163.

Sims A.P., Folkes B.F. 1964. A kinetic study of the assimilation of (^{15}N)-ammonia and the synthesis of amino acids in an exponentially growing culture of *Candida utilis*. - Proc. R. Soc. Lond. B. 159: 479-502.

Sohlenkamp C., Shelden M., Howitt S., Udvardi M. 2000. Characterization of *Arabidopsis* AtAMT2, a novel ammonium transporter in plants. - FEBS Lett. 467: 273-278.

Somerville C.R., Ogren W.L. 1980. Inhibition of photosynthesis in *Arabidopsis* mutants lacking in leaf glutamate synthase activity. - Nature 286: 257-259.

Sonoda Y., Ikeda A., Saiki S., von Wirén N, Yamaya T., Yamaguchi J. 2003. Distinct expression and function of three ammonium transporter genes (*OsAMT1;1-1;3*) in rice. - Plant Cell Physiol. 44: 726-734.

Srivasankar S., Oaks A. 1996. Nitrate assimilation in higher plants: The effect of metabolites and light. - Plant Physiol. Biochem. 34: 609-620.

Sivastava H.S., Singh R.P. 1987. Role and regulation of L-glutamate dehydrogenase activity in higher plants. - Phytochemistry 26: 597-610.
Stewart G.R., Mann A.F., Fentem P.A. 1980. Enzymes of glutamate formation: Glutamate dehydrogenase, glutamine synthetase and glutamate synthase. - In Miflin B.J. (Ed.) The biochemistry of plants. - New York, USA, Academic Press, Vol.5, pp. 271-327.
Stitt M., Müller C., Matt P., Gibon Y, Carillo P., Morcuende R., Scheible W.-R., Krapp A. 2002. Steps towards an integrated view of nitrogen metabolism. - J. Exp. Bot. 53: 959-970.
Suenaga A., Moriya K., Sonoda Y., Ikeda A., von Virén N, Hayakawa T. Yamaguchi J., Yamaya T. 2003. Constitutive expression of a novel-type ammonium transporter *OsAMT2* in rice plants. - Plant Cell Physiol. 44: 206-211.
Suzuki A., Gadal P. 1984. Glutamate synthase: Physiochemical and functional properties of different forms in higher plants and other organisms. - Physiol. Veg. 22: 471-486.
Suzuki A., Rothstein S. 1997. Structure and regulation of ferredoxin-dependent glutamate synthase from *Arabidopsis thaliana* - cloning of cDNA, expression in different tissues of wild-type and *glts* mutant strains, and light induction. - Eur. J. Biochem. 243: 708-718.
Suzuki A., Gadal P., Oaks, A. 1981. Intracellular distribution of enzymes associated with nitrogen assimilation in plants. - Planta 151:457-461.
Syntichaki K.M., Loulakakis K.A., Roubelakis-Angelakis K.A.. 1996. The amino acid sequence similarity of plant glutamate dehydrogenase with the extremophilic archaeal enzyme conforms to its stress related function. - Gene 168: 87-92.
Takei K., Takahashi T., Sugiyama T., Yamaya T., Sakakibara H. 2002. Multiple routes communicating nitrogen availability from roots to shoots: a signal transduction pathway mediated by cytokinin. - J. Exp. Bot. 53: 971-977.
Tanksley S.D., McCouch S.R. 1997. Seed banks and molecular maps: Unlocking genetic potential from the wild. - Science 277: 1063-1066.
Tempest D.W., Meers J.L., Brown C.M. 1970. Synthesis of glutamate in *Acerobacter aerogenes* by a hitherto unknown route. - Biochem. J. 117: 405-407.
Tingey S.V., Walker E. L., Coruzzi G.M. 1987. Glutamine synthetase genes of pea encode distinct polypeptides which are differentially expressed in leaves, roots and nodules. - EMBO J. 6: 1-9.
Tishner R., Lorenzen H. 1980. Changes in the enzyme pattern in synchronous *Chlorella sorokiniana* caused by different nitrogen sources. - Z. Pflanzenphysiol. 100: 333-341.
Tjaden G., Edwards J.W., Coruzzi G.M. 1995. *cis*-Elements and trans-acting factors affecting regulation of a nonphotosynthetic light-regulated gene for chloroplast glutamine synthetase. - Plant Physiol. 108: 1109-1117.
Tobin A.K., Yamaya T. 2001. Cellular compartmentation of ammonium assimilation in rice and barley. - J. Exp. Bot. 52: 591-604.
Vance C.P., Miller S.S., Gregerson R.G., Samac D.A., Robinson D.L., Gantt J.S. 1995. Alfalfa NADH-dependent glutamate synthase: structure of the gene and importance in symbiotic N_2 fixation. - Plant J. 8: 345-358.
Vincent R., Fraisier V., Chillou S., Limami M.A., Deleens E., Phillipson B., Douat C., Boutin J.P., Hirel B. 1997. Overexpression of a soybean gene encoding cytosolic glutamine synthetase in shoots of transgenic Lotus corniculatus L. plants triggers changes in ammonium assimilation and plant development. - Planta 201: 424-433.
von Wirén N., Bergfeld A., Ninnemann O., Frommer W.B. 1997. OsAMT1-1. A high-affinity ammonium transporter from rice (*Oryza sativa* L. cv. Nipponbare). - Plant Mol. Biol. 3: 681.
von Wirén N., Lauter F.-R., Ninnemann O., Gillissen B., Walch-Liu P., Engels C., Jost W., Frommer W.B. 2000. Differential regulation of three functional ammonium transporter genes by nitrogen in root hairs and by light in leaves of tomato. - Plant J. 21: 167-175.
Walker E.L., Coruzzi G.M. 1989. Developmentally regulated expression of the gene family for glutamine synthase in *Pisum sativum*. - Plant Physiol. 91: 702-708.
Wallsgrove R.M., Lea P.J., Miflin B.J. 1979. Distribution of the enzymes of nitrogen assimilation within the pea leaf cell. - Plant Physiol. 63: 232-236.

Wallsgrove R.M., Turner J.C., Hall N.P., Kendall A.C., Bright S.W.J. 1987. Barley mutants lacking chloroplast glutamate synthetase - biochemical and genetic analysis. - Plant Physiol. 83: 155-158.

Wang M.Y., Siddiqi M.Y., Ruth T.J., Glass D.M. 1993. Ammonium uptake by rice roots. II. Kinetics of $^{13}NH_4^+$ influx across the plasmalemma. - Plant Physiol. 103: 1249-1258.

Wang R., Guegler K., LaBrie S.T., Crawford N.M. 2000. Genomic analysis of a nutrient response in *Arabidopsis* reveals diverse expression patterns and novel metabolic and potential regulatory genes that are induced by nitrate. - Plant Cell 12: 1491-1510.

Watson A.T., Cullimore J.V. 1996. Characterization of the expression of the glutamine synthetase *Gln-α* gene of *Phaseolus vulgaris* using promoter-reporter gene fusions in transgenic plants. - Plant Sci. 120: 139-151.

Yamaya T., Matsumoto H. 1985. Influence of NH_4^+ on the oxygen uptake of mitochondria isolated from corn and pea shoots. - Soil Sci. Plant Nutr. 31: 513-520.

Yamaya T., Oaks A. 1987. Synthesis of glutamate by mitochondria: An anapleurotic function of glutamate dehydrogenase. - Physiol. Plant. 70: 749-756.

Yamaya T., Oaks A. 1988. Distribution of two isoforms of glutamine synthetase in bundle sheath and mesophyll cells of corn leaves. - Physiol. Plant. 72: 23-28.

Yamaya T., Oaks A., Matsumoto H. 1984. Characteristics of glutamate dehydrogenase prepared from corn shoots. - Plant Physiol. 76: 1009-1013.

Yamaya T., Oaks A., Rhodes D., Matsumoto H. 1986. Synthesis of (^{15}N)glutamate from (^{15}N)H_4 and (^{15}N)glycine by mitochondria isolated from pea and corn shoots. - Plant Physiol. 81: 754-757.

Yamaya T., Hayakawa T., Tanasawa K., Kamachi K., Mae T., Ojima K. 1992. Tissue distribution of glutamate synthase and glutamine synthetase in rice leaves. Occurrence of NADH-dependent glutamate synthase protein and activity in the unexpanded non-green leaf blades. - Plant Physiol. 100: 1427-1432.

Yamaya T., Tanno H., Hirose N., Watanabe S., Hayakawa T. 1995. A supply of nitrogen causes increase in the level of NADH-dependent glutamate synthase protein and in the activity of the enzyme in roots of rice seedlings. - Plant Cell Physiol. 36: 1197-1204.

Yamaya T., Obara M., Nakajima H., Sasaki S., Hayakawa T., Sato T. 2002. Genetic manipulation and quantitative-trait loci mapping for nitrogen recycling in rice. - J. Exp. Bot. 53: 917-925.

Yoneyama T., Kumazawa K. 1974. A kinetic study of the assimilation of ^{15}N-labelled ammonium in rice seedling roots. - Plant Cell Physiol. 15: 655-661.

Chapter 3

ATMOSPHERIC NITROGEN - POLLUTANT OR FERTILISER?

Lucy J. Sheppard and Håkan Wallander

INTRODUCTION

The nitrogen cycle operates on a pan-continental scale, and the issues concerned with maintaining homeostasis are global (Thomas 1998). Nature functions as a balancing act. Human activities disrupt this balance by uncoupling the availability of N from the users of N, both temporally and spatially. Long-distance transport of N from industrial, agricultural and densely populated conurbations has the potential to transform biological communities in remote areas that have evolved under low N availability, through eutrophication. Locally the increasing intensity of farming activities can pollute forests within close proximity. This chapter is concerned with the effects of human manipulation of the N cycle on forest productivity and vitality, and on the environmental conditions that modify the impact of N.

Understanding the role of N as a pollutant has paralleled the rise in wealth and demand for energy in western societies. Humans, domesticated/farm animals and crop plants of the developed world metabolise protein for energy on a scale comparable with the increased mobility and use of combustion energy (Thomas 1998). Public perception is that N is a fertiliser, and thus N deposition must be beneficial. In many cases the positive or neutral impacts of acid rain on forests were considered to be due to the ameliorating fertiliser effect of N (Nihlgård 1985, Sheppard *et al*. 1993). N deposition, except at very high concentrations/doses, will tend to retain or improve the overall lush "healthy" visual impression. The main threats from N deposition are associated with the combination of its fertilising and acidifying effects (Skeffington and Wilson 1988, van der Eerden *et al*. 1998, Cannell and Thornley 2000, Sverdrup *et al*. 2002). However, the consequences for forest resilience and stability arising from the loss of biodiversity *e.g.* forest floor species (Kellner and Redbo-Torstenssen 1995) in response to eutrophication, which favours faster growing nitrophilic species, are unknown (UNECE 2003). Because understorey species are effective competitors for N in the early stages of forest growth they do influence the internal N cycling of an ecosystem (Buchmann *et al.* 1996).

Today the N threat to forest health and yields in Europe is widely believed to equal or exceed that from acid rain (Aber *et al.* 1989, 1998). Most temperate and boreal forests were historically N limited (Aerts and Chapin 2000, Körner 2003) and are particularly at risk from N eutrophication. The indigenous species are adapted to low resource supply and thus have inherently low growth rates even under conditions of high resource supply (Chapin *et al.* 1990). Low growth rates can compromise the ability to utilise N in high concentrations. This means that there is the potential for overload *i.e.* accumulation and ultimately damage. We have focused on such forests because they are the principal recipients of the N fertiliser effect (UNECE 2003) although, if experimental evidence is to be believed, the enhanced growth is not sustainable. We examine the conditions that determine the direction of the N deposition impact and why growth enhancement may slow down, stop or reverse.

Unsurprisingly, given that the demand for N outstrips that for other nutrients, almost 10-fold for P and 2- to 4-fold for K, trees like most plants have evolved mechanisms (physiological, biochemical, developmental) and life histories in order to optimise N acquisition. N as a fertiliser is a nutrient that underpins autotrophic C fixation. Thus one of the main responses to N is to accelerate ontogeny and ageing and complete the life cycle pre-senescence more quickly.

N acquisition presents a major but necessary C expense for plants. The largest proportion of assimilated C is utilised for N uptake, translocation and assimilation: 25-45% is used for NH_4^+, 20-50% for NO_3^-, and 25-50% for mycorrhizal formation (Chapin *et al.* 1987). N is a constituent of amino acids and thus protein, nucleic acids, enzymes and energy transfer materials. Trees need to take up more N than is needed for dry matter accumulation, to fuel the annual growth flush (Millard and Proe 1993). However, and despite their tendency to grow in low N environments, trees have evolved a highly integrated system for regulating N uptake. In this world of increasing N deposition trees face a challenge: how to link long-term demand to short-term uptake strategies? We shall explore some of these contradictory aspects of N nutrition in an effort to understand how and why N deposition can be both good and bad for forests.

The impact of N on crops has dominated research. Only experience will show how useful an insight they provide into how natural forests respond to N deposition. Crop plants have been bred to respond positively to N supplements and other growth limitations are corrected by human intervention (P, K and micro nutrient fertilisers, irrigation, pathogen and pest control, matching plants or cultivars to climates). By contrast, for most forests the increase in N availability by N deposition is variable, and may cause, indirectly, a decrease in availability of other nutrients. Thus it may or may not stimulate growth. Secondary effects of enhanced growth such as the increased demand for light and water are also less likely to be met by comparison with the agricultural cropping situation. In addition interactions with stresses, *e.g.* frost and winter desiccation may be heightened together with pathogen and pest attacks, which too are unlikely to be mitigated or controlled (NEGTAP 2001). Finally, coniferous forests grow in cold climates where temperature may, but not always (Bergh *et al.* 1999) exert a ceiling on growth and the potential to use/detoxify N. Although, their slow growth rates will increase the time taken to exhaust other nutrients.

NITROGEN DEPOSITION

Sources of N pollution

Most oxidised N is generated through fossil fuel combustion (Dignon and Hameed 1989). The main sources of oxidised N are linked to human transportation. The majority of NO is formed by the combination of N and O from the air within the flames (Palmer and Seery 1973). Oxidised N emissions are dominated by NO and NO_2, highly reactive gases that are subject to transformation to nitric and/or nitrous acid respectively (Levine *et al.* 1991). NO_2 produced in the urban environment also contributes to the N load of rural systems and may be deposited wet or dry. Annual average concentrations of NO_2 in Europe range from less than 0.5 to 2 µg m^{-3} and for NO are below 0.2 µg m^{-3}.

Atmospheric reduced N (NH_y) is predominantly an agricultural by-product, increasing in direct proportion to animal numbers and body mass (Asman *et al.* 1998). NH_y dominates N deposition in Europe and dry deposition of NH_3 dominates the NH_y inputs close to emission sources. Annual average NH_3 concentrations range from 20 to 63 µg m^{-3} down to less than 1 µg m^{-3} (Fangmeier *et al.* 1994, Fowler *et al.* 1998a,b, Krupa 2003). Emissions are linked to farming practices and may vary seasonally and between countries. The fitting of catalytic converters to vehicle exhausts has provided an additional source of NH_y. NH_4^+ ions can travel as aerosols of $(NH_4)_2SO_4$ that have an atmospheric residence time of 4 to 15 days and contribute significantly to the N deposition load to remote ecosystems. NH_3 and NO_x have relatively short half-lives of less than 3 days and for HNO_3 the half-life is less than a few hours (Harrison *et al.* 2000).

Expected N deposition

Anthropogenic N can be deposited wet, as in rain, cloud droplets or aerosols or dry as in gaseous form or solid particles. N is removed from the atmosphere and deposited to the plant/soil surfaces in precipitation through the process of nucleation scavenging whereby aerosols of the pollutants provide condensation nuclei on which cloud droplets form. Forests, especially older ones, tend to be tall and aerodynamically rough, facilitating efficient exchange of heat and momentum with the atmosphere. Old forests may be more vulnerable than young forests, because they receive higher N loads, including the largest deposition rates of gaseous N and N in aerosols and cloud droplets of all vegetation types. Forests effectively filter the atmosphere for N and are therefore vulnerable to excessive uptake from gaseous sources. Forests growing at high altitude sites also see enhanced N deposition from orographic cloud and via occult deposition, where the vegetation is capped by cloud for long periods (Fowler et al. 1998a,b).

N deposition in Europe has changed little over the last 20 years. Locally elevated N deposition levels, however, are now a feature of countries in Western Europe (Asman et al. 1998). Historically N inputs to European forests were only 10-20% of the 1990 values. Deposition loads vary across Europe with the highest values in the Netherlands (ca. 50 kg NH_y-N ha^{-1} yr^{-1}) compared to ca. 1 kg NH_y-N ha^{-1} yr^{-1} in northern Sweden. The average N deposition to forests between 1995 and 1999, measured at 234 plots was 19 kg N ha^{-1} yr^{-1} with the lowest inputs in Scandinavia and the highest (more than 22.4 kg N ha^{-1} yr^{-1}) in Central Europe (UNECE 2003). Wet deposited NH_4^+ and NO_3^- ions, are the major N source experienced by forests except where these occur within 1 or 2 km of localised sources of reduced N. Europe is typified by having a patchy landscape, so that agricultural areas and forests occur in close proximity. This means that some forests in the Netherlands, Germany and Denmark can receive up to 60 kg N ha^{-1} yr^{-1} (Asman et al. 1998). Trees growing downwind within 100 m of intensive livestock units may be exposed to maximum average exposure concentrations of 90 µg m^{-3} NH_3 (Fowler et al. 1998c).

Annual deposition of NO_2 to forests has been estimated between 0.08 and 1.9 kg N ha^{-1} yr^{-1} rising to 12 kg N ha^{-1} yr^{-1} in the vicinity of urban areas (Hanson and Lindberg 1991). The rate of dry deposition of nitric acid to the vegetation is fast, quadruple that of wet acidic deposition (Dollard et al. 1987). However, conifer cuticles provide effective barriers that minimise uptake of HNO_3 (Marshall and Candle 1989).

Cautionary note on quoted N deposition values

We have avoided quoting too many concentration and dose values because experience has shown these to be clouded in uncertainty. Gaseous concentrations mostly represent monthly averages, collected by passive sampling and provide no indication of actual concentrations, ambient conditions, duration and frequency of exposure and length of the potential recovery intervals, all factors that influence plant response. Dry deposition velocities that attempt to quantify the dose are also subject to modification by meteorological conditions in the ambient environment and at the leaf surface. With respect to the annual wet dose, wet deposition is often quoted as the N concentration in throughfall but for conifers this can be greatly modified by dry deposition and reactions on or with the leaf. Lastly, it is not always clear how authors deal with ambient deposition: is it based on actual measurements or is it derived from model studies? For more discussion see Sutton *et al.* (2003).

Critical Loads

Irrespective of the eutrophication and toxicity (see below) associated with N deposition, the contribution of N to acidification in Western Europe is increasing. In 1990 it was 53% and by 1999 it had risen to 72% (Tarrason and Hjellbrekke 2001). Fortunately this increase mostly reflects a drop in the contribution from S based pollutants in response to effective abatement strategies. However, despite projected falls in N emissions by 2010, recent estimates show little change (NEGTAP 2001). NO_x levels peaked in Europe *ca.* 1990 but the reduced N budget (NH_y) is proving difficult to describe, with the variations in farming practice contributing to large uncertainties (NEGTAP 2001). Cheaper options for N abatement have been exhausted and the more expensive ones may require government legislation to implement.

In order to protect European forests from the polluting effects arising from N deposition Critical Loads (CL) have been set. A CL represents a quantitative estimate of exposure to a pollutant below which significant harmful effects on specified sensitive components of the ecosystem do not occur according to present knowledge (UNECE 2003). CLs are not exceeded as long as an ecosystem remains in ecological equilibrium. With respect to acidification by N however, the level of N accumulation has reached levels that indicate CLs have been exceeded for a large proportion of the 6000 monitoring sites laid out in a systematic 16x16 km grid over Europe. CLs aimed at preventing further N accumulation are exceeded in 92% of monitoring plots and represent an annual deposition of 8 kg N ha^{-1} yr^{-1}. CLs for acidity are calculated from steady state soil models based on a mass balance approach, though appreciation of the impact of acidity on forest vitality, an empirical approach, is now under consideration (Tech Report UNECE 2003). At the recent Empirical Critical Loads Expert Workshop in Berne 2002 a CL of 10-15 kg N ha^{-1} yr^{-1} was set to protect coniferous forests in Europe (UNECE 2003).

TREE GROWTH RESPONSES TO NITROGEN

Assessing the impacts of N deposition on forests has drawn on long-term fertiliser studies (Sweden particularly), gradients of N deposition, N manipulations, including removal of N and latterly measurement of deposition in relation to forest vitality indicators. Oden (1968) suggested that while N inputs should enhance forest growth in northern Europe, effects might be counteracted by the acidifying impact of these compounds. Tamm and co-workers put this hypothesis to the test. In the short-term, acidification can increase both base cation and N availability, possibly as a result of the depressed activities of microorganisms in response to increasing acidity. Effects of N on tree growth were generally neutral or positive (Tamm and Popovic 1995). Height growth is sensitive to K availability, so if height or volume increments are the measure of growth they are less likely to register significant effects compared to basal area, which is more coupled to N (Tamm and Popovic 1995, Sheppard *et al.* 2000, 2001). Values for Norway spruce reported in Germany between 1958 and 1987 showed an increase in volume increments in response to 100 kg N ha^{-1} yr^{-1} of 2-22% and even higher for basal area increments (Kenk and Fischer 1988).

N saturation experiments (NITREX, Wright and Tietema 1995) and experimental manipulation studies (EXMAN, Kreutzer *et al.* 1998) throughout northern and central Europe have provided limited evidence that simulated N deposition can drive increases in growth. The magnitude and nature of the effects of N deposition were shown to be strongly dependent on site type (soil chemistry and climate) and the developmental stage of the stand. In Nordic countries responses to N have been mixed depending on the availabilities of P, K and micronutrients. An 11-year old project for forest research in these countries has concluded that forest ecosystems show a relatively slow response and high resilience to elevated N inputs (Sigurgeirsson 2003). Most of the N was retained by the soil so that inputs of 50-55 kg N ha^{-1} yr^{-1} did not significantly increase growth though positive effects could be seen at one of the two sites.

Emmett (1999) summarised tree growth responses to known inputs. One important observation was that inputs below 60 kg N ha^{-1} yr^{-1} took several years before changes in growth were recorded and given the short-term nature of experimental programmes often failed to register significant growth effects. In southern Germany average deposition of 25 kg N ha^{-1} yr^{-1} appears to have almost eliminated N deficiency and increased yields (Kenk and Fischer 1988). Binkley and Högberg (1997) concluded, from a review of Swedish literature, that N deposition had not caused any negative effects on forest productivity, or vitality, and that there was no evidence of damage from N induced acidification. Spiecker *et al.* (1996) reported an overall improvement in forest growth rates throughout Europe based on a comprehensive survey. More recently UNECE (2003) published data from their comprehensive monitoring network (International Cooperative Programme, ICP) showing significant correlations between N deposition and forest growth. Atmospheric inputs when assessed with meteorological data and soil and foliar chemistry strongly suggest that N deposition has fuelled this increased growth. However, other growth enhancing factors have also changed over the same time period, namely the increase in temperature and CO_2 emissions (Briffa *et al.* 1998). Thus, it is almost impossible to state categorically that N deposition alone has caused the growth enhancements seen over the latter half of the 20th century.

Limitations and lag time

N driven growth enhancement may not be sustainable because other nutrients or resources required for growth are not enhanced in relation to N. Indeed P, and base cations are often reduced as result of these inputs (see later). Increment cores from more than 31,000 Norway spruce are smaller now than those recorded 50 years ago in Norway, except where modelled inputs indicate N deposition is less than 7 kg N ha^{-1} yr^{-1} (Nellmann and Thomsen 2001). At forest sites with a recent history of high N deposition, cf. forests in the Netherlands and Denmark, removal of N from the precipitation, using roofs, increased growth (Beier 1998). However, predicting the amount of N that forests can absorb, and process without detrimental effects is providing quite a challenge (Näsholm 1998).

In contrast to pot experiments, conducted in controlled environments where temperature and water are generally more optimal than in the forest, and pests and diseases are kept to a minimum, responses to N inputs under field conditions are slow (Tamm and Popovic 1995). The short 3-5 year experimental period, representing less than 5% of the life span of most forests, may be insufficient to realise a significant growth response. Tamm and Popovic (1995) concluded that while field manipulations afford the opportunity to evaluate N deposition effects they must be run for many years before meaningful results can be derived. One of the main findings from the NITREX experiments was the small response of diameter increment to N inputs (Gunderson *et al.* 1998). Aber *et al.* (1999) did not measure growth increases for at least 5 years. The reasons for this are many, but one fundamental difference between pots and the field is the ratio of plant to soil so that in the field the plant is competing with the complexity of the soil, its chemistry and biota. In addition, the biomass in seedling studies is relatively small but deposition is expressed on a unit area basis, irrespective of the size and thus dilution capacity of the receiver. Sheppard *et al.* (1997) concluded this was an overlooked factor when making extrapolations from pots to the field.

At the soil level additional N must satisfy the "immobilisation" capacity before N concentrations in soil solution can increase (Aber 1992). A cascade effect following on from increased soil water N leading to higher foliar N, which in turn stimulates growth, describes the N effect. Gunderson *et al.* (1998) concluded that response times were strongly related to soil N pool sizes and that soil solution N, which reflects the dissolved and adsorbed inorganic N pool, responded within one year whereas needles and litter could take 4 years and changes in the forest floor C/N ratio might take more than 5 years. A recent summary from the European monitoring sites confirms a strong positive correlation between N deposition and foliar N for Scots pine that has now fed through to yield. The relationship between yield and N deposition is strongly temperature and stand age dependent. In Sweden Sverdrup *et al.* (2002) showed a strong north south axis between the amount of N used and temperature.

EFFECTS ON SOILS AND N UPTAKE BELOWGROUND

Soil acidification

N deposition can acidify the soil directly through inputs of HNO_3 in rain, or as wash off of dry deposited NO_2, which can be subsequently oxidised on plant surfaces. Inputs of H^+ ions associated with oxides of N have fallen by 30-50% over the last 20 years in line with improvements in fossil combustion (NEGTAP 2001). However, acidification can also occur via base cation leaching following cation displacement in response to inputs of H^+ and NH_4^+. Soil acidity reflects the relative proportions of base cations (Ca^{2+}, Mg^{2+}, K^+ and Na^+) to H^+ and Al species occupying the soil exchange complex. Most forest soils supporting conifers tend to be acidic. If they are organic their exchange sites are dominated by H^+ ions whereas mineral soils tend to be dominated by Al species (Al^{3+}, Al $(OH)^{2+}$ and Al $(OH)_2^+$).

N deposition in rain or throughfall (precipitation that has interacted with the canopy and is modified by bi-directional exchange) generally contains both NO_3^- and NH_4^+ in varying proportions depending on the source of the air. When NH_4^+ ions adsorb to exchange sites they displace base cations, which are then carried down the profile. Acidification occurs when exchangeable base cations are removed faster than they can be replaced by weathering (Sverdrup and Warfvinge 1995, Jönsson *et al*. 2003). In Wales, N transformations contribute 1 keq ha^{-1} yr^{-1} acidity compared to the 0.37 keq ha^{-1} yr^{-1} generated from SO_4^{2-} (Reynolds 1997). The significance of N deposition for base cation cycling in forests should not be underestimated since NO_3^- and NH_4^+ inputs affect both the sink for base cations, tree growth, their source, cation exchange sites and their availability (mobile anion effect, see later). Too much N will increase the sink size while also reducing the source. Removal of base cations occurs when the incoming N stimulates tree growth and concomitantly the demand for, and uptake of, base cations. Thus whether N acts as a fertiliser or pollutant will largely be determined by interactions with base cations and pH regulation both inside and outside the tree.

In organic soils uptake by the forest is met through litterfall recycling and precipitation. The atmosphere can also supply base cations. For forests growing close to the sea, deposition can be large, so that deficiencies are less likely than they would be in forests growing in continental Europe (Kreutzer *et al*. 1998). Large inputs of a single growth-limiting nutrient must eventually perturb the system because of the demand made on other nutrients and the law of the limiting (Liebig 1840) and more recently demonstrated for conifers by Ingestad (1979). In natural systems the N cycle is closed and nutrients cycle in balance with each other, with minimal losses via denitrification or NO_3^- leaching (Garner 1984). N deposition perturbs this balance but the extent and implications will depend on soil type *i.e.* geology and weathering rates.

N-stimulated growth is normally matched by the production of additional litter (Berg and McClaugherty 2003). This can increase the cation exchange capacity of the system, so that it exceeds the amount of base cations returned in the additional litter. This is because organic matter consists largely of weak acids, which provide cation exchange sites (Berg and Matzner 1997). Younger forests, with a high nutrient demand, may be particularly sensitive to increases in cation exchange capacity, depending on the rate of organic matter breakdown. Increasing amounts of N in the litter have different effects on breakdown processes dependent on the decomposition stage (Berg and Matzner 1997, Schulze et al. 2000). Effects on the decomposers appear mixed especially for encytraeids, minuscule pale worms that physically and chemically break down the size and composition of the litter (Standen 1982). Boxman et al. (1998) concluded that the microfaunal decomposers were generally not very sensitive to N deposition. Further stages of breakdown involving saprophytic fungi, ectomycorrhiza and other microorganisms are initially stimulated in response to the decline in C/N ratio but thereafter breakdown is retarded (Berg and Staaf 1981). This is because higher available N represses the formation of lignolytic enzymes by white-rotting fungi and lignin retards the rate of decomposition (Berg and McClaugherty 2003). N can also be incorporated into lignin during the humification process to form recalcitrant compounds that further retard decomposition. At very high N loads disintegration of humus may occur. This is probably the effect of the replacement of white rotting fungi by soft rotting fungi which tolerate high N levels resulting in increased production of soluble organic matter (Berg and McClaugherty 2003).

Nitrification

The most important consequence of N deposition for forest soil acidification occurs via the mobile anion effect associated with NO_3^-. Nitrate ions are negatively charged and as such are highly mobile in the soil, being "repelled" from exchange sites. Unless they are removed by uptake or chemical transformation these mobile anions migrate through the soil profile taking base cations with them. Depending on soil pH, nitrification can also generate significant amounts of H^+ so much so that the process can become self-limiting. Nitrification is temperature and moisture dependent, which is important in its own right and via its effect on oxygen availability. In the cold acid soils with a pH lower than 4 where many conifers grow only a varying proportion of the deposited NH_4^+ may be nitrified to NO_3^-. Thus in acid soils, pH < 4, negative impacts of N are more likely to arise from the effects of NH_4^+ than via the impact of nitrification.

Nitrification occurs mainly via the activities of autotrophic nitrifying bacteria that can fix CO_2 (dark fixation) or heterotrophic bacteria that rely on external C sources (Sverdrup et al. 2002). Nitrification rates also depend on the NH_4^+

concentration, which reflects tree demand for N and the ease of uptake. In forests with high N input the tree N demand will decline, as will the amount of ectomycorrhizal mycelia (Nilson and Wallander 2003), which results in less NH_4^+ uptake into trees and more for nitrifiers. The capacity of the microbial pool to immobilise NH_4^+ (incorporate into microbial tissue) may also decline in soils saturated in N since soil microbes are usually C limited (Magill et al. 2000). A decline in their activities will slow down nitrification and mineralisation. Thus the N cycling processes become self-limiting, homeostasis breaks down and the forests start to leak N either into stream water or in gaseous form (NO. N_2O) via denitrification (Gunderson et al. 1998, Tietema 1998, Skiba et al. 1999).

Changes in soil N pools - fertilisation effect

The main sink for N deposition is the soil. In an N deposition simulation Buchmann et al. (1996) found that 87% of reduced N and 79% of oxidised N deposited, remained in the soil. The availability of the deposited N and its potential to immediately fertilize the forest is mediated through the chemical, physical and biological environment, a product of geology, climate and landuse history. These factors influence the speed at which N becomes available via 1) effects on the microbial population involved in N transformations (Skiba et al. 1999), 2) influence over the soil N reserves (Emmett et al. 1998), and 3) previous crops and management may have removed a large proportion of the base cations (Spiecker 1999). Thus the fate of N deposited to soil cannot be predicted on the basis of inputs alone.

N inputs to soil under forests are significantly modified by interaction with the canopy (Cape et al. 2001). Rainfall N content can be significantly enhanced from the removal of dry deposited N from the canopy surface. Inorganic N ions may be absorbed via foliage and bark or leaf microflora and transformations between organic and inorganic N occur on the leaf surface (Chiwa et al. 2003). These biological transformations vary from day to day, so that the chemistry of throughfall and stemflow may be very different from bulk precipitation. The potential for these N inputs to fertilise the forest will depend on the strength of the relative sinks for N, the forest itself and competing organisms as leaf and soil microorganisms and ground-layer vegetation (Aber et al. 1998). The strength of trees to act as sink is a function of growth potential (light, water, temperature, length of growing season and the availability of other macro and micro nutrients). Young forests provide the largest sink strength and thus repository for N, having the smallest internal reserves and largest demands for N and other nutrients in order to realise their growth potential.

The fate of NH_4^+ inputs is largely determined by soil pH and organic matter content, both organic soils and organic horizons - the O horizon or forest floor have large capacities to take up NH_4^+ and effectively immobilise it. At Aber (N. Wales), one of the NITREX experiments, 40-50% of the NH_4^+ was retained in the organic layer (Gunderson *et al.* 1998). In Sweden it is estimated that more than 95% of incoming N is immobilised (Sverdrup *et al.* 2002). Immobilisation prevents loss from the system through leaching but does not ensure its availability as a nutrient. Deposition and availability are often temporarily uncoupled and during this time the N form may be transformed. The death of microorganisms will release mineral N, often in flushes. Cell lysis can occur in response to rapid changes in temperature (winter freezing) or moisture (summer droughts). A large proportion of NH_4^+ is utilised to support microbial growth and activities. In this instance NH_4^+ can increase the capacity of the system for N mineralisation. However, too much NH_4^+ can reduce mineralisation via toxic effects on soil microbes. The "binding" of N by interaction with lignin can also increase the potential for soils to immobilise N without immediate consequences for living systems (Sverdrup *et al.* 2002). The time taken for N deposition to significantly enhance soil water N concentrations can be rapid (Gunderson *et al.* 1998) or extend into years depending on the capacity of the soil to absorb N.

Our understanding of N immobilisation is incomplete particularly with respect to the role of fungi (see below) and the significance of C availability (Aber *et al.* 1998). Suffice to say in many of the N manipulation and fertiliser studies the soil has, and continues, to provide the largest sink for N, buffering forests against the effects except where the aerial uptake pathway is significant.

Ectomycorrhiza

In north temperate forests tree roots are usually colonised by ectomycorrhizal (EM) fungi, which supplement nutrient uptake. Globally this relationship evolved when organic matter began to accumulate in certain soils 200 million years ago, in the Ordovician period and increased toward current levels globally in the late Devonian/Carboniferous, (John Raven, personal communication). The tree host benefits through improved nutrient acquisition, since the fungus explores the soil efficiently for nutrients (especially N and P) in return for host C (Smith and Read 1997).

EM mycelia can contribute up to 33% of soil microbial biomass (Högberg and Högberg 2002). There are more than 6,000 known EM species (Smith and Read 1997) with a range of growth strategies in the soil. The mycelia of some types extend far out from the root tips (long distance exploration types) while other types extend very small distances (contact exploration types) both, increasing the absorptive surface (Agerer 2001). Root systems of adult trees may be associated with numerous EM fungi possessing a range of abilities to use different N forms (Eltrop and Marschner 1996), providing the host with the capacity to capture many froms of N and maximising the potential for N uptake (Martin and Lorillou 1997). The distance between individual EM hyphae in forest soil may be as little as 20 μm (Olsson *et al.* 2002). EM fungi have a large capacity for sequestering N in their extensive external mycelia. Nitrogen can also be stored in the fungal mantles that enclose the mycorrhizal roots in osmiophilic vacuolar bodies, whose occurrence increases after N additions (Wallenda and Kottke 1998). These N stores can be mobilised rapidly and utilised by the plant under conditions of high demand, *e.g.* during bud burst in the spring (Wallander *et al.* 1997).

Most stored fungal N is allocated to the host plant (Smith and Read 1997) but a significant proportion is retained in structural components such as chitin, where the C/N ratio exceeds that in protein, in the EM cell wall. Glutamate is the preferred transport medium but sometimes NH_4^+ ions are passed to the root (Finlay *et al.* 1988). France and Reid (1983) suggested that plants exercise a considerable influence over the N assimilating enzymes of their fungal partners. The different steps in the N assimilation pathways are spatially compartmented in the different tissues making up the EM, but a description of regulation is some way off (Martin and Lorillou 1997). Down-regulation of amino acid uptake in response to the host's improved N status and by high endogenous NH_4^+ levels has been seen, with up-regulation of uptake in response to increased carbohydrates (Persson and Näsholm 2002). Uptake of N via mycorrhizal roots appears to be less acidifying for the rhizosphere than uptake by non-mycorrhizal roots (Martin and Lorillou 1997). This may follow because 1) different N forms are accessed by EM *cf.* tree roots, and/or 2) the same N forms are used but the symbiosis generates less H^+ during growth and fewer organic anion salts in the end product than does the tree alone (John Raven, personal communication). Little is known about the role of EM in the plant acid base balance but given that acidification is already an issue with N, symbiosis with EMs may have "hidden" benefits for the tree.

Changes in EM communities and mycelia

N additions both in the form of N fertilisation and N deposition influence the composition of EM communities. Several studies have shown a reduction in sporocarp (fruit body) formation in response to N addition, both as N deposition (Termorshuizen and Schaffers 1991, Lilleskov *et al.* 2001) and in N fertilised plots (Wiklund *et al.* 1994, Brandrud 1995). The magnitude of the effect is greatest for sporocarp numbers, in particular for specialist fungi *e.g. Tricholoma, Cortinarius* and *Suillus,* which decline. Inputs of $(NH_4)_2SO_4$ appear to be more detrimental than NH_4NO_3 (Brandrud 1995). Brandrud and Timmermann (1998) postulated three phases of N enrichment effects on mycorrhizal flora: 1) rapid decrease in diversity and production of sporocarps of most species except *Lactarius* spp., 2) loss of sensitive species-rich key groups *e.g. Cortinarius* and the retreat of more stress tolerant groups *e.g. L. rufus,* and 3) near complete disappearance of sporocarp production with decreased density of mycorrhiza and fine roots. Morphotype diversity is restricted with loss of thick mantled types formed by sporophore producing macro-fungi. Arnebrant and Söderström (1994) showed that changes in EM communities after N fertilisation could persist for more than 13 years.

Investigation of EM fungi belowground is complicated because extramatrical mycelia of EM fungi are difficult to distinguish from those of saprotrophs. Boxman et al. (1998) observed from the NITREX experiments on more than 35-year old trees that N reduced mantle thickness and the amount of external mycelium although inter-annual differences confounded effects. In laboratory experiments it has been shown that the growth of EM mycelia is severely inhibited by N additions (Wallander and Nylund 1992) with differences in species sensitivity (Arnebrant 1994). Most EM species that have been used in laboratory experiments are rather tolerant of N addition and this may have led to underestimates of the negative influence of N on growth of external EM mycelia that occur in the field.

New methods for studying mycelia involving their molecular identification (Dickie et al. 2002, Landeweert et al. 2003) and the use of ingrowth mesh bags to quantify EM mycelia (Wallander et al. 2001) have greatly added to our understanding of belowground effects. Using the mesh bag method, Nilsson and Wallander (2003) showed that growth of external mycelia decreased by ca. 50% with the addition of 100 kg N ha^{-1} yr^{-1} for 11 years in the Skogaby experimental forest in southern Sweden.

Effects of N additions on the EM communities on root tips have been studied in N fertilisation experiments and along N deposition gradients in the field (Kårén and Nylund 1997, Wallenda and Kottke 1998, Fransson et al. 2000, Jonsson et al. 2000, Taylor et al. 2000, Peter et al. 2001, Erland and Taylor 2002). The general conclusion from such studies is that almost all root tips remain colonised by mycorrhizal fungi after N additions but the species composition is usually altered and morphotype diversity is reduced (Taylor et al. 2000). Belowground communities are less N sensitive than is the production of above-ground sporocarps (Peter et al. 2001).

The consequence of such changes in species composition for nutrient cycling and tree growth is difficult to predict because little is known about nutrient uptake capabilities in different species. EM species with a high capacity to utilise complex organic N sources have decreased in abundance in ecosystems exposed to N additions along an N gradient in Europe (Taylor et al. 2000) and along an N deposition gradient in Alaska (Lilleskov et al. 2002). This type of adaptation probably occurs because the production of N mobilising enzymes uses assimilate and would be of less benefit when inorganic N sources increase in the soil.

High N inputs to forest ecosystems can lead to a reduction in the amount of C allocated to the roots (Vogt et al. 1982) and thus to reduced carbohydrate concentrations in the roots, which may have a negative effect on the growth of EM fungi (Wallenda et al. 1996). Increased N supply causes a switch from gluconeogenesis to glycolysis in the host and a reduction in the C allocation to

EM, which leads to a fall in the amount of living hyphae via starvation (Hampp et al. 1999). In addition, Wallander (1995) suggested that increasing N availability resulted in a larger proportion of available carbohydrate in the fungal tissue being used for NH_4^+ assimilation rather than growth of the external mycelium or sporocarp production. Assimilation of NH_4^+ into amino acids consumes large amounts of C (Martin and Canet 1986), the majority of which is returned to the host as amino acids. High C consumption during uptake of NH_4^+ was also demonstrated as increased respiration in microcosms (Ek 1997, Bidartondo et al. 2001) when it was shown that the N tolerant *Paxillus involutus* had a much larger capacity to obtain C from the host tree than N sensitive *Rhizopogon* and *Suillus* species. Growth of the EM mycelia appears to respond to host C allocation rather than soil N status.

Consequences of changes in EM communities and mycelia

The reduced capability of EM infected root tips to utilise organic N sources commensurate with N additions may be dramatic. Lilleskov et al. (2002) calculated that the capability was 70% with low N deposition falling to 7% at the highest N deposition along an N deposition gradient in Alaska. The impact however, may not be instantaneous for nutrient dynamics and tree growth because the increasing availabilities of inorganic N will reduce the benefits of using organic N sources by the trees. But, if the capability to utilise complex organic N sources is not re-established and N deposition declines it may have serious consequences for the N cycling of the forest in the longer term (Näsholm et al. 2000). Preferential use of inorganic N sources consumes more C which may be detrimental for the growth of EM mycelia and consequently for tree growth.

Reduced growth of the EM mycelia can reduce uptake of elements other than N that could be in low supply. Since it is likely that other nutrients will supersede N as the limiting element in forest ecosystems exposed to high N input this will accelerate the speed at which these other nutrients start to restrict the N-driven growth response. Hagerberg et al. (2003) showed that EM fungi could compensate for a P deficiency situation that developed in a Norway spruce stand under high N availability. The EM fungi intensified the colonisation of apatite (P containing mineral) added in mesh bags, which in turn led to dissolution of the mineral and a higher availability of P. Colonisation of deeper soil layers with higher amounts of weatherable minerals by mycorrhizal roots may be one way to compensate nutrient deficiencies that develop in N-saturated stands (Landeweert et al. 2001).

Non-mycorrhizal roots

Although N uptake by EM exceeds that by roots, most trees also have non-mycorrhizal roots that may contribute to N uptake (Marschner 1991, Eltrop and

Marschner 1996). Effects of N deposition have generally started as positive changing to negative only when inputs exceed demand and soil water N concentrations increase (Boxman et al. 1998). A threshold of more than 2 mg N L^{-1} (NO_3^- and NH_4^+) has been identified (Matzner and Murach 1995). Reduction in N inputs to Scots pine, which caused a fall in soil water N, saw a recovery in root growth (Persson and Ahlstrom 2002). Loss of EMs from roots will increase the risk from heavy metal and Al^{3+} toxicity via adsorption to the fungal cell (Jentsche and Godbold 2000) and such ions will be increased by N inputs (see below). Negative effects on root growth will have serious implications for drought tolerance, since N additions also increase the aboveground transpiring surface.

NITROGEN UPTAKE ABOVEGROUND

Wet deposition

N deposition offers trees many routes for N uptake. Wet N (NH_4^+ more than NO_3^- ions) can be absorbed through the aerial tree parts as well as the roots. In the 80's Abrahamsen (1984) observed that Norwegian forests supplemented their poor N supply from the atmosphere. Canopy uptake via foliage is well documented for spruce trees (Skiba et al. 1986, Lovett and Lindberg 1993, Macklon et al. 1996). Bark (Katz et al. 1989) is estimated to have a 5-fold higher uptake potential than needles possibly reflecting the higher contact time (Boyce and McCune 1992), though it represents a smaller overall surface area. Lovett and Lindberg (1993) estimated that Spruce-fir forests in the US satisfy less than 15%, approximately 1-12 kg N ha^{-1} yr^{-1} of their N requirement from canopy uptake. Boyce et al. (1996) using ^{15}N-labelled NH_4NO_3 estimated 2-8% of tree N demand was met via the foliage, which took up 5 times more NH_4^+ than NO_3^- ions. Chiwa et al. (2003) showed that up to 40% of NH_4NO_3 at 48 kg N ha$^-$ yr^{-1} sprayed 50 times at 2mm precipitation equivalent, over a 10-m high Sitka spruce canopy at Deepsyke in the Scottish Borders could be retained during the growing season. The fate of this N was not followed using ^{15}N and some may have been utilised by the phylloplane microflora. Three years of throughfall measurements under this spruce canopy have also shown that canopy uptake varies unpredictably and significantly from year to year ranging from 15 to 40% of the added N.

Uptake of NH_4^+ via the foliage and bark appears to respond to diffusion gradients (Wilson 1992) but in addition electrical gradients are also active. These, and the "negativity of the cuticle", demand that uptake of positive (+ve) ions be accompanied by the equivalent charge loss. They also explain how ions of similar charge compete with each other for uptake sites. At Deepsyke the presence of H^+ ions modified the proportion of NH_4^+ to NO_3^- that was absorbed from this equimolar solution favouring NO_3^- uptake. Leaching of base cations from cuticular exchange sites may offer the potential for recycling by root uptake from the soil. The efficiency of this mechanism compared with internal recycling is not known.

Ionic strength does not appear to influence uptake (Macklon *et al.* 1996). This may be due to fact that as liquid evaporates from the leaf surface the ionic strength increases. N concentrations measured in throughfall suggest that uptake of N by the canopy decreases with age (unless dry deposition increases with age). If this is the case it suggests foliar uptake is regulated and related to growth. In the study by Cape *et al.* (2001) N uptake was not significantly enhanced when the trees were supplied twice as frequently. The absence of growth enhancement appears to suggest that uptake through foliage/bark may be linked to growth via N assimilation. Young trees, having a high requirement for N would be expected to assimilate the N more quickly and thus increase the potential for diffusion. Uptake rates are tree species specific (van der Eerden *et al.* 1992) and variable.

Uptake is dictated by the charge on the ion and the cuticular material and most importantly the ionic strength of the apoplastic free space. Klemm *et al.* (1989) concluded that the N concentration gradient from outside the leaf to the interior favours uptake. Transport across the cuticle is the rate-limiting step for NH_4^+ (Macklon and Armstrong 1994). It appears that trees can influence foliar uptake of N via the ionic strength of the apoplastic space and coupling to N assimilation. However, while N uptake may be linked to the availability of reductant and C skeletons these alone are insufficient to promote growth and thus utilisation of N.

Dry deposition

The impact of gas phase NH_3 depends on the physical and chemical properties of the leaf surface that influence the rate of diffusion through the stomata. Thus uptake is influenced by stomatal density, environmental factors that control opening, outside gas concentration and internal gas concentration. The internal concentration will reflect pollutant solubility and decomposition rate (potential for interaction with the cell wall enzymes) and the rate of removal *i.e.* N assimilation. The physical properties of the leaf may be auto correlated thus, high conductance is incompatible with a low stomatal density, and transpiration rates are coupled to the radiation regime and vapour pressure deficits. Trees cannot prevent gaseous uptake *per se* (Gebauer and Schulze 1997). Assimilation rate is probably the most important step, regulating uptake. Thus young foliage with its high potential for assimilation exhibits the greatest uptake potential (Bruckner-Schatt *et al.* 1995). Tree N status also affects uptake, which is highest in N poor trees, although this must be balanced against their poor growth, assimilation rates.

NH$_3$ can be both absorbed and emitted by foliage, depending on the concentration gradient between the leaf interior and ambient air and the pH of the substomatal fluid. NH$_3$ gas shows alkaline tendencies in a protogenic solvent but its dissolution effect on pH is counterbalanced by H$^+$ ions extruded from cytoplasmic assimilation (Husted and Schjoerring 1995). The gas phase NH$_3$ concentration is in equilibrium with the NH$_4^+$ concentration in the substomatal apoplastic fluid, which forms a physical, albeit not chemical, continuum with the liquid phase of the leaf apoplast and the xylem. NH$_3$ will be deposited as long as this internal equilibrium concentration, known as the compensation point, which is associated with a particular partial pressure, is below that of ambient air. The partial pressure controls the solubility of NH$_3$ in the apoplast in accordance with Henry's law.

The size of the apoplastic NH$_y$ pool is regulated via the balance between uptake and metabolism. Deposition is the highest in summer when assimilation keeps reducing the internal N concentration. Trees growing in N enriched soil will have higher compensation points and potentially absorb more NH$_3$ (Gessler *et al.* 2002). However, these trees are also more likely to emit NH$_3$ when the atmospheric concentrations fall, such as with a change in wind direction from a local NH$_3$ source. The ability to emit NH$_3$ provides forests with a mechanism for restricting N accumulation, were this not the case uptake via stomata could be difficult to control (Gessler *et al.* 2002). Re-emission prevents toxic accumulations of NH$_3$/NH$_4^+$ in the needle apoplast and can increase in response to root N uptake (Husted and Schjoerring 1995). Detoxification of NH$_3$ has been reviewed by Givan (1979). When the humidity is high NH$_3$ can also be deposited to water films, so long as the pH is acidic, on the leaf surface and access the sub stomatal cavity by this route, which is largely independent of stomatal opening (Krupa 2003).

It has been suggested that uptake of gaseous NH$_3$ occurs more readily than that of other N forms that use the combined cuticular and root uptake pathways. Pérez-Soba and van der Eerden (1993) showed that Scots pine seedlings exposed to 50 µg m^{-3} NH$_3$ or fertilised with 65 kg (NH$_4$)$_2$SO$_4$ had a 40% higher needle N concentration with NH$_3$ compared to 8%, with root applied fertiliser. Given the potential uptake pathways of stomata, water films, cuticle, bark and roots, trees receiving their N as NH$_3$ should respond to N deposition more quickly than those supplied with wet NH$_y$. Trees that rely on belowground uptake will derive their N more slowly because they have to compete with the powerful soil/microorganism sink. This means that effects of dry deposition of N should be observed more quickly than if the same amount of N had been deposited wet.

N DEPOSITION AND PHYSIOLOGICAL PROCESSES

N uptake across the plasmalemma

Regulation of N uptake across the plasmalemma requires information on N status of individual organs to be conveyed bi-directionally so that physically separate functions *e.g.* photosynthesis and mineral uptake can be co-ordinated and integrated. Mechanisms have evolved that enable uptake rates to be linked to "demand" which changes with growth and developmental stage, rather than to the N concentration in the growth medium (chapter 1). N uptake kinetics display seasonality linked to the temperature dependence of enzyme activities. In addition trees can alter the amount and the morphology of the root system, and the extent and type of EM association via changes in C partitioning. Trees also need to be able to integrate and coordinate N acquisition from several sources *i.e.* roots and shoots. However, short-term NO_2 fumigation experiments (Muller *et al.* 1996) suggest the internal control operating in response to demand is not perfect, *i.e.* does not operate beyond certain limits and probably would not offer protection from an "overload" situation.

NO_3^- uptake represents a thermodynamic challenge; being negatively charged it has to move against an electrical potential gradient (Crawford 1995, chapter 1). Uptake is active and coupled to metabolism. Most northern conifers growing in acid soils will have to process less than 10% (approximately) of their N in the oxidised state (Stewart *et al.* 1989). NH_4^+ moves in response to the negative electrical gradient generated from the hydrolysis of ATP by the plasmalemma H^+ ATPase pump (Näsholm 1991, chapter 2). In Scots pine amino acid uptake was down regulated by NH_4^+ and up regulated by the presence of amino acids, suggesting that the organic N form is the preferred because it consumes the least energy in assimilation (Persson and Näsholm 2002). N is also taken up as amino acids. There is an excellent review of N acquisition and assimilation in EM by Martin and Lorillou (1997). However, the physiological processes involved in the transfer of N across the fungal plasmalemma or the interfacial matrix and the root plasmalemma, are still poorly understood as is their regulation (Chalot *et al.* 2002). Chalot and Brun (1998) suggested that amino acids may move from the EM to the host by diffusion, but such a route in the absence of membrane channels would be slow.

Molecular studies, predominantly undertaken with crop plants, have increased our understanding of N uptake and regulation. In such crops NH_4^+ and NO_3^- uptake by roots appears to be carefully controlled by several high affinity transporters (HATS) (chapters 1 and 2), one or two of which have been demonstrated in a few forest trees (Kronzucker *et al.* 1995, 1997, Suárez *et al.* 2002). Glutamine is an important signalling factor, which for crops, though still to be irrefutably demonstrated in trees, can serve to down- regulate N uptake via effects on genes encoding the respective HATS (chapters 1 and 2). HATS for

NO_3^- operate over a wide concentration range from μM to mM, compared to those for NH_4^+. The ability to operate over a wide range provides flexibility to up or down regulate the flux, in relation to demand and the concentration in the soil solution. Communication of foliar N status to the roots is well demonstrated (chapter 1). In Norway spruce one mole of NH_3 uptake by foliage led to the reduction of root NO_3^- uptake by one mole (Rennenberg *et al.* 1998). Root to shoot signalling by long distance transport of amino acids is now starting to gain credibility as mechanism for regulating the N nutrition of mature trees (Weber *et al.* 1998).

N functions and demand

The photosynthetic capacity of leaves is related to their N content mostly because the proteins of the Calvin cycle and thylakoids represent the greatest fraction of the leaf N content. Amounts of chlorophyll and the major enzyme of photosynthesis (RuBP carboxylase) are proportional to the leaf N content (Evans 1989). Because N is a component of amino acids and protein it is both structurally and metabolically important (Marschner 1995). In addition N can act as a signalling agent, *e.g.* for stomatal opening (Raven 2003). Signals derived from NO_3^- are involved in triggering widespread changes in gene expression that affect the intimate programming of C and N metabolism. Internal and external NO_3^- concentrations signal adjustments in root growth and architecture to the physiological state of the plant in relation to external N sources (Zhang *et al.* 1999, chapter 1) Excess or insufficiency will compromise any or all of the functions (see above). N demand, which tends to reflect the potential for assimilation/detoxification, is determined by species, age and environmental conditions that affect growth (light > temperature > water availability). Thus demand varies both seasonally and diurnally. At night or in winter N assimilation, which prevents accumulation and the potential for toxicity, will be slower although stomatal closure will restrict gaseous uptake.

Nitrogen assimilation

The N and C cycles are intimately linked so that pollutant N effects on N cycling will feed back to C assimilation and use (Lawlor 1994). NH_3 absorbed via the stomata or trans-cuticular route is assimilated in the foliage and corresponds to an increase in glutamine synthetase activity (Pérez-Soba *et al.* 1994). Light is a pre-requisite as it provides the C skeletons via photosynthesis and is the electron donor for reducing ferredoxin. Light is also the source of reducing power for NO_3^- reduction and modulates nitrate reductase by enzyme phosphorylation (chapter 6). NO_3^- assimilation can utilise up to 25% of the reductant used in photosynthesis (Gebauer and Schulze 1997). Both NO_3^- and

NH_4^+ levels will tend to accumulate under low light intensities increasing the likelihood of toxic effects (see below).

In the roots N assimilation rate is influenced by soil temperature and the rate of supply of C from the shoots, which may be relatively slow in northern conifers. Glutamine, glutamic acid and aspartic acid occupy central positions in N metabolism, but the amino acid most associated with N deposition is arginine (Näsholm 1991). Arginine is known to accumulate at high N concentrations and its production was seen as a mechanism for sequestering NH_4^+. Now, it is accepted that arginine accumulation is indicative of nutrient stress, imbalance, which lowers growth rates but does not inhibit nutrient uptake and may or may not lead to a surplus of NH_4^+ (Chapin 1980, Rabe and Lovatt 1986, Näsholm and Ericsson 1990). Arginine has the lowest C/N ratio (6C/4N) so that storing N in this form requires the least C to remove the potentially toxic NH_4^+ ion (Marschner 1995). Arginine concentrations in the root can serve to down regulate NH_4^+ uptake (Näsholm *et al.* 1997). The likelihood of measuring elevated arginine levels increases with decreasing site productivity (Edfast *et al.* 1990). Arginine represented 4% of total N in response to elevated N but the proportion was 27% when nutrient imbalance was evident. Scots pine at Norrliden, Sweden, that had received 5 to 8 times ambient N deposition (max. 20 kg N ha^{-1} yr^{-1}) accumulated arginine, up to 10-27% of the 1.7-2.2% N total dry $weight^{-1}$, except when P and K were added, when accumulation represented only 7-16% (Näsholm and Ericsson 1990). Not withstanding this the correlation between N deposition and vacuolar arginine accumulation can be high (Huhn and Schulz 1996). N deposition can stimulate nutrient deficiency by enhancing growth and concomitantly reducing P and base cation availability via effects of acidification (Jonsson *et al.* 2003). Growth reduction is associated with arginine levels exceeding 30% of the total foliar N (Krause *et al.* 1986). In pristine areas of northern Scandinavia arginine measures less than 0.06% foliar dry weight, whereas in N polluted areas of the Netherlands values of 0.65-0.9% can occur.

CONSEQUENCES OF NITROGEN DEPOSITION

Foliar nitrogen

Foliar N concentrations (usually reported for current year) in Scots pine from the ICP forest plots (De Vries *et al.* 2002) and in Norway spruce from Switzerland (Fluckiger and Braun 1998) are correlated with N deposition and have increased over recent decades. Responses in foliar N concentrations to the N deposition dose may be neither linear, nor immediate. Foliar N concentrations are influenced by tree age, reflecting the balance between metabolic activity and assimilation rate, needle age, via sequestration and carbohydrate status at the time of sampling in respect to non-structural carbohydrate levels (Linder 1995) phenology and branch position in the canopy, which affects sink strength and chlorophyll-N content (Rosengren-Brinck 1994). Responses are affected by the source of N, air or soil, and whether it is wet or dry. The N status of the tree affects how quickly foliar N changes and thus the potential for accumulation. For NH_3 deposition this is implemented via the compensation point (Sutton et al. 1993a).

At Aber (N. Wales), changes in foliar N in more than 30 year-old Sitka spruce were generally insignificant, despite additions of up to 75 kg N ha^{-1} yr^{-1} irrigated at 1 m above the forest floor. NO_3^- leached but almost no N was absorbed through the roots and the % N dry weight was about 1.8% (Gunderson et al. 1998). There were no growth responses but the N was not taken up by the roots to accumulate above the optimum for Spruce (Innes 1995). Seven years of treating a Sitka spruce canopy approximately twice weekly with N, with or without acidity, at 48 kg N ha^{-1} yr^{-1} where significant amounts of foliar uptake were observed (Cape et al. 2001) failed to significantly increase foliar N status in the 15-year old trees (at canopy closure) even though foliar N at 1.2% dry weight (Sheppard et al. 2001) was verging on deficient. In fertiliser studies where N applications as a solid to the surface of the soil are in the range of 75-150 kg N ha^{-1} yr^{-1}, irregular and infrequent increases in foliar N tend to be large in the first year but thereafter show little change as the trees readjust to their earlier homeostatic state (Dutch 1995). Many of the NITREX N manipulation experiments have failed to change foliar N (Gunderson et al. 1998). However at Skogaby, (Sweden) Norway spruce, post canopy closure growing on an N deficient site, exhibited very large increases in growth which were soon accompanied by increases in foliar N, which are now at or exceed the recommended value of 1.8% N for optimum tree growth (Nilsson and Persson 2001). These results appear to confirm links between growth responses and the likelihood of foliar N changing in response to wet N deposition but it is not always clear why foliar concentrations have responded in the ways they have. In response to dry N deposition as NH_3, foliar N concentrations appear to increase more per unit N (Pérez-Soba 1995); however, extrapolation from concentration to deposition dose is quite difficult (Sutton et al. 1993b).

While we have some of the explanations for changes, or lack of them, in foliar N status in response to N deposition or N manipulation experiments, many of the changes are still unpredictable and inexplicable. Foliar N concentrations higher than 3% dry weight are rarely measured except in the Netherlands in senescing needles. Symptoms of N effects, white/yellow needle bases on young Scots pine that were associated with % N dry weight of 2.3-3.1 are restricted to damaged trees on acidic, nutrient poor heathland (van Dijk and Roelofs 1988). Conifers would appear to be relatively successful at matching uptake to demand so damage is uncommon except where N deposition is very high or other factors are weakening the trees.

Nutrient imbalance

It is generally believed that the imbalance of nutrients with respect to N is more likely to impair growth than high foliar N *per se* (Linder 1995). Thus we tend to discuss nutrients as ratios rather than concentrations, which also means that inter-annual comparisons are not biased by differences in stored carbohydrate. Target ratios for Norway spruce are 10, 4, 35 and 25 for P/N, K/N, Mg/N and Ca/N respectively (Sternquist *et al.* 2002). Lower values indicate N excess with the potential for reductions in stem growth, which are usually accompanied by needle loss and changes in C balance not withstanding increases in susceptibility to stress, pests and disease. Young trees are less affected by high N inputs since they are less sensitive than older trees to suboptimal P and K status because the strength of their growing sinks maximises the likelihood of internal remobilisation and retranslocation of nutrients (Marschner 1995, Thelin 2000). However, once these needles have been shed, and if defoliation is severe (more than 50%) such trees will suffer (Marschner 1995). High concentrations of soil water NH_4^+ can interfere with K^+ uptake via antagonism, as the K channel is approximately the same size as the NH_4^+ ion (Lips *et al.* 1987). Uptake of other cations may also be directly impaired in the presence of high NH_4^+, thus exacerbating the likelihood of nutrient deficiency (Krupa *et al.* 2003).

Potential for N toxicity

The uptake kinetics of nutrients other than N can be rapidly adjusted to function at very low concentrations to maintain supply commensurate with the demand imposed by N (Sverdrup *et al.* 2002). However, N can accumulate as the supply of other nutrients fails, because these restrict growth but do not appear to restrict N uptake. This might reflect past evolutionary strategies. N has historically been in short supply, so that strategies for restricting N uptake were probably less desirable than the evolution of strategies for maximising uptake. Where deficiencies restrict anabolic processes, *i.e.* protein synthesis and growth or restrict glucose production or enhance its depletion *e.g.* in cold stress, there is the potential for toxic NH_3 to accumulate (Rabe and Lovatt 1986, Rabe 1990). Tolerance of high NH_3 levels appears to be quite good under optimum growth conditions (Pérez-Soba *et al.* 1994), but these rarely occur in forests. Direct injury, attributed to NH_3, has been observed in the proximity of intensive animal units (Fangmeier *et al.* 1994, Krupa 2003) and N fertiliser factories (De Temmerman *et al.* 1988). Damage was observed in Flanders to Austrian pine where deposition was unprecedented at 300-700 kg N ha^{-1} yr^{-1}. High atmospheric NH_3 concentrations have also led to enhanced foliar N, which increases the risk from pests and pathogens. One of the highest foliar N concentrations (3%) for pine needles occurred in response to monthly means between 45-65 µg m^{-3} NH_3 (Heinsdorf and Krauss 1991). Direct foliar damage is

not usually observed with wet deposition, which suggests toxicity may be linked to increased uptake of N in gaseous form. Summarising, NH_3 can:
- adversely affect acid base regulation, and uncouple photo phosphorylation (Raven 1985),
- inhibit starch synthesis and glucose accumulation (Hampp *et al*. 1999),
- saturate membrane lipids and increase permeability (Näsholm 1991)
- inhibit protein synthesis and increase the proportion of free amino acids (Näsholm 1991),
- inhibit chlorophyll synthesis,
- attack and degrade the cuticle, which can adversely affect water loss and ion uptake (van der Eerden *et al*. 1992).

Damage is most common in winter when the potential for assimilation (detoxification) is low and the potential for interaction with cold stress is high. In many cases the damage is transient and restricted to some needle shedding.

Sensitivity to stress

N eutrophication has been linked to increased sensitivity to stress, particularly frost hardiness (Aronsonn 1980). While it is known that N can extend the growing season in the autumn and thus increase the risk of frost injury, mechanisms by which N increase winter damage are controversial (DeHayes *et al*. 1989) and experimental data from manipulation studies are inconclusive (Sheppard *et al*. 2003). N is more likely to improve winter hardiness in shoots (Sheppard 1994, Pfanz and Sheppard 2001). In spring the risk of frost injury to newly flushed shoots is real, through earlier budburst, brought on by accelerated carbohydrate assimilation in response to the improved foliar N status (Pfanz and Sheppard 2001). More recently N deposition has been linked to secondary stresses following freezing damage (Jönsson 2002).

There does appear to be a link between local NH_3 emissions and frost damage (Dueck *et al*. 1991, van der Eerden *et al*. 1992). Frost damage and fungal attacks were associated with foliar N concentrations of 1.6-2.0% in Scots pine, 1.5-1.8% in Douglas fir and 1.1-2.0% in Austrian pine, trees where the "norm" is between 1.0-1.5% N (De Temmerman *et al*. 1988). However, when Clement (1996) investigated the impact of NH_3 on frost hardiness in Scots pine he found no effects on current year shoots. Chlorophyll concentrations, photosynthesis and protein contents were increased, while carbohydrate content was decreased. One-year old needles did show reduced winter hardiness and Clement (1996) concluded that NH_3 *per se* had not reduced the hardiness but rather the imbalance between N and K and subsequent effects on cryoprotectants were to blame.

Excess N has also been linked to reduced water use efficiency and increased risk from drought, but our understanding is far from clear, especially for trees. For further reading on the effects of different N forms and the interaction with pH regulation and water use see Raven (1985), Yin *et al.* (1996) and Yin and Raven (1997). Simply, N increases the shoot/root ratio so that the uptake area is reduced at the expense of the increasing transpiring area (van Dijk *et al.* 1990). High NH_4^+ can also restrict root growth and restrict K uptake with implications for osmoregulation (Marschner 1995).

Growth of epiphytic green algae on leaf surfaces protected from frequent desiccation, in response to N eutrophication is a cause for concern for forests as light penetration may be impeded (Poikolainen *et al.* 1998). Evidence for increased vulnerability to pests, disease and herbivores, both small and large grazers, is convincing in response to nutrient imbalances with respect to N (Fluckiger and Braun 1998, 1999). This is probably because of lower concentrations of phenolics and higher concentrations of free amino acids. N fertilisation increased aphid population numbers and litterfall in Sitka spruce (Parry 1974). In spruce, both N and S treatments, especially in combination increased aphid numbers (Stadler *et al.* 2001). Damage to pines from fungal pathogens (*Brunchorstia pinea* and *Sphaeropsis sapinea*) associated with increased N content has increased in the Netherlands (Roelofs *et al.* 1985, van Dijk *et al.* 1990).

CONCLUDING REMARKS

Conifers, particularly those growing in northern latitudes face a dilemma: if they control N uptake too tightly they may have insufficient N to sustain the annual growth flush. Northern conifers have mainly evolved under conditions of restricted N supply so it seems unlikely they will have evolved mechanisms to restrict N storage. Root N uptake is regulated, foliar uptake of gaseous NH_3 is controlled through the compensation point phenomenon and cuticular pathways operate control via concentration gradients. So, can trees fine-tune N uptake sufficiently to prevent accumulation or is the fate of our forests outside the trees' control? Field manipulations suggest trees cannot cope when the availability of N exceeds that of other nutrients. Nutrient imbalance restricts growth and N starts to accumulate in the tissue, increasing susceptibility to stress, pests and diseases.

Successful trees expand their canopy quickly, as light can soon limit the potential for C fixation. In response to N trees invest their new assimilate aboveground in light capture and partition less to the roots. This would seem to be counter-intuitive in our increasingly N rich environment, given that N is not the only nutrient required. Thus trees face a paradox, at a time when they need to maximise nutrient uptake to meet the N driven growth demand they partition less C to roots and mycorrhizas. The restricted ability to forage for nutrients and loss of mycorrhizas will leave roots vulnerable to Al^{3+} in the acidifying soils, and pathogens. In addition reduced root growth means reduced water uptake - a decoupling of root and shoot balance so that their increased evaporative canopy surface is supported by a smaller water absorbing capacity.

The extent to which trees are affected by N deposition would appear to be mostly outside the trees' control since the availability of N for uptake is predominantly under the control of soil microbes via retention and transformation (Garner 1984, Bengtsson and Bergwall 2000). It is the characteristics of the soil and its biota that determine if and when a forest ecosystem will saturate in response to N deposition (Garner 1984). Näsholm (1991) concluded that estimating the potential of a soil to retain N presents the largest uncertainty in the setting of Critical Loads (CL) for N. The activities of the soil microflora control the capacity for N immobilisation and the speed of release. Soils, particularly those rich in organic material, have a high capacity to immobilise N. However, this capacity is not exhaustible and many forests are already N saturated. When this occurs N moves away from being a fertiliser to becomming a pollutant.

The speed at which the above happens is also influenced by the availability of nutrients other than N. NH_4^+ and H^+ released during nitrification temporarily enhance base cation availability, ultimately leading to their removal via the mobile anion effect. Where weathering rates fail to match the losses the system will no longer be able to sustain the accelerated demand for nutrients. In agricultural systems enhanced growth is met through fertilisers processed to match the demands of the growing crop. In the case of anthropogenic N deposition to forests only N increases.

Experiments have revealed that at N additions, 5-10 times higher than current N deposition, many soils can buffer a large proportion of the incoming N. They also show that younger forests generally benefit from these N inputs. However, there are already soils that because of their low levels of available nutrients cannot sustain the N driven growth and in these soils N deposition has exceeded the CL and will need to be reduced before forests can recover.

Direct injury from N would appear unlikely. However, the balance of literature suggests that forests growing near local sources of NH_3 will be negatively affected more quickly than forests receiving wet N, especially under dark and cold conditions. Damage to such forests is most likely to be indirect via interaction with cold, pests or pathogens.

The answers lie belowground

- N availability and loss is regulated exclusively through belowground microbial processes.
- The potential for N to act as a fertiliser or pollutant will depend on its availability in relation to the relative availability of other nutrients especially P and base cations. The speed at which N moves from being a fertiliser to pollutant will depend on forest age and growth rate.

ACKNOWLEDGEMENTS

John Raven, Peter Millard, Marcel van Oijen, Neil Cape and Torgny Näsholm are thanked for their feedback on the text and answering unanswerable questions. Eleanor MacDonald has painstakingly typed in all the references.

REFERENCES

Aber J.D. 1992. Nitrogen cycling and nitrogen saturation in temperate forest ecosystems. - Trends Ecol. Evol. 7: 220-224.
Aber J.D., Nadelhoffer K.J., Steudler P., Melillo J.M. 1989. Nitrogen saturation in northern forest ecosystems. - Bioscience 39: 378-386.
Aber J., McDowell W., Nadelhoffer K., Magill A., Berntson G., Kamakea M., McNulty S., Currie W., Rustad L., Fernandez I. 1998. Nitrogen saturation in temperate forest ecosystems. - Bioscience 48: 921-934.
Abrahamsen G. 1984. Effects of acidic deposition on forest soil and vegetation. - Philos. Trans. R. Soc. Lond. B 305: 360-382
Aerts R., Chapin F.S. III. 2000. The mineral nutrition of wild plants revisited: a re-evaluation of processes and patterns. - Adv. Ecol. Res. 30: 2-69.
Agerer R. 2001. Exploration types of ectomycorrhizae. A proposal to classify ectomycorrhizal mycelial systems according to their patterns of differentiation and putative ecological importance. - Mycorrhiza 11: 107-114.
Arnebrant K. 1994. Nitrogen amendments reduce the growth of extrametrical mycelium. - Mycorrhiza 5: 7-15.
Arnebrant K., Söderström B. 1994. Effects of different fertiliser treatments on ectomycorrhizal colonisation potential in two Scots pine forests in Sweden. - For. Ecol. Man. 53: 77-89.
Aronsson A. 1980. Frost hardiness in Scots pine (*Pinus silvestris L.*) II. Hardiness during winter and spring in young trees of different mineral nutrient status. - Stud. For. Suec. 155: 14-50.
Asman W.A., Sutton M.A., Schjørring J.K. 1998. Ammonia: emission, atmospheric transport and deposition. - New Phytol. 139: 27-48.
Beier C. 1998. Water and element fluxes calculated in a sandy forest soil taking spatial variability into account. - For. Ecol. Man. 101: 269-280.
Bengtsson G., Bergwall C. 2000. Fate of ^{15}N labelled nitrate and ammonium in a fertilized forest soil. - Soil Biol. Biochem. 32: 545-557.
Berg B., Staaf H. 1981. Leaching, accumulation and release of nitrogen in decomposing forest litter. - In: Clark F.E., Rosswall T. (Eds.) Terrestrial nitrogen cycles, processes, ecosystem strategies and management impacts. - Ecol. Bull. 33: 163-178.
Berg B., Matzner E. 1997. Effect of N deposition on decomposition of plant litter and soil organic matter in forest systems. - Environ. Rev. 5: 1-25.
Berg B., McClaugherty C. 2003. Plant litter, decomposition, humus formation, carbon sequestration. - Berlin, Heidelberg, Germany, Springer Verlag.
Bergh J., Linder S., Lundmark T., Elving B. 1999. The effect of water and nutrient availability on the productivity of Norway spruce in northern and southern Sweden. - For. Ecol. Man. 119: 51-62.
Bidartondo M.I., Ek H., Wallander H., Söderström B. 2001. Do nutrient additions alter carbon sink strength of ectomycorrhizal fungi? - New Phytol. 151: 543-550.
Binkley D., Högberg P. 1997. Does atmospheric deposition of nitrogen threaten Swedish forests? - For. Ecol. Man. 92: 119-152.
Boxman A.W., Blanck K., Brandrud T.-E., Emmett B.A., Gundersen P., Hogervorst R.F., Kjønaas O.J., Persson H., Timmermann V. 1998. Vegetation and soil biota response to experimentally-

changed nitrogen inputs in coniferous forest ecosystems of the NITREX project. - For. Ecol. Man. 101: 65-79.

Boyce R.L., McCune D.C. 1992. Water holdup capacity and residence time of red spruce and balsam fir branches. - Trees 6: 19-27.

Boyce R.L., Friedland A.J., Chamberlain C.P., Poulson S.R. 1996. Direct canopy nitrogen uptake from ^{15}N-labeled wet deposition by mature red spruce. - Can. J. For. Res. 26: 1539-1547.

Brandrud T.E. 1995. The effects of experimental N addition on the ectomycorrhizal flora in an oligotrophic spruce forest at Gåardsjön, Sweden. - For. Ecol. Man. 71: 111-122.

Brandrud T.E., Timmermann V. 1998. Ectomycorrhizal fungi in the NITREX site at Gårdsjön, Sweden; below and above-ground responses to experimentally-changed nitrogen inputs 1990-1995. - For. Ecol. Man. 101: 207-214.

Briffa K.R., Schweingruber F.H., Jones P.D., Osborn T.J., Harris I.C., Shiyatov S.G., Vaganov E.A., Grudd H. 1998. Trees tell of past climates: but are they speaking less clearly today? - Philos. Trans. R. Soc. Lond. B. 353: 65-73.

Bruckner-Schatt G., Peters K., Bauer G.A., Schulze E.D. 1995. Reduced nitrogen: emission, immission deposition and above-ground uptake in a spruce ecosystem (in German). - In: Gregor H.-D. (Ed.) Wirkungscomplex Stickstoff und Wald. - Berlin, Germany, Umweltbundesambt, pp. 30-43.

Buchmann N., Gebauer G, Schulze E.-D. 1996. Partitioning of ^{15}N-labeled ammonium and nitrate among soil, litter, below- and above-ground biomass of trees and understory in a 15-year-old *Picea abies* plantation. - Biogeochemistry 33: 1-23.

Burkhardt J., Eiden R. 1994. Thin water films on coniferous needles. - Atmos. Environ. 28: 2001-2017.

Cannell M.G.R., Thornley J.H.M. 2000. Nitrogen states in plant ecosystems: a viewpoint. - Ann. Bot. 86: 1161-1167.

Cape J.N., Dunster A., Crossley A., Sheppard L.J., Harvey F.J. 2001. Throughfall chemistry in a Sitka spruce plantation in response to six different simulated polluted mist treatments. - Water Air Soil Poll. 130: 619-624.

Chalot M. Brun A. 1998, Physiology of organic nitrogen acquisition by ectomycorrhizal fungi and mycorrhizas. - F.E.M.S. Micro Rev. 22: 21-24.

Chalot M., Javell A., Blaudez D., Lambilliote R., Cooke R., Sentenac H., Wipf D., Botton B. 2002. An update on nutrient transport processes in ectomycorrhizas. - Plant Soil 244: 165-175.

Chapin F.S. 1980. The mineral nutrition of wild plants. - Annu. Rev. Ecol. Syst. 11: 233-60.

Chapin F.S., Bloom A.J., Field C.B., Waring R.H. 1987. Plant responses to multiple environmental factors. - Bioscience 37: 49-57.

Chapin F.S., Schulze E.D., Mooney H.A. 1990. The Ecology and economics of storage in plants. - Annu. Rev. Ecol. Syst. 21: 423-447.

Chiwa M., Crossley A., Sheppard L.J., Sakugawa H., Cape J.N. 2003. Throughfall chemistry and canopy interactions in a Sitka spruce plantation sprayed with six different simulated polluted mist treatments. - Environ. Pollut. 1-25.

Clement H. 1996. Interaction of atmospheric ammonia pollution with frost tolerance of plant. A study on winter wheat and Scots pine. - Ph.D. Thesis, University of Groningen, The Netherlands.

Crawford N.M. 1995. Nitrate: nutrient and signal for plant growth. - Plant Cell 7: 859-868.

DeHayes D.H., Ingle M.A., Waite C. E. 1989. Nitrogen fertilization enhances cold tolerance of red spruce seedlings. - Can. J. For. Res. 19: 1037-1043.

De Temmerman L., Ronse A., van den Cruys K., Meeus-Verdinne K. 1988. Ammonia and pine tree dieback in Belgium. - In: Mathy P. (Ed.) Air pollution and ecosystems. - Dordrecht, The Netherlands, Reidel Publishing Company, pp. 774-779.

De Vries W., Reinds G. J., van Dobben H., De Zwart D., Aamlid D., Neville P., Posch M., Auee J., Voogd J.C.H., Vel E. 2002. Intensive monitoring of forest ecosystems in Europe. - Geneva and Brussels, Technical Report 2002, UNECE and EC.

Dickie I.A., Xu B., Doide R. T. 2002. Vertical niche differentiation of ectomycorrhizal hyphae in soil as shown by T-RFLP analysis. - New Phytol. 156: 527-535.
Dignon J., Hameed S. 1989. Global emissions of nitrogen and sulfur-oxides from 1860 to 1980. - J. Air Waste Man. 39: 180-186.
Dollard J.G., Davies T.J., Lunstrom J.P. 1987. Measurements of the dry deposition rates of some trace gas species. - In: Angeletti G., Restilli G. (Eds.) Physico-chemical behaviours of atmospheric pollutants. - Dordrecht, The Netherlands, Kluwer Academic Publishers, pp. 470-479.
Dueck T.A., Dorel F.G., ter Horst R., van der Eerden L.J. 1991. Effects of ammonia, ammonium sulphate and sulphur dioxide on the first sensitivity of Scots pine (*Pinus sylvestris* L.). - Water Air Soil Pollut. 54: 35-49.
Dutch J. C. 1995. The role of nutrition in the decline of Sitka spruce. - In: Coutts M. P. (Eds.) Decline in Sitka spruce on the South Wales coalfield. - Forestry Commission Technical Paper 9.
Edfast A.-B., Näsholm T., Ericsson A. 1990. Free amino acid concentrations in needles of Norway spruce and Scots pine trees on different sites in areas with two levels of nitrogen deposition. - Can. J. For. Res. 20: 1132-1136.
Ek H. 1997. The influence of nitrogen fertilization on the carbon economy of *Paxillus involutus* in ectomycorrhizal association with *Betula pendula*. - New Phytol. 135: 133-142.
Eltrop L, Marschner H. 1996. Growth and mineral nutrition of non-mycorrhizal and mycorrhizal Norway spruce (*Picea abies*) seedlings grown in semi-hydroponic sand culture. - New Phytol. 133: 469-478.
Emmett B.A. 1999. The impact of nitrogen on forest soils and feedbacks on tree growth. - Water Air Soil Poll. 116: 65-74.
Emmett B.A., Reynolds B., Silgram M., Sparks T.H., Woods C. 1998. The consequences of chronic nitrogen additions on N cycling and soilwater chemistry in a Sitka spruce stand, North Wales. - For. Ecol. Man. 101: 165-175.
Erland S., Taylor A.F. S. 2002. Diversity of ecto-mycorrhizal fungal communities in relation to the abiotic environment. - In: van der Heijden M. G. A., Sanders I. (Eds.) Mycorrhizal ecology - Ecol. Studies 157: 163-224.
Evans J.R. 1989. Photosynthesis and nitrogen relationships in leaves of C_3 plants. - Oecologia 78: 9-19.
Fangmeier A., Hadwiger-Fangmeier A., van der Eerden L.J.M., Jäger H.-J. 1994. Effects of atmospheric ammonia on vegetation - a review. - Environ. Pollut. 86: 43-82.
Finlay R.D., Ek H., Odham G., Söderström B. 1988. Mycelial uptake, translocation and assimilation of nitrogen from ^{15}N-labelled ammonium by *Pinus sylvestris* plants infected with four different ectomycorrhizal fungi. - New Phytol. 110: 59-66.
Flückiger W., Braun S. 1998. Nitrogen deposition in Swiss forests and its possible relevance for leaf nutrient status, parasite attacks and soil acidification. - Environ. Pollut. 102: 69-76.
Flückiger W., Braun S. 1999. Nitrogen and its effect on growth, nutrient status and parasite attacks in beech and Norway spruce. - Water Air Soil Poll. 116: 99-110.
Fowler D., Cape J.N., Leith I. D., Choularton T.W., Gay M.J., Jones A. 1988a. The influence of altitude on rainfall composition at Great Dunfell. - Atmos. Environ. 22: 1355-1362.
Fowler D., Flechard C., Skiba U., Coyle M., Cape J.N. 1998b. The atmospheric budget of oxidized nitrogen and its role in ozone formation and deposition. - New Phytol. 139: 11-23.
Fowler D., Pitcairn C.E.R., Sutton M.A., Flechard C., Loubet B., Coyle M., Munro R.C. 1998c. The mass budget of atmospheric ammonia in woodland within 1 km of livestock buildings. - Environ. Pollut. 102: 343-348.
France R.C., Reid C.P.P. 1983. Interactions of nitrogen and carbon in the physiology of ectomycorrhizae. - Can. J. Bot. 61: 964-984.
Fransson P.M.A., Taylor A.F.S., Finlay R.D. 2000. Effects of continuous optimal fertilization upon a Norway spruce ectomycorrhizal community. - Tree Physiol. 20: 599-606.

Garner J.H.B. 1984. Nitrogen oxides, plant metabolism and forest ecosystem response. - In: Alscher R. G., Wellburn A. R. (Eds.) Plant Responses to the gaseous environment. - Dordrecht, The Netherlands, Kluwer Academic Publishers, pp. 301-314.

Gebauer G., Schulze E.-D. 1997. Nitrate nutrition of central European forest trees. - In: Rennenberg H., Eschrich W., Ziegler H. (Eds.) Trees - contributions to modern tree physiology. - Leiden, The Netherlands, Backhuys Publishers, pp. 273-291.

Gessler A., Rienks M., Rennenberg H. 2002. Stomatal uptake and cuticular adsorption contribute to dry deposition of NH_3 and NO_2 to needles to adult spruce (*Picea abies*) trees. - New Phytol. 156: 179-194.

Givan C.V. 1979. Metabolic detoxification of ammonia in tissues of higher plants. - Phytochemistry 18: 375-382.

Gundersen P., Emmett B.A., Kjønaas O.J., Koopmans C.J. Tietema A. 1998. Impact of nitrogen deposition on nitrogen cycling in forests: a synthesis of NITREX data. - For. Ecol. Man. 101: 37-55.

Hagerberg D., Thelin G., Wallander H. 2003. The production of ectomycorrhizal mycelium in forests: Relation between forest nutrient status and local mineral sources. - Plant Soil 252: 279-290.

Hampp R., Wiese J., Mikolajewski S., Nehls U. 1999. Biochemical and molecular aspects of C/N interaction in ectomycorrhizal plants: an update. - Plant Soil 215: 103-113.

Hanson P.J., Lindberg S.E. 1991. Dry deposition of reactive nitrogen compounds: a review of leaf, canopy and non-foliar measurements. - Atmos. Environ. 25A: 1615-1634.

Harrison A.F., E.-D., Gebauer G., Bruckner G. 2000. Canopy uptake and utilization of atmospheric pollutant nitrogen. - In: Schulze E.-D. (Ed.) Carbon and nitrogen cycling in European forest ecosystems. - Ecol. Studies 142: 171-88.

Högberg M.N., Högberg P. 2002. Extramatrical ectomycorrhizal mycelium contributes one-third of microbial biomass and produces, together with associated roots, half the dissolved organic carbon in a forest soil. - New Phytol. 154: 791-795.

Huhn G., Schulz H. 1996. Contents of free amino acids in Scots pine needles from field sites with different levels of nitrogen deposition. - New Phytol. 134: 95-101.

Husted S., Schjoerring J.K. 1995. Apoplastic pH and ammonium concentration in leaves of *Brassica napus* L. - Plant Physiol. 109: 1453-1460.

Ingestad T. 1979 Mineral nutrient requirements of *Pinus sylvestris* and *Picea abies* seedlings. - Physiol. Plant. 45: 373-380.

Innes J.L. 1995. Influence of air pollution on the foliar nutrition of conifers in Great Britain. - Environ. Pollut. 88: 183-192.

Jönsson AM. 2000 Bark lesions and sensitivity to frost in beech and Norway spruce. Ph.D. Thesis, University of Lund, Sweden.

Jonsson L., Dahlbert A., Brandrud T.E. 2000. Spatiotemporal distribution of an ectomycorrhizal community in an oligotrophic Swedish *Picea abies* forest subjected to experimental N addition: above and below-ground views. - For. Ecol. Man. 132: 143-156.

Jönsson U., Rosengren U., Thelin G., Nihlgård B. 2003. Acidification-induced chemical changes in coniferous forest soils in southern Sweden 1988-1999. - Environ. Pollut. 123: 75-83.

Kåren O., Nylund J.E. 1997. Effects of ammonium sulphate on the community structure and biomass of ectomycorrhizal fungi in a Norway spruce stand in South West Sweden. - Can. J. Bot. 75: 1628-1643.

Katz C., Oren R., Schulze E.-D. Milburn J. 1989. Uptake of water and solutes through twigs of *Picea abies* (L.) Karst. - Trees 3: 33-37.

Kellner O., Redbo-Torstensson P. 1995. Effects of elevated nitrogen deposition on the field-layer vegetation in coniferous forests. - Ecol. Bull. 44: 227-237.

Kenk G., Fischer H. 1988. H Evidence from nitrogen fertilisation in the forests of Germany. - Environ. Pollut. 54: 199-218.

Klemm O. *et al.* 1989. 3-C Leaching and uptake of ions through above-ground Norway spruce tree parts. - In: Schulze E.-D., Lange O. L., Oren R. (Eds.) - Ecol. Studies 77: 210-237.

Körner C. 2003. The carbon charging of pines at the tree line: a global comparison. - Oecologia 135: 10-20.
Kronzucker H.J., Siddiqi M.Y., Glass A.D.M. 1995. Compartmentation and flux characteristics of nitrate in spruce. - Planta 196: 674-686.
Kronzucker H.J., Siddiqi M.Y., Glass A.D.M. 1997. Root discrimination against soil nitrate and the ecology of forest succession. - Nature 385: 59-61.
Krause G. H. M., Arndt U., Brandt C. J., Bucher J., Kenk G., Matzner E. 1986. Forest decline in Europe: development and possible causes. - Water Air Soil Pollut. 31: 647-668.
Kreutzer K. *et al.* 1998. Atmospheric deposition and soil acidification in five coniferous forest ecosystems: a comparison of the control plots of the EXMAN sites. - For. Ecol. Man. 101: 125-142.
Krupa S.V. 2003. Effects of atmospheric ammonia (NH_3) on terrestrial vegetation: a review. - Environ. Pollut. 124: 179-221.
Landeweert R., Hoffland E., Finlay R., van Breemen N. 2001. Linking plants to rocks: ectomycorrhizal fungi mobilize nutrients from minerals. - Trends Ecol. Evol. 16: 248-254.
Landeweert R., Leefland P., Kuyper T.W., Hoffland E., Rosling A., Wernars K, Smit E. 2003. Molecular identification of ectomycorrhizal mycelium in soil horizons. - Appl. Environ. Microbiol. 69: 327-333.
Lawlor D.W. 1994. Relation between carbon and nitrogen assimilation, tissue composition and whole plant function. - In: Roy J., Garnier E. (Eds.) A whole plant perspective on carbon-nitrogen interactions - The Hague, The Netherlands, SPB Academic Publishing, pp. 47-60.
Levine J.S., Cofer W.R., Winstead E.L., Rhinehart R.P., Cahoon D.R., Sebacher D.I., Sebacher S., Stocks B.J. 1991. Biomass burning: combustion emissions, satellite imagery, and biogenic emissions. - In: Levine J.S. (Ed.) Global biomass burning: atmospheric, climate and biospheric implications. - Cambridge, MA, The MIT Press, pp. 264-271.
Liebig J.v. 1840. Die Chemie in ihrer Anwendung auf Agrikultur und Physiologie. - Vieweg, Braunschweig, Germany.
Lilleskov E., Fahey T.J., Lovett G.M. 2001. Ectomycorrhizal fungal aboveground community change over an atmospheric nitrogen deposition gradient. - Ecol. Appl. 11: 397-410.
Lilleskov E.A., Hobbie E.A., Fahey T.J. 2002. Ectomycorrhizal fungal taxa differing in response to nitrogen deposition also differ in pure culture organic nitrogen use and natural abundance of nitrogen isotopes. - New Phytol. 154: 219-231.
Linder S. 1995. Foliar analysis for detecting and correcting nutrient imbalances in Norway spruce. - Ecol. Bull. 44: 178-190.
Lips S.H., Soares M.I.M., Kaiser J.J., Lewis O.A.M. 1987. K^+ modulation of nitrogen update and assimilation in plants. - In: Ullrich, W.R., Aparicio P.J., Castillo F. (Eds.) Inorganic nitrogen in plants and microorganisms. - Berlin, Germany, Springer Verlag, pp. 233-239.
Lovett G.M., Lindbert S. E. 1993. Atmospheric deposition and canopy interactions of nitrogen in forests. - Can. J. For. Res. 23: 1603-1616.
Macklon A.E.S., Armstrong J.A. 1994: Uptake of ammonium and nitrate ions from acid mist applied to Sitka spruce [*Picea sitchensis* (Bong.) Carr.]. - Trees 10: 261-267.
Macklon A.E.S., Sheppard L.J., Sim A., Leith I.D. 1996. Uptake of ammonium and nitrate ions from acid mist applied to Sitka spruce [*Picea sitchensis* (Bong.) Carr.] grafts over the course of one growing season. - Trees 10: 261-267.
Magill A.H., Aber J.D., Berntson G.M., McDowell W.H., Nadelhoffer K.J., Melillo J.M., Steudler P. 2000. Long-term nitrogen additions and nitrogen saturation in two temperate forests. - Ecosystems 3: 238-253.
Marschner H. 1991. Mechanisms of adaptation of plants to acid soils. - Plant Soil 134: 1-20.
Marschner H. 1995. Mineral nutrition of higher plants. 2nd edition. - San Diego, Ca, Academic Press Inc.
Marschner H., Häussling M, George E. 1991. Ammonium and nitrate uptake rates and rhizosphere pH in non-mycorrhizal roots of Norway spruce [*Picea abies* (L.) Karst.]. - Trees 5: 14-21.

Marshall J.D., Candle S.H. 1989. Evidence for trans-cuticular uptake of HNO_3 vapour by foliage of Eastern white pine (*Pinus strobes* L.). - Environ. Pollut. 60: 15-28.

Martin F., Canet D. 1986. Biosynthesis of amino acids during [13C] glucose utilization by the ectomycorrhizal ascomycete *Cennococcum geophilum* monitored by 13C nuclear magnetic resonance. - Physiol. Veg. 24: 209-218.

Martin F., Lorillou S. 1997. Nitrogen acquisition and assimilation in ectomycorrhizal systems. - In: Rennenberg H., Eschrich W., Ziegler H. (Eds.) Trees - Contributions to modern tree physiology. - Leiden, The Netherlands, Backhuys Publishers, pp. 423-439.

Matzner E., Murach D. 1995. Soil changes induced by air pollutant deposition and their implication for forests in central Europe. - Water Air Soil Poll. 85: 63-76.

Millard P., Proe M.F. 1993. Nitrogen uptake, partitioning and internal cycling in *Picea sitchensis* (Bong.) Carr. as influenced by nitrogen supply. - New Phytol. 125: 113-119.

Muller B., Touraine B., Rennenberg H. 1996. Interaction between atmospheric and pedospheric N nutrition in spruce (*Picea abies* L. Karst) seedlings. - Plant Cell Environ. 19: 345-355.

Näsholm T. 1991. Aspects of nitrogen metabolism in Scots pine, Norway spruce and birch as influenced by the availability of nitrogen in pedosphere and atmosphere. - Ph.D. Thesis, Swedish Univ. Agric. Sci., Umeå, Sweden.

Näsholm T. 1998. Qualitative and quantitative changes in plant nitrogen acquisition induced by anthropogenic nitrogen deposition. - New Phytol. 139: 87-90.

Näsholm T., Ericsson A. 1990. Seasonal changes in amino acids, protein and total nitrogen in needles of fertilized Scots pine trees. - Tree Physiol. 6: 267-281.

Näsholm T., Nordin A., Edfast A-B., Högberg P. 1997. Identification of coniferous forests with incipient nitrogen saturation through analysis of arginine and nitrogen-15 abundance of trees. - J. Environ. Qual. 26: 302-309.

NEGTAP 2001. Transboundary air pollution: Acidification, eutrophication and ground-level ozone in the UK. DEFRA Contract EPG 1/3/153.

Nellmann C., Thomsen M.G. 2001. Long-term changes in forest growth: Potential effects of nitrogen deposition and acidification. - Water Air Soil Poll. 128: 197-205.

Nihlgård B. 1985. The ammonium hypothesis - an additional explanation to the forest dieback in Europe. - Ambio 14: 2-8.

Nilsson L.-O., Persson T. 2001. The Skogaby experiment - effect of N and S deposition to a forest ecosystem. - In: SNV Naturvårdsverket Report 5173 (In Swedish with English abstract).

Nilsson L.-O., Wallander H. 2003. The production of external mycelium by ectomycorrhizal fungi in a Norway spruce forest was reduced in response to nitrogen fertilization. - New Phytol. (*in press*).

Oden S. 1968. The acidification of air and precipitation and its consequences for the natural environment. - Bull. Ecol. Comm. 1: 1-86.

Olsson P.-A., Jacobsen I., Wallander H. 2002. Foraging strategies and resource allocation of mycorrhizal fungi in a patchy environment. - In: van der Heijden M.G.A., Sanders I. (Eds.) Mycorrhizal ecology. - Ecol. Studies 157: 93-115.

Palmer H.-B., Seery D.-J. 1973. Chemistry of pollutant formation in flames. - Annu. Rev. Phys. Chem. 24: 235-262.

Parry W.H. 1974. The effects of nitrogen levels in Sitka spruce needles on *Elatobium abietinum* (Walker) populations in north-east Scotland. - Oceol. 15: 304-20.

Pérez-Soba M. 1995. Physiological modulation of the vitality of Scots pine trees by atmospheric ammonia deposition. - Ph.D. Thesis, University of Groningen, The Netherlands.

Pérez-Soba M., van der Eerden L. J. M. 1993. Nitrogen uptake in needles of Scots pine (*Pinus sylvestris* L.) when exposed to gaseous ammonia and ammonium fertilizer in the soil. - Plant Soil 153: 231-242.

Pérez-Soba M., Stulen I., van der Eerden L.J. 1994. Effect of atmospheric ammonia on the nitrogen metabolism of Scots pine (*Pinus sylvestris*) needles. - Physiol Plant. 90: 629-636.

Persson P., Näsholm T. 2002. Regulation of amino acid uptake in conifers by exogenous and endogenous nitrogen. - Planta 215: 639-644.

Peter M., Ayer F., Egli S. 2001. Nitrogen addition in a Norway spruce stand altered macromycete sporocarp production and below-ground ectomycorrhizal species composition. - New Phytol. 149: 311-325.
Poikolainen J., Lippo H., Hongisto M. Kubin E., Mikkola K., Lindgren M. 1998. On the abundance of epiphytic green algae in relation to the nitrogen concentrations of biomonitors and nitrogen deposition in Finland. - Environ. Pollut. 102: 85-92.
Rabe E. 1990. Stress physiology: The functional significance of the accumulation of nitrogen containing compounds. - J. Hortic. Sci. 65: 235-243.
Rabe E., Lovatt C.J. 1986. Increased arginine biosynthesis during phosphorus deficiency. - Plant Physiol. 81: 774-779.
Raven J.A. 1985. Regulation of pH and generation of osmolarity in vascular plants: a cost-benefit analysis in relation to efficiency of use of energy, nitrogen and water. - New Phytol. 101: 25-77.
Raven J.A. 2003. Can plants rely on nitrate? - Trends Plant Sci. 8: 314.
Rennenberg H., Gessler A. 1999. Consequences of N deposition to forest ecosystems - recent results and future research needs. - Water Air Soil Pollut. 116: 47-64.
Rennenberg H., Kreutzer K., Papen H., Weber P. 1998. Consequences of high loads of nitrogen for spruce (*Picea abies*) and beech (*Fagus sylvatica*) forests. - New Phytol. 139: 71-86.
Reynolds B. 1997. Predicting soil acidification trends at Plynlimon using the SAFE model. - Hydrol. Earth Syst. Sci. 1: 717-728.
Roelofs J.G.M., Kempers A. J., Houdijk A.L.F.M. 1985. The effect of airborne ammonium sulphate on *Pinus nigra* var. *maritime* in the Netherlands. - Plant Soil 84: 45-56.
Rosengren-Brinck U. 1994. The influence of nitrogen on the nutrient status in Norway spruce (*Picea abies* L. Karst). - Ph.D. Thesis, University Lund, Sweden.
Schulze E.-D., Högberg P., van Oene H., Persson T., Harrison A. F. Read D., Kjoller A., Matteucci G. 2000. Interactions between the carbon and nitrogen cycles and the role of biodiversity: A synopsis of a study along a north-south transect through Europe. - In: Schulze E.-D. (Ed.) Carbon and nitrogen cycling in European forest systems. - Ecol. Studies 142: 468-491.
Sheppard L.J. 1994. Causal mechanisms by which sulphate, nitrate and acidity influence frost hardiness in red spruce: review and hypothesis. - New Phytol. 127: 69-82.
Sheppard L.J., Crossley A. 2000. Responses of a Sitka spruce ecosystem after 4 years of simulated wet N deposition: effects of NH_4NO_3 supplied with and without acidity (H_2SO_4 pH 2.5). - Phyton 40: 169-174.
Sheppard L.J., Pfanz H. 2001. Impacts of air pollutants on cold Hardiness. - In: Bigras F. J., Colombo S. J. (Eds.) Conifer cold hardiness. - Dordrecht, The Netherlands, Kluwer Academic Publishers, pp. 335-366.
Sheppard L.J., Cape J.N., Leith I.D. 1993. Influence of acidic mist on frost hardiness and nutrient concentrations in red spruce seedlings. - New Phytol. 124: 607-615.
Sheppard L.J., Crossley A., Harvey F.J., Wilson D., Cape J.N. 1997. Field application of acid mist to a single clone of Sitka spruce: Effects on foliar nutrition and frost hardiness. - Environ. Pollut. 98: 175-184.
Sheppard L.J., Crossley A., Parrington J., Harvey F.J., Cape N. 2001. Effects of simulated acid mist on a Sitka spruce forest approaching canopy closure: significance of acidified versus non-acidified nitrogen inputs. - Water Air Soil Poll. 130: 953-958.
Sheppard L.J., Rosengren U., Emmett B. A. 2003. Do nitrogen additions change the sensitivity of detached shoots from Sitka and Norway spruce to freezing temperatures? evidence from three field manipulation studies. - Scand. J. For. Res. 18: 487-498.
Sigurgeirsson A. 2003. Final reports from SNS projects. - Scand. J. For. Res. 18: 290.
Skeffington R.I., Wilson E. J. 1988. Excess nitrogen deposition: Issues for consideration. - Environ. Pollut. 54: 159-184.

Skiba U., Peirson-Smith T.J., Cresser M.S. 1986. Effects of simulated precipitation acidified with sulphuric and/or nitric acid on the throughfall chemistry of Sitka spruce *Picea sitchensis* and heather *Calluna vulgaris*. - Environ. Pollut. B 11: 255-270.

Smith S.E., Read D.J. 1997. Mycorrhizal Symbiosis. 2nd edition. San Diego, Ca, Academic Press.

Spiecker H. 1999. Overview of recent growth trends in European Forests. Water Air Soil Poll. - 116: 33-46.

Spiecker H., Mielikainen K., Kohl M., Skovsgaard J.P. (Eds.) 1996. European forest institute report 5. - Berlin, Germany, Springer Verlag.

Stadler B., Müller T., Sheppard L., Crossley A. 2001. Effect of *Elatobium abietinum* on nutrient fluxes in Sitka spruce canopies receiving elevated nitrogen and sulphur deposition. - Agric. For. Ent. 3: 253-261.

Standen V. 1982. Associations of *Enchytraeidae* (*Oligochaeta*) in experimentally fertilized grasslands. - J. Anim Ecol. 51: 501-522.

Stewart G.R., Pearson J, Kershaw J.L., Clough E.C.M. 1989. Biochemical aspects of inorganic nitrogen assimilation by woody plants. - Ann. Sci. For. 46: 648s-653s.

Stjernquist I., Selldén G. 2002. Forest vitality and stress implications. - In: Sverdrup H., Stjernquist I. (Eds.) Developing principles and models for sustainable forestry in Sweden. - Dordrecht, The Netherlands, Kluwer Academic Publishers, pp. 197-271.

Suárez M.F., Avila C., Gallardo F., Cantón F. R., Garcia-Gutiérrez A., Claros M.G., Cánovas F.M. 2002. Molecular and enzymatic analysis of ammonium assimilation in woody plants. - J. Exp. Bot. 53: 891-904.

Sutton M.A., Pitcairn C.E.R., Fowler D. 1993a. The exchange of ammonia between the atmosphere and plant communities. - Adv. Ecol. Res., 24: 301-93.

Sutton M.A., Fowler D., Moncrieff J.B. 1993b. Exchange of atmospheric ammonia with vegetated surfaces: Unfertilised vegetation. - Quart. J. R. Meteorol. Soc. 119: 1023-45.

Sutton M.A., Cape J.N., Rihm B., Sheppard L.J., Smith R.I., Spranger T., Fowler D. 2003. The importance of accurate background atmospheric deposition estimates in setting critical loads for nitrogen. - In: Achermann B. (Ed.) Empirical critical loads for nitrogen, expert workshop Berne, 2002. - Environ. Doc. No. 164: 231-257.

Sverdrup H., Hagen-Thorn A., Holmquist J., Wallman P., Warfvinge P., Walse C., Alveteg M. 2002. Biogeochemical processes and mechanisms. - In: Sverdrup H., Stjernquist I. (Eds.) Developing principles and models for sustainable forestry in Sweden – Dordrecht, The Netherlands, Kluwer Academic Publishers, pp. 91-196.

Sverdrup H., Warfvinge P. 1995. Past and future changes in soil acidity and implications for growth under different deposition scenarios. - Ecol. Bull. 44: 335-352.

Tamm C.O., Popovic B. 1995. Long-term field experiments simulating increased deposition of sulphur and nitrogen to forest plots. - Ecol. Bull. 44: 301-321.

Tarrason L., Hjellbrekke A.-G. 2001. Mapping of concentrations in Europe combining measurements and acid deposition models. - Water Air Soil Poll. 130: 1529-1534.

Taylor A.F.S., Martin F., Read D.J. 2000. Fungal diversity in ectomycorrhizal communities of Norway spruce (*Picea abies* (L) *Karst*.) and Beech (*Fagus sylvatica* L.) in forests along north-south transects in Europe. - In: Schulze E.-D. (Ed.) Carbon and nitrogen cycling in European forest ecosystems. Ecol Studies 142: 343-365.

Termorshuizen A.J., Schaffers A. 1991. The decline of carpophores of ectomycorrhizal fungi in stands of *Pinus sylvestris L.* in the Netherlands: possible causes. - Nova Hedwigia 53: 267-289.

Thelin G. 2000. Nutrient imbalance in Norway spruce. - Ph.D. Thesis, University of Lund, Sweden.

Thomas H. 1998. Air today – gone tomorrow? - New Phytol. 139: 225-229.

Tietema A. 1998. Microbial carbon and nitrogen dynamics in coniferous forest floor material collected along a European nitrogen deposition gradient. - For. Ecol. Man. 101: 29-36.

Tischner R. 2000. Nitrate uptake and reduction in higher and lower plants. - Plant Cell Environ. 23: 1005-1024.

UNECE 2003. Empirical Critical Loads for nitrogen 2003. - Expert Workshop Berne 2002 Proceedings SAEFL Berne.
Van der Eerden L.J., Lekkerkerk L.J.A., Smeulders S.M., Jansen A.E. 1992. Effects of atmospheric ammonia and ammonium sulphate on Douglas fir (*Pseudotsuga menziesii*). - Environ. Pollut. 76: 1-9.
Van der Eerden L.J., de Vries W., van Dobben H. 1998. Effects of ammonia deposition on forests in the Netherlands. - Atmos. Environ. 32: 525-532.
Van Dijk H.F.G., de Louw M.H.J., Roelofs J.G.M., Verburgh J.J. 1990. Impact of artificial, ammonium-enriched rainwater on soils and young coniferous trees in a greenhouse. Part II - Effects on the Trees. - Environ. Pollut. 63: 41-59.
Van Dijk H.F.G., Roelofs J.G.M. 1988. Effects of excessive ammonium deposition on the nutritional status and condition of pine needles. - Physiol. Plant. 73: 494-501.
Vogt K.A., Grier C.C., Meier C.E., Edmonds R.L. 1982. Mycorrhizal role in net primary production and nutrient cycling in *Abies amabilis* ecosystems in western Washington. - Ecology 63: 370-380.
Wallander H. 1995. A new hypothesis to explain allocation of dry matter between ectomycorrhizal fungi and pine seedlings. - Plant Soil 168-169: 243-248.
Wallander H., Nylund J.-E. 1992. Effects of excess nitrogen and phosphorous starvation on extrametrical mycelium in Scots pine seedlings. - New Phytol. 120: 495-503.
Wallander H., Massicotte H., Nylund J.-E. 1997. Seasonal variation in ergosterol, chitin and protein in ectomycorrhizal roots collected in a Swedish pine forest. - Soil Biol. Biochem. 29: 45-53.
Wallander H., Nilsson L.-O., Hagerberg D., Baath E. 2001. Estimation of the biomass and seasonal growth of external mycelium of ectomycorrhizal fungi in the field. - New Phytol. 151: 753-760.
Wallenda T., Schaeffer C., Einig W., Wingler A., Hampp R., Seith B., George E., Marschner H. 1996. Effects of varied soil nitrogen supply on Norway spruce (*Picea abies* [L] *Karst.*) II Carbon metabolism in needles and mycorrhizal roots. - Plant Soil 186: 361-369.
Wallenda T., Kottke I. 1998. N deposition and ectomycorrhizas. - New Phytol. 139: 169-187.
Weber P., Stoermer H., Gessler A., Schneider S., Vonsengbusch D., Hanemann U., Rennenberg H. 1998. Metabolic responses of Norway spruce (*Picea abies*) trees to long-term forest management practices and acute $(NH_4)_2SO_4$ fertilisation: transport of soluble non-protein nitrogen compounds in xylem and phloem. - New Phytol. 140: 461-485.
Wiklund K., Nilsson L.-O., Jacobsson S. 1994. Effect of irrigation, fertilization, and artificial drought on basidiomycete production in a Norway spruce stand. - Can. J. Bot. 73: 200-208.
Wilson E.J. 1992. Foliar uptake and release of inorganic nitrogen compounds in *Pinus sylvestris* L. and *Picea abies* (L.) Karst. - New Phytol. 120: 407-416.
Wilson E.J., Tiley C. 1998. Foliar uptake of wet deposited nitrogen by Norway spruce: an experiment using ISN. - Atmos. Environ. 32: 513-518.
Wright R.F., Tietema A. 1995. NITREX – special issue. - For. Ecol. Man. 71: 1-169.
Yin Z-H., Raven J.A. 1997. A comparison of the impacts of various nitrogen sources on acid-base balance in C_3 *Triticum aestivum* L. and C_4 *Zea mays* L. plants. - J. Exp. Bot. 48: 315-324.
Yin Z-H., Kaiser W.M., Heber U., Raven J.A. 1996. Acquisition and assimilation of gaseous ammonia as revealed by intracellular pH changes in leaves of higher plants. - Planta 200: 380-387.
Zhang H., Jennings A., Barlow P.W., Forde B.G. 1999. Dual pathways for regulation of root branching by nitrate. - Proc. Natl. Acad. Sci. USA 96: 6529-6534.

Chapter 4

SYMBIOTIC NITROGEN FIXATION

Javier Ramos and Ton Bisseling

INTRODUCTION

Nitrogen is one of the most abundant elements on earth and occurs in three main reservoirs: lithosphere, atmosphere and biosphere. The complex relationships between these reservoirs constitute the nitrogen cycle. The lithosphere contains the largest reserves of nitrogen, but only a small part is accessible to plants *e.g.* as NO_3^- and NH_4^+. The N content of most surface mineral soils is about 0.02-0.5%. However, most of the soil N is in organic form associated with humus and silicate clays and only about 2-3% of this is mineralised each year (Brady 1990). Thus, the amount of readily available N in the form of NO_3^- and available NH_4^+ compounds is generally only about 1-2% of the total soil N, with the exception to areas where large amounts of fertiliser have been added.

About 80% of the atmosphere is composed of N_2. The latter is the substrate of nitrogen fixation and the products of this process are the primary source of fixed N, in fact for all living organisms including humans. The quantity of nitrogen fixed in nature is about 230×10^6 Tm annually. Of this, approximately 13% is fixed by electric storms (into nitric acid), whereas 87% is the result of biological fixation of nitrogen. Of the latter, 80% is fixed via symbiotic associations and 20% by free living organisms (Gutschick 1980, Vance 1988). The biological costs of fixing N_2 is substantial, being equivalent to about 2.5% of the energy generated by primary photosynthesis on the earth (Gutschick 1980).

In agriculture, in addition to N sources created by biological nitrogen fixation, N-fertilisers are used. The current annual world industrial production of ammonia exceeds 130×10^6 Tm per year (Bakemeier *et al.* 1997), mostly produced by the Haber-Bosch process. The production of artificial fertiliser is expensive due to high energy costs. Further, the excessive application of fertilisers, contributes to environmental hazard by leaching of nitrate into the ground. Thus biological fixation of N_2 has considerable advantages from an ecological as well as an economical point of view (Galloway et al. 1995, Socolow 1999, Sainju et al. 2003).

DIAZOTROPHIC ORGANISMS

The nitrogenase enzyme complex is at the core of biological nitrogen fixation. Only some microorganisms, diazotrophic prokaryotes, possess this enzyme complex and are able to fix nitrogen. Several microbes are able to fix nitrogen in a free-living form.

The biological fixation of nitrogen is the most important way by which organisms have access to N sources. This process requires a fairly high proportion of the ATP, which is generated by oxidative phosphorylation. The nitrogenase enzyme complex, however, is sensitive to O_2, that irreversible inactivates the enzyme. Therefore diazotrophs must employ mechanisms, which permit the supply of O_2 required for energy regeneration and simultaneously nitrogenase has to be protected from the deleterious effect of O_2. The most efficient nitrogen fixing systems are those in which the microbe has established a symbiosis with higher plants. The symbiosis between *Gunnera* and *Nostoc*, legumes and rhizobia and actinorhizal plants and *Frankia* have in common that the nitrogen fixing prokaryote becomes internalised in the cells of their angiosperm host. Cyanobacteria and gram-negative rhizobia are related to the microbes that are endosymbiotic ancestors of chloroplasts and mitochondria, respectively (Andersson et al. 1998, Douglas 1998). However, in contrast to these organelles they are genetically completely independent from the host. On the other hand *Frankia* is a gram-positive bacterium without such relationship with ancient endosymbionts (Allen and Raven 1996).

Plant endosymbioses are characterised by microbes that are hosted by the plant in intracellular manner. However the microorganisms are separated from the host cytoplasm by a plant membrane, the symbiosome membrane (SM) (see below for details). Thus, the microsymbiont may be intracellular, but it is always extracytoplasmic because of the integrity of the SM. In the following paragraphs of this chapter the actinorhizal and cyanobacteria-*Gunnera* symbiosis will briefly be summarized whereas the rhizobial-legume interaction will be described in more detail.

In addition to these intimate and specialised symbiotic associations, there are several free-living nitrogen-fixing bacteria that grow in close association with plants. For example, *Azotobacter*, *Acetobacter* and *Azospirillum* species have been shown to fix nitrogen when growing in the rhizosphere of tropical grasses, as root surface-colonizing bacteria, or as endophytic diazotrophs (Steenhoudt et al. 2000). Often they are found in association with economically important Gramineae as sugarcane (Elbeltagy et al. 2001, Sevilla et al. 2001).

Gunnera hosting cyanobacteria

Cyanobacteria have co-evolved with the changing oxidation state of the ocean and atmosphere to accommodate the machinery of oxygenic photosynthesis and oxygen-sensitive N_2 fixation. Therefore they are unique in the sense that they produce oxygen and have to protect the nitrogenase from oxygen damage (Berman-Frank *et al.* 2003). In the most advanced cases this is solved by the synthesis of nitrogenase in specialized cells, the so-called heterocysts. These cells have a thicker cell wall, that contains polysaccharides and glycolipids, which restricts O_2 diffusion. Heterocysts import disaccharides from vegetative cells and export fixed N as glutamine (Adams 2000). Nitrogen-fixing cyanobacteria can be either free-living or as partners in a variety of symbioses and hetrocysts can be formed by both (Rai *et al.* 2000). Some cyanobacteria (particularly *Nostoc*) are able to establish symbiotic nitrogen fixation associations with lichens, bryophytes (notably leafy liverworts, Adams 2002), aquatic ferns (Braun-Howland and Nierzwicki-Bauer 1990), gymnosperms (coralloid tips in cycads, Chang *et al.* 1988) and the angiosperm *Gunnera*. *Gunnera* sp. can contain nitrogen-fixing cyanobacteria in specialized gland organs located at the base of petioles (Bergman *et al.* 1992). *Gunnera* is often referred to as the only angiosperm that forms a nitrogen-fixing symbiosis with cyanobacteria, however, this is not strictly true since some tropical angiosperms have cyanobacterial films on their leaf surfaces. Nevertheless *Gunnera* is unique since the bacteria (*Nostoc*) enter the *Gunnera* cells, where they form many heterocysts (Meeks and Elhai 2002). *Nostoc* strains are compatible with several *Gunnera* species and further cyanobacterial isolates from the same *Gunnera* species are genetically diverse (Rassmussen and Svenning 2001). This suggests that the host specificity of the *Nostoc/Gunnera* symbiosis is not very strict. In the case of *Gunnera/Nostoc* symbiosis no new structures are developed for supply and protection from oxygen, as the microbes are hosted by an already existing gland of the plant and nitrogen fixation takes place in heterocysts. In legumes and actinorhizal plants, new organs, *e.g.* root nodules, are formed.

Actinorhizal nodules

Another group of microbes able to establish endosymbiotic associations are actinomycetes belonging to the genus *Frankia*. *Frankia*, normally grows in hyphal form. They are slow-growing, aerobic heterotrophs (Schwencke 1991, Benson and Sylvester 1993). They are able to form nitrogen-fixing root structures known as nodules (sometimes called actinorhizae) with more than 200 woody plants species, belonging to eight different non-legume angiosperm families (Bond 1983, Baker and Mullin 1992, Pawlovski and Bisseling 1996, Wall 2000). In contrast to the rhizobial legume interaction the host range is rather broad. These host plants are named actinorhizal plants. Actinorhizal plants are generally bushes or trees of ecological relevance (examples are *Alnus*, *Casuarina*, *Coriaria*, *Myrica*), and further they are grown together with other trees to improve soil fertility and forestry production. Alder and the other woody hosts of *Frankia* are typical pioneer species, important in the colonization of poor soils or coastal dunes. These plants probably benefit from the nitrogen-fixing association, while supplying the bacterial symbiont with photosynthetic products (Pawlowski and Bisseling 1996).

Molecular phylogenetic studies revealed that all actinorhizal plants belong to a single clade named Rosid I. Further, legumes as well as the only non-legume rhizobia-host, *Parasponia*, belong to this clade. The phylogenetic relationship of the species belonging to this clade and their (dis)ability to nodulate has led to the hypothesis that the ability to establish a nitrogen fixing symbiosis has evolved (and got lost) several times within this clade. Although predisposition/ability to establish a nodule symbiosis is restricted to plant species belonging to a single clade it is unlikely that species belonging to this clade have a large set of symbiosis specific genes. This idea is supported by studies showing that genes and processes underlying nodulation are in fact recruited from processes that are widespread among the higher plants, so there are perhaps a few symbioses specific genes that determine the predisposition to nodulation (Gualtieri and Bisseling 2000). Maybe this set of genes has already been identified by genetic screens and (part) of these will be discussed below (Figure 2).

Actinorrhizal nodules are composed of lobes, which are modified lateral roots. Woronin described in 1866 the microbe in these nodules as a parasitic fungus having hyphae. Later, the microbe was correctly identified as belonging to the order Actinomycetales, family *Frankiaceae*, genus *Frankia* (Becking *et al.* 1964, Lechevalier and Lechevalier 1979). Like rhizobia, some *Frankia* species infect the root system by penetrating root hairs (Berry and Torey 1983, Berry *et al.* 1986). In these cases root hairs deform, getting twisted and branched (van Ghelue *et al.* 1997). When *Frankia* enter the root hair, they become embedded in a layer of cell wall-like material (Berry and Torey 1983) surrounded by pectinaceous matrix, creating a tubular structure comparable to the rhizobial infection thread (see below). In general, several root hair infections are involved in the infection of one nodule (Diem *et al.* 1983). In case *Frankia* enters the plant by infection thread like structures, a pre-nodule, is formed by cortical cell division near the infection site, before *Frankia* actually enters the plant root (Berry *et al.* 1989).

A second mode of infection involves intercellular penetration of the root epidermis, and colonization of the intercellular spaces of the root cortex. In these cases a prenodule is not formed. After initial extra-cellular growth, *Frankia* eventually penetrates hosts cells. This mode of infection occurs among others in hosts of the *Eleagnaceae* (Miller and Baker 1985). Which type of infection used is primarily determined by the host plant, as particular *Frankia* strain can use both modes of infection, depending on the host species.

Ultimately *Frankia* invade lateral root primordia which development is subsequently altered and a nodule like structure is formed (Angulo *et al.* 1975). Since actinorhizal nodules are formed from root primordia they posses a central vascular bundle. Around the vascular tissue *Frankia* infects the cortical cells. The expansion of those cells produces the swelling of the root nodule. Within these cells *Frankia* hyphae branch and multiply, while the hosts cells surround them with a plasma membrane and cell wall material. Eventually in several cases terminal swellings are formed at the hyphae extremes (vesicles) where nitrogen fixation will take place. Microscopic studies showed that the cell envelope surrounding the vesicles resembles the glycolipid membranes of cyanobacteria heterocysts (Torrey and Callaham 1982), forming a diffusion barrier to oxygen and therefore protecting the nitrogenase complex. In some species the composition of the plant cell wall (*Myrica*, Zeng and Tjepkema 1994) as well as a haemoglobin (*Casuarina*, Gherbi *et al.* 1997) also contribute to the protection against oxygen.

NITROGEN FIXATION OF THE RHIZOBIAL - LEGUME SYMBIOSIS

Rhizobia is the common name given to a group of small, rod-shaped, gram negative bacteria which collectively have the ability to infect the roots (or, in some cases, the stems) of leguminous plants and produce nodules (van Berkum and Eardly 1998). Once the nodule is formed, the differentiated bacteria (bacteroids) living in the infected plant cells, reduce atmospheric nitrogen to ammonia which is excreted to the plant cell and is, in turn, assimilated to organic nitrogen by the plant. The plant provides the bacteroid with carbon skeletons (photosynthate) which are needed by *Rhizobium*, a strict aerobe, to provide the energy that is needed for nitrogen fixation.

Because of their great diversity legumes are widely adapted to different climates and soils. Many legume species are able to establish a root nodule symbiosis with rhizobial bacteria, which enables the plant to fix atmospheric nitrogen and become less dependent on nitrogen sources from the soil. This property has been studied in about 60% of legume species, and the percentage of species that can be nodulated in the three legume families is: 23% in *Ceasalpinioideae*, 90% in *Mimosoideae* and 97% in *Papilionoideae* (de Faria *et al.* 1989, van Berkum and Eardly 1998). Legumes are among the world's most important agricultural crops, being a major source of protein, oil, and forage for human and animal consumption (Phillips 1993). Therefore from the start of agriculture legumes have been used as rotation crop together with cereals (Lev-Yadun *et al.* 2000). Humans also indirectly benefit from nitrogen fixation as legumes are cultivated as forage and pasturage for cattle, that assure the supply of meat, milk and other products (Harwood 1990). Furthermore, legumes are beneficial to soil enrichment, contributing to about 80% of biological nitrogen fixation in agriculture (Vance 1996).

With one notable exception, namely *Parasponia* of the elm family (Trinick 1979), the ability to form nitrogen-fixing symbiosis with rhizobia is restricted to the legume family. The nodulation process has been studied in detail in a few legume species, therefore it can not give a complete picture on the existing diversity (chemical communication, invasion routes etc.). Nevertheless the observed mechanisms are most likely involved in most associations of rhizobia and legumes.

Molecular studies of the symbiosis are greatly facilitated by selecting *Medicago truncatula* and *Lotus japonicus* as model legumes, respectively, for indeterminate or determinate nodulation (see below; Cook, 1999, Stougaard, 2001, Udvardi, 2001). Both legume species have a small diploid genome, are autogamous, have a short generation time and large seed production, and are amenable for transformation and mutants have been generated. In addition, the chloroplastic genome of *L. japonicus* and the genomes of *Sinorhizobium meliloti* and *Mesorhizobium loti* (the bacterial partners of these model legumes) have been entirely sequenced. Further the nuclear genomes of *M. truncatula* and *L. japonicus* are being sequenced at a fast pace and will be most likely available by 2006 (VandenBosch and Stacey 2003).

Taxonomy of rhizobia

In 1888, Beijerinck first isolated a bacterium from legume root nodules, which he named *Bacillus radicicola*. Since then, the taxonomy and nomenclature of the root nodule bacteria, has been in constant review. After Frank (1889) named the bacterium *Rhizobium leguminosarum*, it was observed that no strain could nodulate all legumes, but that each could only nodulate some legumes. This led to the concept of cross-inoculation groups, with organisms grouped according to the hosts they nodulate. Some were described as having a narrow host range whereas others have a broad host range. Species description was then primarily based on the cross-inoculation groups. Fred *et al.* (1932) first recognized the taxonomic diversity of root nodule bacteria and classified them based on growth rates, and also established their relationship with agrobacteria.

The taxonomic classification based on cross-inoculation groups was maintained till 1980 (Skerman, 1980) and *Rhizobiacea* were divided in 8 groups: those nodulating *Medicago*, *Melilotus* and *Trigonella* (*R. meliloti*); clover (*R. trifolii*); pea, lentil, *Vicia* (*R. leguminosarum*); bean (*R. phaseoli*); lupine (*R. lupini*); soybean (*R. japonicum*); cowpea, acacia, peanuts (*Rhizobium spp.*) and *Lotus* (*Rhizobium spp.*). Thus, rhizobia isolated from one species of clover would usually nodulate other species of clover, and were therefore collectively called *R. trifolii*, while rhizobia isolated from lucern would also nodulate *Melilotus* and *Trigonella* and were called *R. meliloti*. Changes in the methods used in bacterial taxonomy (Graham *et al.* 1991) and advances in molecular biology have currently resulted in a rhizobial taxonomy based on DNA sequences and a wide range of other characteristics This has resulted in six genera that can nodulate legumes (Young and Haukka 1996, van Berkum and Eardly 1998).

– The genus *Rhizobium* (Jordan 1984), forms nodules on legumes from areas with temperate climates. They are rather fast growing bacteria with a generation time of about 2 hours. They have a narrow host range and nodulation and nitrogen fixation genes are located on rather large plasmids. *R. leguminosarum* has been subdivided in biovars, bv. *viciae, trifolii*.

– *Allorhizobium* (de Lajudie *et al.* 1998) efficiently nodulates *Neptunia natans*, a stem nodulated tropical legume found in waterlogged areas of Senegal. It is debated if it should be part of *Rhizobium* (Tighe *et al.* 2000).

– *Bradyrhizobium* (Jordan 1982) can be distinguished from other rhizobia by their slow growth rate in culture and the alkalinisation of the media in which they grow. Their genes related to symbiosis are located on the chromosome. They have a broad host range and form nodules on legumes of tropical and subtropical origin. The symbionts of soybean (formerly *Rhizobium japonicum*) are included in this group. They also nodulate the non-leguminous *Parasponia* (*Ulmacea*). The taxonomic relationships inside the group are still little studied, but it seems very heterogenous (Jordan 1982).

– *Mesorhizobium* (Jarvis *et al.* 1997) is a genus recently described (formerly *Rhizobium*). The intermediate growth speed compared to *Rhizobium* and *Bradirhizobium* has been the basis for its name. Their symbiotic genes are located in their genome. Hosts include *Lotus, Astragalus* and chickpea.

– *Sinorhizobium* (de Lajudie *et al.* 1994) has phenotypic characteristics similar to *Rhizobium*, but it is genetically distant. This genus includes the microsymbiont of *Medicago, Trigonella* and *Melilotus*, which is named *S. meliloti*, (Crow *et al.* 1981). This genus also includes species that nodulate *Sesbania, Acacia* and *Leucaena*.

– *Azorhizobium* (Dreyfus *et al.* 1988) is able to form nodules on the stems of *Sesbania rostrata*. Molecular data clearly demonstrated that this genus is distinct from other members of the rhizobia (Willems and Collins 1993) and a close relative of *Xantobacter*.

The publication of the full sequence of the symbiotic plasmid of *Sinorhizobium* sp. NGR234 (Freiberg *et al.* 1997), and the full genome sequence of *Mesorhizobium loti* (Kaneko *et al.* 2000) and *Sinorhizobium meliloti* (Galibert *et al.* 2001) has provided a wealth of data that most likely will lead to further refinement of the rhizobial taxonomy (Downie and Young 2001).

The above named genera that are able to form nodules on legumes will be named rhizobia in the rest of this chapter. All rhizobia are gram-negative aerobic bacteria, they have usually a rod-shaped form and do not form spores. Most rhizobia only fix nitrogen within nodules. However a few are known to fix nitrogen as free living bacteria. Rhizobia can actively move by the use of a polar or subpolar flagella, or 2 to 6 peritrichous flagella's (Yang and Lin 1998). In cultures, bacterial colonies have a mucilaginous nature because of extracellular polysaccharides. Rhizobia live as saprophytes in soil, competing with chemiheterotrophic bacteria. Root exudates stimulate growth of specific rhizobia and other microbes by which the host "optimises" the composition of the microbe population in the rhizosphere (Bladergroen and Spaink 1998).

Early events in legume nodule formation

Colonization of legume roots by *Rhizobium* starts when the bacteria move chemotactically to the roots and there compounds present in the root exudate stimulate bacterial growth. Legume roots secrete flavonoids, such as luteolin and naringenin (Charrier *et al.* 1995, Shirley 1996, Bladergroen and Spaink 1998), that serve as chemoattractant to rhizobia (Caetano-Anolles *et al.* 1988). This is interesting as these compounds also activate a set of rhizobial genes that are involved in the synthesis of a compound molecule that sets in motion the process of root nodule formation (Zuanazzi *et al.* 1998). This set of rhizobial genes is named nodulation (*nod*) genes and the secreted signal molecule is the so-called Nod factor. Rhizobia produce Nod factors with a similar basic structure. This basic structure is made up of a backbone of three to five β 1,4-linked N-acetyl-D-glucosamine residues and a fatty acyl chain attached to the non-reducing sugar residue. Based on their structure Nod factors are also named lipochitooligosaccharides (Dénarié *et al.* 1996, Downie 1998). The variation in Nod factor structure is caused by specific substitutions at the terminal sugar residues as well as by the structure of the acyl chain. Some *nod* genes have a regulatory function, like *nodD* that is constitutively active and encodes the protein that together with a plant factor as coinducer, often a flavonoid, will activate the transcription of the other *nod* genes (Schlaman *et al.* 1998). Thus, NodD acts as a sensor of a plant signal and also as an activator of transcription of *nod* loci. Some *nod* genes are involved in the synthesis of the chitooligosaccharide (*nodA*, *nodB*, *nodC*) which is acylated by the products of the *nodF* and *nodE* genes (Carlson *et al.* 1994). Other *nod* genes are responsible for specific substitutions on the reducing or non-reducing terminal sugar residue, respectively. For example, the major Nod factor of *S. meliloti*, contains a backbone of four glucosamine units, an acyl chain of 16 C-atoms in length with two unsaturated bonds (determined by *nodE* and *nodF*), an acetyl group at the non-reducing terminal sugar residue (determined by *nodL*), and a sulfate group at the reducing terminal sugar residue (determined by *nodH*, *nodP* and *nodQ*) (Lerouge *et al.* 1990, Carlson *et al.* 1994, Figure 1).

As described above host specificity is a typical characteristic of the rhizobial legume interaction. Nod factor structure is one of the levels at which host specificity is regulated (Perret *et al.* 2000). A comparison of Nod factor structure and host and microbe phylogeny provided interesting clues about the mechanism by which rhizobia obtained the ability to establish a symbiosis with legumes. Rhizobia interacting with members of the *Leguminosae* are taxonomically very diverse and do not form a discrete family/genus (Berkum and Eardly 1998). Furthermore, several closely related legume species are infected by distantly related bacteria and conversely closely related rhizobia can interact with legume species belonging to different tribes. However a comparison of the structure of rhizobial Nod factors shows that the phylogeny of Nod factor structure coincides with that of host phylogeny. In other words, if closely related legumes interact with unrelated rhizobia these bacteria produce Nod factors with a similar structure. This suggests that rhizobia obtained *nod* genes by horizontal gene transfer, which is consistent with the fact that in most cases *nod* genes are present on a plasmid (Farrand 1998).

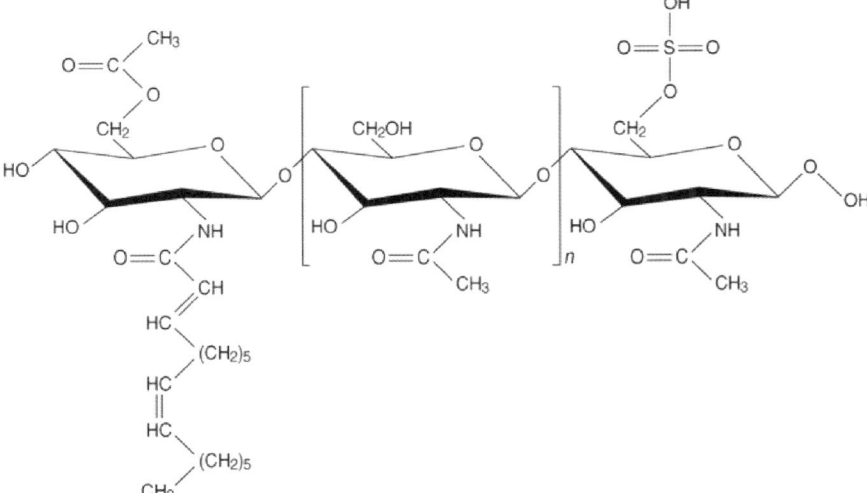

Figure 1. The major Nod factor produced by *Sinorhizobium meliloti*. It is characteristic the specific decoration at the reducing terminal sugar unit and the structure of the acyl chain (Lerouge *et al.* 1990). *n* can be 4 or 5. Pentameric Nod factors can be partially acetylated at their reducing terminal sugar residue.

INFECTION ACTIVITY AND MODE OF ACTION OF NOD FACTORS

The availability of purified Nod factors made it possible to answer the question to what extent these compounds induce responses in the host that normally are induced by the rhizobia. Such studies have been done with several legume species. Legumes turn out to be very sensitive to Nod factors as they can recognize very low concentrations, in some cases at a picomolar level. Therefore, it is often suggested that Nod factors are recognized by a high affinity receptor (Dénarié et al. 1996, Heidstra and Bisseling 1996). Further, the amphiphilic nature of Nod factors, with their hydrophobic lipid tail and hydrophilic sugar backbone, suggests that Nod factor receptors are located in the plasma membrane. The latter is supported by *in vitro* studies showing that Nod factors rapidly insert into membranes (Goedhart et al. 1999).

When roots are treated with Nod factors this induces responses in epidermis, cortex and pericycle. Like rhizobia, purified Nod factors induce *ENOD40* expression in the pericycle within a few hours after "inoculation" (Yang et al. 1993). Further, in all legume species tested so far Nod factors are sufficient to induce cortical cell division that develop into primordia. In some species, *e.g.* alfalfa, complete nodules are formed (Truchet et al. 1991, Hadri and Bisseling 1998). The latter have the same tissue organization as rhizobia induced nodules although, of course, cells are not infected by bacteria. When rhizobia induce nodules these are preferentially formed opposite protoxylem poles. Strikingly, when roots are submerged in a solution containing Nod factors primordia are still preferentially formed opposite protoxylem poles. This implies that the plant provides positional cues by which the primordia are formed at specific places in the roots. This is most likely achieved by the local production of ethylene (Heidstra et al. 1997) and uridine (Smit et al. 1993) that act as a negative and positive regulator, respectively (Hadri and Bisseling 1998).

Nod factors also induce several responses in the root hairs and other epidermal cells. For example Nod factors at picomolar concentrations are sufficient to induce root hair deformations. (Heidstra et al. 1994). Purified Nod factors are in general not sufficient to induce root hair curling or infection thread formation. The latter require the presence of the bacteria. However, as rhizobial mutants that do not produce Nod factors are unable to induce root hair curling or infection thread formation it seems probable that Nod factors are essential for the induction of these processes (Heidstra et al. 1994). In addition to root hair deformation Nod factors induce various other responses in the epidermis. Examples are alkalinisation of the medium, calcium influx at the root hair tip, calcium spiking, and membrane depolarisation (Ehrhardt et al. 1992, Felle et al. 1995, 1996, Kurkdijan 1995, Shaw and Long 2003). Also genes are induced in the root epidermis within a few hours after of which the best studied example is ENOD12 and ENOD11 (Horvath et al. 1993, Journet et al. 1994, Heidstra et al. 1997, Catoira et al. 2000). In most cases the biological significance of the response is not clear, nevertheless such responses turned out to be very good assays to unravel Nod factor perception and transduction mechanisms.

At cortical cells, purified Nod factors induce cell division and preinfection thread formation (see below), but infection threads are not formed unless rhizobia are present (van Brussel et al. 1992). So cortical cells re-enter the cell cycle and form a primordium but cortical cells preparing for infection thread penetration (preinfection thread formation) do not divide, and get arrested in G2 phase (Yang *et al.* 1993 and see below).

In temperate legumes, the inner cortical cells, specially those located opposite prototype poles, divide (Kijne 1992). In tropical legumes the outer cortical cells are the ones mitotically activated. The Nod factors that induce responses in the first group have highly unsaturated fatty acyl groups, while the Nod factors of rhizobia that nodulate tropical legumes generally contain a C18:1 acyl group.

Nod factor perception

After Nod factors had been identified and it became clear that these molecules were active at very low concentrations and in a structure dependent manner it seemed probable that Nod factors are recognized by receptors. Several studies have been initiated aiming to identify such receptors. First biochemical approaches were used to identify Nod factor binding proteins. More recently these biochemical studies have been complemented with genetic approaches in model legumes aiming to dissect the Nod factor perception/signalling cascade. First we will summarize the biochemical approaches to identify Nod factor binding proteins (Cullimore *et al.* 2001, Oldroyd *et al.* 2001, Geurts and Bisseling 2002).

Biochemical approach

Radiolabelled Nod factors were used to identify binding proteins in *Medicago* species. Two different Nod factor binding sites, NFBS1 and NFBS2, have been identified (Cullimore *et al.* 2001). These molecules efficiently bind Nod factors. However, they do not discriminate between biologically active *S. meliloti* Nod factors that contain the *O*-sulfate at the reducing end or an inactive molecule that lacks this substitution. The finding that biological activity and binding properties do not perfectly match makes it unlikely that NFBS1 and/or NFBS2 are specific Nod factor receptors. However, it is still well possible that they are part of a Nod factor binding complex in combination with other components (Cullimore *et al.* 2001).

To identify Nod factor binding proteins educated guesses have been made as well. A promising candidate is a lectin-nucleotide phosphohydrolase (LNP) first identified in *Dolichos biflorus* that can bind Nod factors. *DbLNP* does not show any sequence homology to conventional lectins and has an apyrase activity (Etzler *et al.* 1999). Support for its involvement in Nod factor perception came from studies in which roots where pre-treated with an antibody against LNP. This pre-treatment blocked root hair deformation as well as nodulation by rhizobia (Etzler *et al.* 1999).

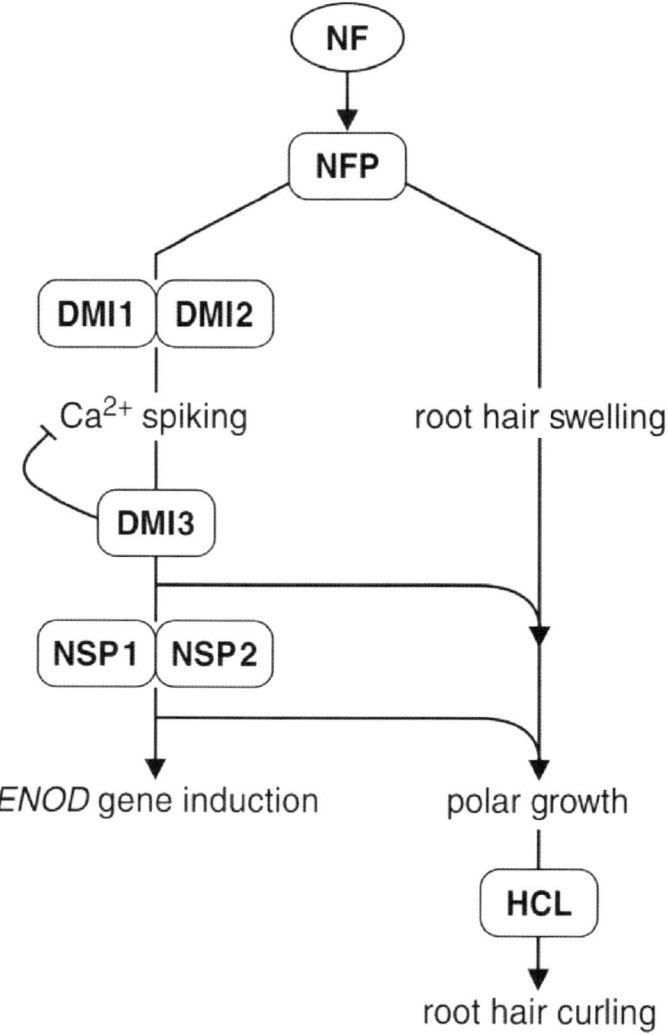

Figure 2. Genetic dissection of the Nod factor signaling pathway. *MtNFP* is blocked in all Nod factor induced responses, but the AM symbiosis is not affected. NFP is a good candidate to be directly involved in Nod factor perception. *Mtdmi1,2* and *3* mutants still respond to Nod factors with root hair swelling but they fail to reestablish tip growth. Since *Mtdmi1* and *2* mutants can not respond with calcium spiking they are placed upstream of *Mtdmi3*. *Mtdmi1,2* and *3* mutants fail to establish the AM symbiosis and therefore are part of a common signaling pathway activated by both microsymbionts. In *nsp* mutants tip growth can be reestablished but in a rather inefficient manner. Further the induction of gene expression is blocked. HCL acts downstream of NSP and in this mutant curling and infection are disturbed. (From Limpens and Bisseling 2003).

Genetic approach

To unravel the mechanisms underlying Nod factor perception and transduction a genetic approach has recently been applied. Especially in the two model legumes, *Lotus* and *Medicago* (Stougaard 2001, Cook *et al.* 1999) and in pea (Borisov *et al.* 2000), several genetic loci have been identified that are essential for the early steps in nodulation. In most cases it is still unclear whether these are orthologous. Nevertheless we will name such loci according to the *Medicago* nomenclature (for example *DOES NOT MAKE INFECTIONS, DMI*).

A set of mutants (*dmi1, dmi2* and *dmi3* in *Medicago*) is blocked in almost all early steps of nodulation. In addition these mutants are also blocked in their interaction with endomycorrhizal fungi (see below, Catoira *et al.* 2000). The fact that the *dmi* mutants are disturbed in both the *Rhizobium* and arbuscular mycorrhiza (AM) symbiosis, shows that a part of the Nod factor signalling pathway is shared with a pathway induced by a molecule (Myc factor) secreted by AM fungi (Albrecht *et al.* 1999, Catoira *et al.* 2000, Stougaard 2001). This turned out to be a useful property to dissect the signalling cascade The root surface of *dmi1, dmi2* and *dmi3* mutants can be colonized by AM fungi but the interaction is blocked at the appressorium stage. Upon Nod factor treatment or inoculation with rhizobia the only detectable morphological response is the swelling of root hair tips showing that root hair deformation depends only in part on DMI1, DMI2 and DMI3 This conclusion is drawn because swellings are induced, whereas efficient induction of tip growth leading to deformed hairs depends on these proteins. The signal molecule of AM fungi has not yet been characterized, but recent studies demonstrate that they indeed produce a diffusible signal molecule (Myc factor) that can induce responses in the legume host (Kosuta *et al.* 2003).

As described above Nod factors induce calcium spiking in root hairs of legumes (Ehrhardt *et al.* 1996, Long 1996, Wais *et al.* 2000, Walker *et al.* 2000). This response has been very useful to position the *DMI* genes in the signalling pathway. In the *dmi1* and *dmi2* mutants calcium spiking is not induced by Nod factors, whereas this response is induced in the *dmi3* mutant (Wais *et al.* 2000). This indicates that DMI1 and DMI2 act upstream of DMI3 in the signalling pathway.

Since DMI1 and DMI2 act in an early step of both nodulation and mycorrhization, it seems possible that they could play a role in Nod and Myc factor perception. *DMI2*, and its orthologs from pea, alfalfa (*NORK*) and *Lotus* (*SYMRK*), has recently been cloned (Endre *et al.* 2002, Stracke *et al.* 2002). It encodes a receptor-like kinase (RLK), containing a characteristic extracellular domain with three leucine-rich repeats (LRRs), a transmembrane domain and an intracellular serine/threonine kinase domain. This is in line with a function in perception and transduction of an extracellular signal. However, it seems unlikely that a DMI2 like receptor is sufficient for the perception of both Nod and Myc factors. First of all the morphological responses induced by mycorrhizal fungi and rhizobia are different (Albrecht et al. 1999). Second, Nod factors induce *MtENOD11* in the epidermis, whereas Myc factors induce the expression of this gene in the root cortex (Kosuta et al. 2003). These observations strongly suggest that these factors are recognized by different receptors. Therefore it seemed probable that a mutation in a Nod factor receptor gene, active upstream of *DMI1* and *DMI2*, would not affect the mycorrhizal interaction, whereas it would eliminate all Nod factor induced responses including calcium spiking and root hair deformation/swelling. Such loci were found in *Medicago* (*NOD FACTOR PERCEPTION, NFP*) and in pea, and *Lotus* (Kistner and Parniske 2002, Geurts and Bisseling 2002, Stougaard 2001, Walker et al. 2000, Amor et al. 2003). NFP loci have recently been cloned from *Lotus* (*LjSym1* and *LjSym5*) and pea (*Sym10*) and these are receptor kinases containing a LysM domain (Radutoiu et al. 2003, Madsen et al. 2003) and have a similar structure as the Nod factor uptake receptor described below.

Genetic studies have also identified a set of mutants that are blocked in Nod factor signal transduction downstream of the calcium spiking response and *DMI3*. All these mutants play a role in a nodulation specific signal transduction pathway as they are still able to establish a normal mycorrhizal association. These mutants have in common that root hair deformation is induced but proper curling does not take place, and as a result, infection is blocked in all these mutants. An example is the *Lotus NIN* gene which was the first gene involved in Nod factor signalling that was cloned (Schauser et al. 1999, Borisov et al. 2003), *NIN* is a putative transcription factor with a typical DNA binding/dimerisation domain in the carboxy-terminal half of the protein and two putative membrane spanning helices.

Several studies indicate that Nod factor perception involves more than one receptor. Especially the infection process requires Nod factors with a very specific structure. For example in the *S. meliloti-Medicago* interaction the acetate substitution at the non-reducing end, and the specific lipid are essential for the induction of infection whereas responses like root hair curling and early nodulin gene induction are not affected. Based on these observations the existence of a Nod factor signalling and a Nod factor entry receptor has been proposed. The latter has a specific function in the infection process (Ardourel et al. 1994). Similarly, in the pea ecotype Afghanistan a *Sym2* allele has been identified whose presence requires an acetate substitution at the reducing end of the Nod factor. In the absence of this substitution the infection process is specifically blocked. Therefore SYM2 is a good candidate to be a Nod factor entry receptor. Recently, *Medicago* genes that have been cloned from the *Sym2* orthologous region of *M. truncatula* have a similar specific function in the initiation of infection as the pea SYM2. These are LysM domain receptor kinases (Limpens et al. 2003), showing that the Nod factor signalling receptor and the entry receptor have a similar structure.

Infection

The formation of a root nodule requires that two major processes are set in motion. On one hand cortical cells have to be dedifferentiated to form a nodule primordium and further the bacteria have to enter the plant, a process under strict control of the plant.

Rhizobia colonize the surface of plant roots forming microcolonies or biofilms. Adhesins and lectins play a role in this process, as well as the formation of cellulose fibrils as a second step (Matthysse and Kijne 1998). In the best studied infection type, rhizobia invade small emerging root hairs. In some way rhizobia are able to redirect the growth of a root hair by which a curl with a shepherd's crook morphology is formed (van Batenburg *et al.* 1986, Esseling *et al.* 2003). During the curling process, the bacteria become entrapped in the pocket of the curl. Subsequently, this microcolony induces the hydrolysis of the plant cell wall in a very local manner and the plasma membrane invaginates (Bauer 1981). New plant material is deposited and a tube-like structure is produced within the root hair cell, the so-called infection thread (Brewin 1998). The bacteria proliferate in the infection thread, that progresses towards the base of the root hair. The infection thread is external to the cell, as it is surrounded by a plant membrane and it contains a matrix of bacterial origin and glycoprotein of vegetal origin (Brewin 1998). When the infection thread reaches the base of the hair the bacteria are released in the intercellular space and a new infection thread is formed in a cortical cell (Figure 3).

In some legumes infection does not start with the formation of infection threads. Instead the bacteria enter in an intercellular manner by growing through the middle lamella between contiguous epidermal cells or enter the root at a site where a lateral root emerges and the cortex and epidermis are damaged (Chandler *et al.* 1978, de Faria *et al.* 1988). The latter is a rather primitive infection process as the intracellular mode of infection seems to provide the plant a better control on the growth of the microsymbiont.

Prior to infection thread growth into the cortex the outer cortical cells undergo morphological changes by entering the cell cycle which arrests at G2 (Yang *et al.* 1994). It has in common with dividings cells that DNA is duplicated but these cells do not get arrested at G2. In this way cytoplasmic bridges are formed, that provide the path through which the infection thread can grow. These radial aligned cytoplasmic bridges are called pre-infection threads (Rae *et al.* 1992, van Brussel *et al.* 1992).

In addition to Nod factors, rhizobial surface polysaccharides are essential factors in the root infection process. The term exopolysaccharide (EPS) is used for polysaccharides with little or no cell association (Leight and Walker 1994, Becker and Pühler 1998). They contribute to bacteria protection, antigenicity, nutrient gathering and attachment to surfaces. The ability to establish an effective symbiosis is severely affected in many rhizobial mutants deficient in

EPS production, inducing nodules devoid of bacteria due to abortion of infection threads or no infecting at all (Rolfe *et al.* 1996). Another complex bacterial cell wall molecule, lipopolysaccharides (LPS) is also often required for proper infection. For example *S. meliloti* carrying an *lpsB* mutation induces infection thread formation but these are defective in the invasion of plant nodule cells, which in part might be due to an effect on endocytosis (Perotto *et al.* 1994, Campbell *et al.* 2002, Campbell *et al.* 2003).

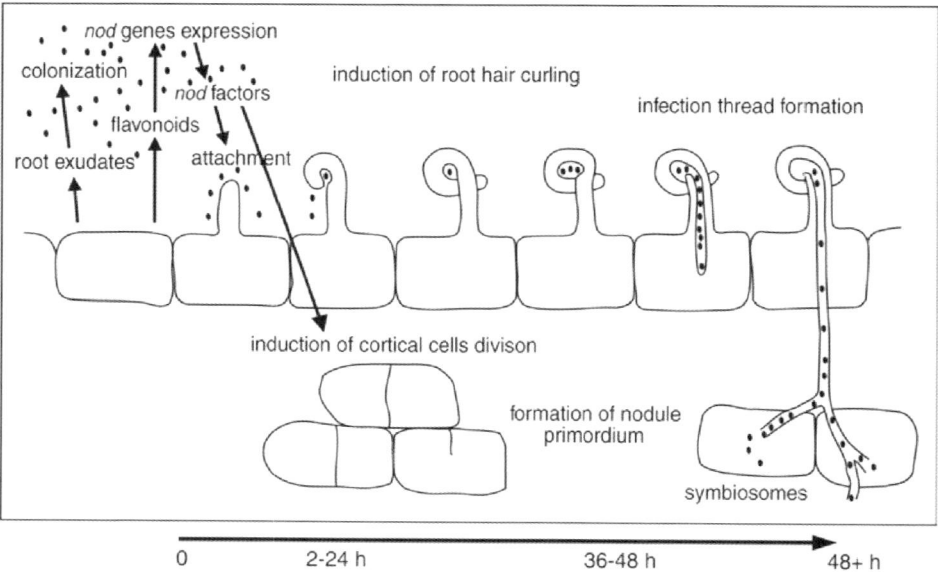

Figure 3. The infection process through the root hairs and the simultaneous formation of the nodule primordium.

Nodule ontogeny and tissue organisation

When the infection thread reaches the nodule primordia, differentiation into a nodule starts. Legume nodules can be divided into two types, determinate and indeterminate nodules. Which nodule type is formed is a characteristic of the host plant (Nap and Bisseling 1990, Mylona *et al.* 1995, Hadri *et al.* 1998). The overall tissue organization of both forms is similar in having a central tissue, with infected cells that harbour the bacteroids and uninfected cells that are interspersed in between the infected cells. These so-called uninfected cells have a specific function in the nitrogen assimilation process. The central tissue is surrounded by a nodule cortex, endodermis and nodule parenchyma. The latter tissue contains the nodule vascular bundles (van de Wiel *et al.* 1990). The difference between the two nodule types is that indeterminate nodules have a persistent meristem at their tip whereas in determinate nodules meristematic activity ceases at an early stage of development. Another major difference concerns their ontogeny as indeterminate nodules originate from primordia formed in the inner cortex. Primordia formed in the outer cortex are the start of determinate nodule development.

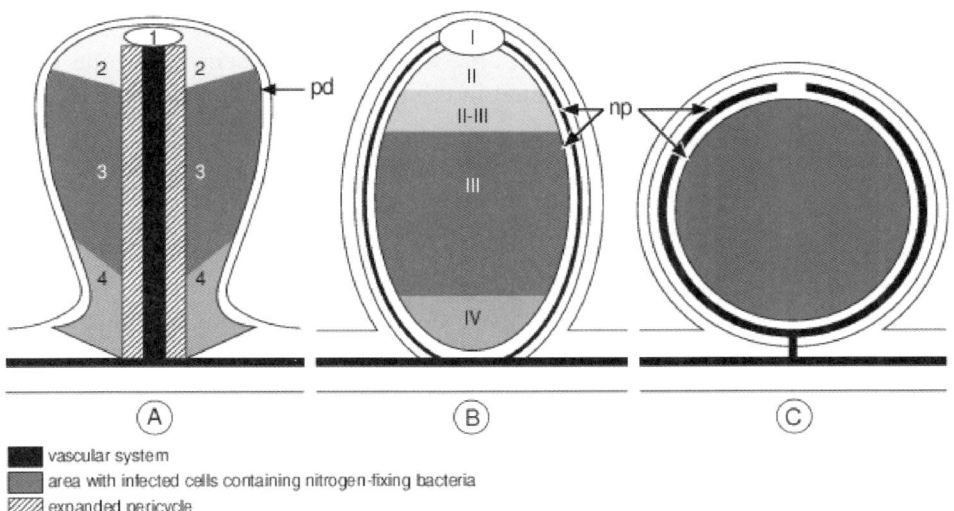

Figure 4. Structures of nodules. A) - Lobe of an actinorhizal nodule. The lobe is surrounded by a periderm (pd). Analogous to indeterminate legume nodules, a zonation of the cortex can be defined (Ribeiro *et al.* 1995). Zone 1 is formed by the meristem. Zone 2 contains cells that become infected by and gradually filled with *Frankia* hyphae, which subsequently form vesicles. In the course of vesicle differentiation, *nif* gene expression and nitrogenase production are induced (Ribeiro *et al.* 1995). The onset of *nif* gene expression marks the shift to nitrogen fixation in zone 3. In the senescence zone 4, nitrogen fixation has ceased, and plant cytoplasm and bacteria are degraded. B) - Indeterminate legume nodule. The central tissue can be divided into five zones (Vasse *et al.* 1990). Directly below the meristem (I), in the prefixation zone (II), cells become infected. Rhizobia are enclosed by peribacteroid membranes (PBMs) and start to differentiate into their symbiotic form, the bacteroids. In the interzone (II-III), bacterial nitrogen fixation starts and takes place throughout the nitrogen fixation zone (III). In the senescent zone (IV), bacteria are degraded. The oxygen diffusion barrier is formed by the nodule parenchyma (np). C) - Determinate legume nodule. The central tissue is surrounded by a nodule parenchyma (in which the vascular bundles are located), an endodermis and the outermost tissue is the nodule cortex. The central tissue is surrounded by the same tissues as in the indeterminate nodules. All cells of the central tissue are more or less in the same developmental stage. This figure was kindly provided by K. Pawlowsky (Pawlovsky and Bisseling 1996).

The presence of a persistent meristem in an indeterminate nodule produces the continuous addition of new cells at the apex, and therefore the tissues are composed of cells at different stages of development with the youngest cells near the meristem and the oldest cells near the root attachment point. The nodule acquires a cylindrical-shape, and it is possible to classify different zones of the central tissue (Vasse *et al.* 1990, Franssen *et al.* 1992). As shown in Figure 4 B the meristem at the nodule apex is called zone I, consisting of non-infected small cytoplasmic rich cells. This is followed by the prefixation (or infection) zone II, there bacteria are released from the infection threads. At the proximal part of this zone, plant cells elongate and bacteria start the differentiation into bacteroids and proliferate. Nitrogen fixation takes place in zone III. A senescent zone IV is present in older nodules and is located in the most proximal part - the base - of the nodule.

In plants with determinate nodules, cells are at similar stage of development. Their mitotic activity stops at an early stage. The nodule grows then only by enlargement of cells due to the absence of an apical meristem. These nodules have a spherical form (Cermola *et al.* 2000, Figure 4 C).

An infected root nodule cell may contain several thousand rhizobial symbionts. To avoid a defence response and to facilitate the communication between the symbionts the bacteroids are located in a new kind of compartment in the plant cell, the symbiosome (SM). (Verma 1992, Verma and Hong 1996). Bacteroids are surrounded by a peribacteroid membrane (PBM) originating from the plant plasma membrane that also shares properties with the vacuolar membrane and contains several nodule-specific proteins (nodulins) performing unique functions for symbiosis (Whitehead and Day 1997). In indeterminate nodules there is a single bacteroid per SM, and this bacteroid can be up to 30-fold larger than free living bacteria (Vasse *et al.* 1990). In determinate nodules, the number of bacteroids per SM varies due to merging of SM rather than by bacterial division (Cermola *et al.* 2000).

Since an infected cell contains many bacteria, endosymbiosis is accompanied by a massive membrane proliferation: an infected root nodule cell harbours a plasmamembrane area of 2,800 μm^2 but an SM area of 21,500 μm^2 (Roth and Stacey 1989). The PBM and the peribacteroid space, the space between the PBM and the bacteroid membrane, mediate the exchange of metabolites between the symbionts (Udvardi and Day 1997). Before discussing nodule metabolism the autoregulatory mechanisms by which nodule number is controlled will be described.

Autoregulation

The legume/rhizobia interaction has a symbiotic nature since both partners profit from this intimate relationship. This requires that the host strictly controls nodule number by which a good balance of costs (*e.g.* carbon sources) and profit (fixed nitrogen) is maintained. When sufficient nodules have been formed on a root system the plant inhibits the formation of new nodules by a systemic response. This mechanism is called auto-regulation (Caetano-Anollés and Greshoff 1991). Also when plants have access to sufficient fixed nitrogen sources in the rhizosphere they do not form root nodules (Zahran 1999, Forde 2002). The mechanisms underlying auto-regulation and inhibition of nodulation by fixed nitrogen are at least in part similar, since legume mutants that have lost the ability to auto-regulate nodule number also form nodules in the presence of fixed nitrogen sources. The molecular mechanism underlying auto-regulation is intriguing since it involves a communication between shoot and root. This was demonstrated by grafting experiments with certain legume mutants that are disturbed in auto-regulation. When a mutated shoot was grafted on a wild type root, this root continued to form new nodules becoming completely covered with nodules as in the mutant.

One of the auto-regulatory genes active in the shoot has recently been cloned from several legume species (*Lotus*, pea, soybean and *Medicago*) (Krusell *et al.* 2002, Penmetsa *et al.* 2003, Searle *et al.* 2003). It is a receptor kinase with an extracellular domain containing LRRs (leucine rich repeats) and an intracellular serine/threonine kinase domain. This auto-regulatory gene has a similar structure as DMI2 (see Nod factor perception) and is homologous to a *Arabidopsis* CLAVATA gene that controls the activity of the shoot meristem (Clark *et al.* 1997). How this receptor like kinase functions within the long-distance communication by which nodule number is regulated remains to be solved.

In actinorhizas, an active feedback control of nodulation also prevents excessive nodulation of the root hair system (Wall and Huss-Danell 1997). However it is unknown whether in this system a CLAVATA-like gene plays a role in auto-regulation.

NODULE METABOLISM

Nitrogenase complex

The reduction of N_2 to NH_3 is catalysed by the nitrogenase enzyme complex. In general this protein complex is not made by free-living rhizobia but instead by the symbiotic bacteroids. The proteins that form this complex are structurally and functionally quite conserved among the different N_2-fixing organisms (Igarashi and Seefeldt 2003, Zehr *et al.* 2003). Although nitrogenase functioning has been studied predominantly in free-living fixing microorganisms due to the structural conservation of its subunits, such studies are of importance to understand the activity of bacteroid nitrogenase (Burris 1991, Lanzilotta and Seefeldt 1997). NH_4^+, the final product of N_2 reduction, is secreted by the bacteroid/symbiosome into the surrounding plant cytoplasm where it is further assimilated by the plant enzymes glutamine synthetase and glutamate synthase (chapter 2). The nitrogenase complex catalyses the reaction:

$$N_2 + 16\ ATP + 10\ H^+ + 8e^- \rightarrow 2\ NH_4^+ + H_2 + 16\ ADP + 16\ Pi$$

Since this reaction uses high amounts of ATP it has high-energy costs (-960 kJ/mol of fixed N_2). The nitrogenase enzyme can use instead of N_2 other substrates like cyanide, azide, NO and acetylene. The latter is used in the acetylene reduction test to quantify the nitrogenase activity.

Nitrogenase is composed of two ferrosulphoproteins. Component I, dinitrogenase or FeMo protein, is an $\alpha_2\beta_2$ heterotetramer of 4 (Fe_4S_4) clusters plus a cofactor containing Fe and Mo, where the catalytic site is located. Its total molecular weight is of about 220 kDa. Component II is a homodimer of about 60 kDa containing a single Fe_4S_4 cluster (Kaminski *et al.* 1998). The α and β apo-subunits of component I are encoded by the *nifD* and *nifK* genes, respectively, whereas component II polypeptides are encoded by *nifH*. Three other genes, *nifB*, *nifE* and *nifN* are required for the biosynthesis of the FeMo cofactor (Fischer 1994), the latter two encoding proteins that are used as scaffold for assembly of the cofactor, that will include the product of *nifB* (Paustian *et al.* 1989, Lanzilotta and Seefeldt 1997, Igarashi and Seefeldt 2003). The Fe protein is reduced *in vivo* by low-potential electron donors, electrons are then transferred one at a time to the FeMo protein, involving the hydrolysis of two ATP molecules at each electron transfer.

nifA and *nifL* have a regulatory function, and control the expression of the other *nif* genes (David *et al.* 1988, Noonan *et al.* 1992). In general the regulatory activity is controlled by oxygen. The Fe-protein and the FeMo-protein are extremely sensitive to O_2, and they become irreversible inactivated in air. Therefore the nitrogenase is only made in the nodule with its microaerobic environment. For the nodule oxygen metabolism see below.

Nitrogen fixation and assimilation

The high-energy costs of the nitrogen fixation process are paid by the plant by photosynthesis products. It has been estimated that nitrogen fixation in the nodules of clover and other leguminous plants may consume as much as about 20% of the total photosynthate (Haystead *et al.* 1980). To provide this energy sacharose is transported from the leaves via phloem to the nodules. There it is hydrolysed to monosaccharides and subsequently phosphoenolpyruvate is formed. The high metabolic activity of the nodule is reflected by the abundance of phosphoenolpyruvate carboxylase, which is about 1 to 2% of total soluble nodule protein (Pathirana *et al.* 1992). It catalyses the formation of oxaloacetate, which is subsequently reduced to malate. This dicarboxylic acid is the major energy source for the bacteroids and plant mitochondria and is used for ammonium assimilation as carbon skeleton in the glutamine synthetase/glutamate synthase pathway (Poole and Allaway 2000, Stitt *et al.* 2002).

Ammonium is a primary product of the nitrogen fixation reaction which is excreted by the bacteroids/symbiosomes. Ammonium is poisonous to plant cells, and is immediately incorporated into amino acids (Day *et al.* 2001). So an essential part of the symbiosis is the suppression of ammonium assimilation by the bacteroids and its accessibility to the host. Recently it has been shown that an amino acid shuttle between the symbionts is essential. Rhizobia have some amino acid transporters on their membrane; a double mutation in these amino acid transporters markedly affects symbiotic nitrogen fixation capacity in the pea *Rhizobium* symbiosis. These mutations do not affect the level of N fixation by the bacteroids, but the fixed N is no longer accessible to the host. These studies suggest that the host provides the bacteria with amino acids (in addition to dicarboxylic acids). These can suppress ammonium assimilation by the rhizobia. Further they cycle amino acids back to the host for *e.g.* asparagine biosynthesis. In this way a mutual dependence of amino acid exchange has evolved (Lodwig *et al.* 2003).

In addition to this amino acid secretion also ammonia is secreted which is assimilated by the host mainly by the GS/GOGAT cycle (Miflin and Habash 2002, chapter 2). The glutamine formed in this cycle can transfer its amide group to aspartate to form asparagine, and several other N containing compounds are made (Lea and Miflin 1980). Part of the ammonium assimilation process takes place in the uninfected cells that are part of the central tissue. For example in soybean, uricase is specifically formed in these uninfected cells and it catalyses the formation of allantoin.

Assimilated ammonium is exported from the nodules to the aerial part of the plant through the xylem. Legumes with indeterminate nodules (temperate legumes; pea, clover, alfalfa) transport N in the form of amides, like glutamine, and asparagine. Legumes with determinate nodules (tropical and subtropical; soybean, cowpea, *Lotus*), transport N as ureides, like allantoine, allantoic acid and citruline (Smith and Atkins 2002).

The oxygen dilemma

Bacteroids produce ATP by oxidative phosphorylation and therefore have a high demand for O_2. However, on the other hand high O_2 concentrations repress the genes encoding nitrogenase and it also inactivates this complex within a minute (Dixon and Wheeler 1986). To solve this dilemma nodules need mechanisms to keep the levels of oxygen in the infected cells at a low concentration, and able to facilitate the supply of O_2 to bacteroids at such a low O_2 concentration. The mechanisms that contribute to this are a high metabolic activity, a controlled oxygen influx and an efficient O_2 transport to the bacteroids.

In legume nodules, the O_2 concentration is controlled by a variable O_2 diffusion barrier (Witty *et al.* 1986). By measuring O_2 concentrations at various positions in the nodule it was shown that the nodule parenchyma forms this barrier to O_2 influx. Further by varying the atmospheric O_2 concentration it was shown that this tissue could adapt rather fast to such changes. Since O_2 diffusion through a water phase is markedly slower than through a gas phase, it is supposed that O_2 diffusion takes place through the intercellular spaces. Therefore it is supposed that the size or content of these intercellular spaces can be regulated in this tissue although the underlining molecular mechanisms are not yet elucidated.

The oxygen concentration in the central infected cells (10-50 nM, Kuzma *et al.* 1993) is about 10^{-4} of that at the nodule surface (250 mM). Although leghemoglobin (see below) may have a role in the fine-tuning of this O_2 concentration (Thumfort *et al.* 1999), the physical barrier to gas diffusion in the nodule parenchyma seems to be the primary mechanism to control gas diffusion (Hunt and Layzell, 1993). At least three proteins have been localized specifically to the nodule parenchyma, which suggests that they may play a role in the regulation of the diffusion barrier. One of these proteins is a glycoprotein (VandenBosch *et al.* 1989), another is an hydroxyproline-rich protein named ENOD2 (Nap and Bisseling 1990) and a third protein is a lectin that has been localized to the intercellular spaces of peanut nodule parenchyma cells (VandenBosch *et al.* 1994).

So how does the nodule provides sufficient O_2 to the bacteroids at such a low concentration? The infected cells of the nodule contain high levels of leghemoglobine (Lb) which is a myoglobin-like protein. Like myoglobin in muscles, it facilitates transfer of O_2 to the bacteroids (Appleby 1992). Lb is a monomeric protein of about 16 kDa with a prostetic protoheme IX group, that binds O_2 in a reversible way. Lb has a high affinity for O_2 (Km = 48-60 nM). Since about 20% of the Lb *in vivo* is bound to O_2 it is calculated that the amount of O_2 bound to Lb is 50,000 fold higher than that of free O_2 in the cytosol of infected cells. This means that almost all O_2 that reaches the bacteroids is carried by Lb. Leghemoglobins were the first nodule-specific proteins identified in legumes, and have also been found in some actinorhizal and *Parasponia* nodules.

The special nodule properties also contribute to the generation of reactive oxygen species (ROS) and since several nodule proteins are sensitive to oxidation special requirements are needed. Despite the fact that the concentration of free O_2 in the infected zone is only of 5-50 nM (Kuzma *et al.* 1993), due the high respiration rate of bacteroids and mitochondria, ROS are formed and their targets are proteins that can easily be oxidized like nitrogenase, leghemoglobine, ferredoxine, hydrogenase and uricase (Dalton 1998,

Matamoros *et al.* 2003). Plant cells have several mechanism for antioxidant protection, and they are present in nodules and have been shown to be essential for maintaining effective nitrogen fixation (Bashor and Dalton 1999).

The nitrogenase enzyme complex is at the core of biological nitrogen fixation. Only some microorganisms, diazotrophic prokaryotes, possess this enzyme complex and are able to fix N. A main feature of nitrogenase is that it is highly sensitive to O_2. Since nitrogen fixing organisms arose more than 3 billion years ago in evolution this oxygen dilemma has not always been a problem. At this stage of evolution the composition of the atmosphere was markedly different in comparison to the current atmosphere. O_2 was present at very low concentrations whereas the toxic HCN was present at rather high levels. Based on the lack of substrate specificity of the current nitrogenase complexes it has been postulated that nitrogenase evolved from enzymes with other functions. An attractive hypothesis is that enzymes involved in detoxification of cyanides are the ancestral enzymes of nitrogenase.

Since the origin of biological nitrogen fixation the abundance of anaerobic environments declined with the oxygenation of atmosphere. However oxygen insensitive forms of nitrogenase did not evolve indicating that oxygen sensitivity is an intrinsic property of this enzyme. As a result a plethora of strategies have evolved to cope with this, the oxygen dilemma.

CONCLUDING REMARKS

Only a small group of prokaryotes have the ability to reduce atmospheric nitrogen into ammonia. This process of biological nitrogen fixation is a primary source of nitrogen for all living organisms on this earth. Especially the diazotrophic microorganisms that can establish a symbiosis with higher plants make a major contribution to global N fixation. The symbiotic interaction of legumes rhizoid and legumes is most important for agriculture and is also the best-studied interaction. The *Rhizobium* legume symbiosis involves several steps from root colonization up to a senescent nodule when rhizobia again become free living organisms. Our knowledge of the contribution of the symbionts is most detailed for the microsymbiont. This is due to very well developed genetic studies by which a rather complete picture has been obtained of the genes involved in *e.g.* nodulation, infection and nitrogen fixation. In contrast research on the host has been hampered for a long time by the lack of plant systems that would allow the cloning of plant genes that by genetic screens were shown to be essential for this symbiotic interaction. The development of the legume model systems *Lotus* and *Medicago* has now markedly changed the perspective of legume research.

The developed molecular tools turn out to be effective for the cloning of legume genes. The last couple of years the international efforts have especially been focused on legume genes involved in Nod factor perception /transduction and early steps of nodulation. Several genes indicated in Figure 2 have in the meanwhile been cloned and it is probable that at the time this book will be published all genes indicated in this figure are cloned. These genes will provide major new opportunities to study the legume-*Rhizobium* interaction. Further, new genetic screens can now be designed by which also host genes involved in other steps of nodulation or other aspects of legume biology can be identified and subsequently cloned. This will create a new area in legume research that could markedly contribute to legume breeding and to a better integration of legumes into future agriculture.

REFERENCES

Adams D.G. 2000. Heterocyst formation in cyanobacteria. - Curr. Op. Microbiol. 3: 618-624.
Adams D.G. 2002. The liverwort-cyanobacterial symbiosis. - Proc. R. Ir. Acad. Sec. B. 102: 27-29.
Albrecht C., Geurts R., Bisseling T. 1999. Legume nodulation and mycorrhizae formation; two extremes in host specificity meet. - EMBO J. 18: 281-288.
Allen J.F., Raven, J.A. 1996. Free-radical-induced mutation vs redox regulation: costs and benefits of genes in organelles. - J. Mol. Evol. 42: 482-492.
Amor B.B., Shaw S.L., Oldroyd G.E., Maillet F., Penmetsa R.V., Cook D., Long S.R., Dénarié J., Gough C. 2003. The NFP locus of *Medicago truncatula* controls an early step of Nod factor signal transduction upstream of a rapid calcium flux and root hair deformation. - Plant J. 34: 495-506.
Andersson S.G., Karlberg O., Canback B., Kurland C.G. 2003. On the origin of mitochondria: a genomics perspective. - Philos.Trans. R. Soc. Lond. B. Biol. Sci. 358: 165-177.
Angulo A.F., van Dijk C., Quispel A. 1975. Symbiotic interactions in non-leguminous root nodules. - In: Nutman P.S. (Ed.) Symbiotic nitrogen fixation in plants. - Cambridge, UK, Cambridge University Press, pp. 475-483.
Appleby C.A. 1992. The origin and functions of haemoglobin in plants. - Sci. Prog. 76: 365–398.
Ardourel M., Demont N., Debellé F., Maillet F., de Billy F., Promé J.C., Dénarié J., Truchet G. 1994. *Rhizobium meliloti* lipooligosaccharide nodulation factors: different structural requirements for bacterial entry into target root hair cells and induction of plant symbiotic developmental responses. - Plant Cell. 6: 1357-1374.
Bakemeier H., Huberich R., Krabetz R., Liebe W., Schunk M., Mayer D. 1997. Ammonia, 5th ed., Ullmann's Encyclopedia of Industrial Chemistry, CD ROM - Weinhein, Wiley -VCH Verlag.
Baker D., Mullin B.C. 1992. Actinorhizal symbioses. - In: Stacey G. R., Burris R.H., and Evans H.J. (Eds.) Biological nitrogen fixation. - New York, USA, Chapman and Hall, pp. 259-292.
Bashor C.J., Dalton D.A. 1999. Effects of exogenous application and stem infusion of ascorbate on soybean (*Glycine max*) root nodules. - New Phytol. 142: 19-26.
Bauer W.D. 1981. Infection of legumes by *Rhizobium*. - Annu. Rev. Plant Physiol. 32: 407-449.
Becker A., Puhler A. 1998. Specific amino acid substitutions in the proline-rich motif of the *Rhizobium meliloti* ExoP protein result in enhanced production of low-molecular-weight succinoglycan at the expense of high-molecular-weight succinoglycan. - J. Bacteriol. 180: 395-399.
Becking J.H. De Boer W.E. Houwink A.L. 1964. Electron microscopy of the endophyte of *Alnus glutinosa*. - Antonie van Leeuwenhoek 30: 343-376.
Beijerinck M.W. 1888. Die Bacterien der Papilionaceenknölchen. - Botanische Zeitung 46: 797-804.

Benson D.R., Silvester W.B. 1993. Biology of *Frankia* strains, actinomycete symbionts of actinorhizal plants. - Microbiol. Rev. 57:293-319.

Bergman B., Johansson C., Söderbäck E. 1992. The *Nostoc-Gunnera* symbiosis. - New Phytol. 122: 379-400.

Berman-Frank I., Lundgren P., Falkowski P. 2003. Nitrogen fixation and photosynthetic oxygen evolution in cyanobacteria. - Res. Microbiol. 154: 157-164.

Berry A.M., Kahn R.K.S., Booth M.C. 1989. Identification of indole compounds secreted by *Frankia* HFPArI3 in defined culture medium. - Plant Soil 118: 205-209.

Berry A.M., McIntyre L., McCully M.E. 1986. Fine structure of root hair infection leading to nodulation in the *Frankia-Alnus* symbiosis. - Can. J. Bot. 64: 292-305.

Berry A.M., Torey J.G. 1983. Root hair deformation in the infection process of *Alnus rubra.* - Can. J. Bot. 61: 2863-2876.

Bladergroen M.R., Spaink H.P. 1998. Genes and signal molecules involved in the rhizobia-leguminoseae symbiosis. - Curr. Opin. Plant Biol. 1998 1: 353-359.

Bond G. 1983. Taxonomy and distribution of non-legume nitrogen-fixing systems. - In: Gordon J.C. and Wheeler C.T. (Eds.) Biological nitrogen fixation in forest ecosystems: foundations and applications. - The Hague, The Netherlands, Martinus Nijhoff, pp 55-87.

Borisov A.Y., Barmicheva E.M., Jacobi L.M., Tsyganov V.E., Voroshilova V.A., Tikhanovich I.A. 2000. Pea (*Pisum sativum* L.) mendelian genes controlling development of nitrogen-fixing nodules and arbuscular mycorrhizae. - Czech. J. Genet. Plant Breeding 36: 106-110.

Borisov A.Y. *et al.* 2003. The *Sym35* gene required for root nodule development in pea is an ortholog of *Nin* from *Lotus japonicus*. - Plant Physiol. 131: 1009-1017.

Brady N.C. 1990. The nature and properties of soils. - New York, USA, Macmillan Publishing Company.

Braun-Howland E., Nierzwicki-Bauer S. 1990. *Azolla-Anabaena* symbiosis: biochemistry, physiology, ultrastructure and molecular biology. - In: Rai A.N. (Ed.) CRC Handbook of symbiotic cyanobacteria. - Boca Raton, Fl, USA, CRC Press, pp. 65-117.

Brewin N.J. 1991. Development of the legume root nodule. - Annu. Rev. Cell Biol 7: 191-226.

Brewin N.J. 1998. Tissue and cell invasion by *Rhizobium*: the structure and development of infection threads and symbiosomes. - In: Spaink H.P., Kondorosi A., Hooykaas J.J. (Eds.) The *Rhizobiaceae*. - Dordrecht, The Netherlands, Kluwer Academic Publishers, pp. 417-429.

Burris R.H. 1991. Nitrogenase. - J. Biol. Chem. 266: 9339-9342.

Caetano-Anolles G., Crist-Estes D.K., Bauer W.D. 1988. Chemotaxis of *Rhizobium meliloti* to the plant flavone luteolin requires functional nodulation genes. - J. Bacteriol. 170: 3164-3169.

Caetano-Anolles G., Gresshoff P.M. 1991. Plant genetic control of nodulation. - Annu. Rev. Microbiol. 45: 345-382.

Callaham D.A., Torrey J.G. 1980. The structural basis for infection of root hairs of *Trifolium repens* by *Rhizobium*. - Can. J. Bot. 59: 1647-1664.

Campbell G.R., Reuhs B.L., Walker G.C. 2002. Chronic intracellular infection of alfalfa nodules by *Sinorhizobium meliloti* requires correct lipopolysaccharide core. - Proc. Natl. Acad. Sci. USA 99: 3938-3943.

Campbell G.R., Sharypova L.A., Scheidle H., Jones K.M., Niehaus K., Becker A., Walker G.C. 2003. Striking complexity of lipopolysaccharide defects in a collection of *Sinorhizobium meliloti* mutants. - J. Bacteriol. 185: 3853-3862.

Carlson R.W., Price N.P.J., Stacey G. 1994. The biosynthesis of rhizobial lipo-oligosaccharide signal molecules. - Mol. Plant-Microbe Interact. 7: 684-695.

Catoira R., Galera C., De Billy F., Penmetsa R.V., Journet E.P., Maillet F., Rosenberg C., Cook D., Gough C., Dénarié J. 2000. Four genes of *Medicago truncatula* controling components of a Nod factor transduction pathway. - Plant Cell 12: 1647-1666.

Cermola M., Fedorova E., Tatè R., Riccio A., Favre R., Patriarca E.J. 2000. Nodule invasion and symbiosome differentiation during *Rhizobium etli-Phaseolus vulgaris* symbiosis. - Mol. Plant-Microbe Interact. 13: 733-741.

Chandler M.R. 1978. Some observations on infection of *Arachis hypogaea* L. by *Rhizobium*. - J. Exp. Bot. 29: 749-755.
Chang D.C.N., Grobbelaar N., Coetzee J. 1988. SEM observations on cyanobacteria-infected cycad coralloid roots. - S. Afr. J. Bot. 54: 491-495.
Charrier B., Coronado C., Kondorosi A., Ratet P. 1995. Molecular characterization and expression of alfalfa (*Medicago sativa* L.) flavanone-3-hydroxylase and dihydroflavonol-4-reductase encoding genes. - Plant Mol. Biol. 29: 773-786.
Clark S.E., Williams R.W., Meyerowitz E.M. 1997. The *CLAVATA1* gene encodes a putative receptor kinase that controls shoot and floral meristem size in *Arabidopsis*. - Cell 89: 575-585.
Cook D.R. 1999. *Medicago truncatula* - a model in the making! - Curr. Opin. Plant Biol. 2: 301–304.
Crow V.L., Jarvis B.D.W., Greenwood R.M. 1981. Deoxyribonucleic acid homologies among acid-producing strains of Rhizobium. - Int. J. Syst. Bacteriol. 31: 152-172.
Cullimore J.V., Ranjeva R., Bono J.J. 2001. Perception of lipo-oligosaccharidic Nod factors in legumes. - Trends Plant Sci. 6: 24-30.
Dalton D.A., Joyner S.L., Becana M., Iturbe-Ormaetxe I., Chatfield J.M. 1998. Enhanced antioxidant defenses in the peripheral cell layers of legume root nodules. - Plant Physiol. 116: 37-43
David M., Daveran M.L., Batut J., Dedieu A., Domergue O., Ghai J., Hertig C., Boistard P., Kahn D. 1988. Cascade regulation of nif gene expression in *Rhizobium meliloti*. - Cell 54: 671-683.
Day D.A., Poole P.S., Tyerman S.D., Rosendahl L. 2001. Ammonia and amino acid transport across symbiotic membranes in nitrogen-fixing legume nodules. - Cell. Mol. Life Sci. 58: 61-71.
deFaria S.M., Hay G.T., Sprent J.L. 1988. Entry of rhizobia into roots as *Mimosa scabrela* Benthan occurs between epidermal cells. - J. Gen. Microbiol. 134: 2291-2296.
deFaria S.M., Lewis G.P., Sprent J.I., Sutherland J.M. 1989. Occurrence of nodulation in the *Leguminoseae*. - New Phytol. 111: 607-619.
deLajudie P., Laurent-Fuelle E., Willems A., Tork U., Coopman R., Collins M.D., Kesters K., Dreyfus B., Gillis M. 1998. Description of *Allorhizobium undicola* gen, Nov., sp. Nov., for nitrogen-fixing bacteria efficiently nodulating *Neptunia natans* in Senegal. - Int. J. Syst. Bacteriol. 48: 1277-1290.
deLajudie P., Willems A., Pot B., Dewettnick D., Maestrojuan G., Neyra M., Collins M.D., Dreyfus B., Kersters K. Gillis M. 1994. Polyphasic taxonomy of rhizobia: emendation of the genus *Sinorhizobium* and description of *Sinorhizobium meliloti* comb. nov., *S. saheli* sp. nov. and *Sinorhizobium terangae* sp. nov. - Int. J. Syst. Bacteriol. 44: 715-733.
Dénarié J., Cullimore J. 1993. Lipo-oligosaccharide nodulation factors: A new class of signalling molecules mediating recognition and morphogenesis. - Cell 74: 951-954.
Dénarié J., Debellé F, Promé J.C. 1996. *Rhizobium* lipo-chitooligosaccharide nodulation factors: signaling molecules mediating recognition and morphogenesis. - Annu. Rev. Biochem. 65: 503-535.
Diem H.G., Gauthier D., Dommergues Y.R. 1983. An effective strain of *Frankia* from *Casuarina sp*. - Can. J. Bot. 61: 2815-2821.
Dixon R.O.D., Wheeler C.T. 1986. - Nitrogen fixation in plants. - New York, USA, Chapman and Hall.
Douglas S.E. 1998. Plastid evolution: origins, diversity, trends. - Curr. Opin. Genet. Dev. 8: 655-661.
Downie J.A. 1998. Functions of rhizobial nodulation genes– In: Spaink H.P., Kondorosi A., Hooykaas J.J. (Eds.) The *Rhizobiaceae*. - Dordrecht, The Netherlands, Kluwer Academic Publishers, pp. 387-402.
Downie J.A., Young J.P.W. 2001. Genome sequencing. The ABC of symbiosis. - Nature 412: 597-598.

Dreyfus B.L., Garcia J.L., Gillis M. 1988. Characterization of *Azorhizobium caulinodans* gen. Nov., sp. Nov., a stem-nodulating nitrogen-fixing bacterium isolated from *Sesbania rostrata*. - Int. J. Syst. Bacteriol. 38: 89-98.
Ehrhardt D.W., Atkinson E.M., Long S.R. 1992. Depolarization of alfalfa root hair membrane potential by *Rhizobium meliloti* Nod factors. - Science 256: 998–1000.
Elbeltagy A., Nishioka K., Sato T., Suzuki H., Ye B., Hamada T., Isawa T., Mitsui H., Minamisawa K. 2001. Endophytic colonization and in planta nitrogen fixation by a *Herbaspirillum sp.* isolated from wild rice species. - Appl. Environ. Microbiol. 67: 5285-5293.
Endre G., Kereszt A., Kevei Z., Mihacea S., Kaló P., Kiss G.B. 2002. Cloning of a receptor kinase gene regulating symbiotic nodule development. - Nature 417: 962-966.
Esseling J.J., Lhuissier F.G., Emons A.M. 2003. Nod factor-induced root hair curling: continuous polar growth towards the point of nod factor application. - Plant Physiol. 132: 1982-1988.
Etzler M.E., Kalsi G., Ewing N.N., Roberts N.J., Day R.B., Murphy J.B. 1999. A Nod factor binding lectin with apyrase activity from legume roots. - Proc. Natl. Acad. Sci USA 96: 5856-5861.
Farrand S.K. 1998. Conjugal plasmids and their transfer. - In: Spaink H.P., Kondorosi A., Hooykaas J.J. (Eds.) The *Rhizobiaceae*. - Dordrecht, The Netherlands, Kluwer Academic Publishers, pp. 199-233.
Felle H.H., Kondorosi E., Kondorosi A., Schultze M. 1995. Nod signal-induced plasma membrane potential changes in alfalfa root hairs are differentially sensitive to structural modifications of the lipochitooligosaccharide. - Plant J. 7: 939–947.
Felle H.H., Kondorosi E., Kondorosi A., Schultze M. 1996. Rapid alkalinization in alfalfa root hairs in response to rhizobial lipochitooligosaccharide signals. - Plant J. 10: 295–301.
Fischer H.-M. 1994. Genetic regulation of nitrogen fixation in rhizobia. - Microbiol. Rev. 58: 352-386.
Forde B.G., 2002. Local and long-range signalling pathways regulating plant responses to nitrate. - Annu. Rev. Plant. Biol. 53: 203-224.
Franssen H.J., Vijn I., Yang W.C., Bisseling T. 1992. Developmental aspects of the Rhizobium-legume symbiosis. - Plant. Mol. Biol. 19: 89-107.
Frank B. 1889. Uber die Pilzsymbiose der Leguminosen. - Berichte der Deutschen Botanischen Gesellschaft 7: 332-346.
Fred E. B., Baldwin I. L., McCoy E. 1932. Root nodule bacteria and leguminous plants. - Madison, USA, University of Wisconsin Press.
Freiberg C., Fellay R., Bairoch A., Broughton W.J., Rosenthal A., Perret X. 1997. Molecular basis of symbiosis between Rhizobium and legumes. - Nature 387: 394-401.
Galibert F. *et al.* 2001. The composite genome of the legume symbiont *Sinorhizobium meliloti*. - Science 293: 668-672.
Galloway J.N., Schlesinger W.H., Levy H., Michaels A., Schnoor J.L. 1995. Nitrogen fixation: atmospheric enhancement-environmental response. - Global Biogeochem. Cycles 9: 235-252.
Geurts R., Bisseling T. 2002. *Rhizobium* Nod factor perception and signalling. - Plant Cell 14: S239-S249.
Gherbi H., Duhoux E., Franche C., Pawlowski K., Berry A., Nassar A., Bogusz D. 1997. Cloning of a full-length symbiotic hemoglobin cDNA and in situ localization of the corresponding mRNA in *Casuarina glauca* root nodule. - Physiol. Plant. 99: 608-616.
Goedhart J., Hink M.A., Visser A.J.W.G., Bisseling T., Gadella T.W.J. Jr. 2000. *In vivo* fluorescence correlation microscopy (FCM) reveals accumulation and immobilization of Nod factors in root hair cell walls. - Plant J. 21: 109–119.
Graham P.H., Sadowsky M.J., Keyser H.H., Y. Barnet Y.M., Bradley J.E., Cooper J.E., D. De Ley D.J., Jarvis B.D.W., Roslycky E.B., Strijdom B.W., Young J.P.W. 1991. Proposed minimal standards for the description of new genera and species of root- and stem-nodule bacteria. - Int. J. System. Bacteriol. 41: 582-587.
Gualtieri G., Bisseling T. 2000. The evolution of nodulation. - Plant Mol. Biol. 42: 181-194.

Gutschick V.P. 1980. Energy flow in the nitrogen cycle, especially in fixation. - In: Newton W.E., Orme-Johnson W.H. (Eds.) Nitrogen fixation, Vol. I. - Baltimore, USA, University Park Press, pp. 17-27.
Hadri A.-E., Bisseling T. 1998 Responses of the plant to Nod factors. - In: Spaink H.P., Kondorosi A., Hooykaas J.J. (Eds.) The *Rhizobiaceae*. - Dordrecht, The Netherlands, Kluwer Academic Publishers, pp. 403-416.
Hadri A.-E., Spaink H.P., Bisseling T., Brewin N.J. 1998. Diversity of root nodulation and rhizobial infection processes. - In: Spaink H.P., Kondorosi A., Hooykaas J.J. (Eds.) The *Rhizobiaceae*. - Dordrecht, The Netherlands, Kluwer Academic Publishers, pp. 347-360.
Harwood R.R. 1990. A history of sustainable agriculture. - In: Edwards C.A., Lal R., Madden P., Miller R.H., House G. (Eds.) Sustainable agriculture systems. - Iowa, USA, Soil and Water Conservation Society, pp. 3-19.
Haystead A., King J., Lamb W.I.C., Marriott C. 1980. Growth and carbon economy of nodulated white clover in the presence and absence of combined nitrogen. - Grass Forage Sci. 35: 123-128.
Heidstra R., Bisseling T. 1996. Nod factor-induced host responses and mechanisms of nod factor perception. - New Phytol. 133: 25-43.
Heidstra R., Geurts R., Franssen H., Spaink H.P., van Kammen A., Bisseling T. 1994. Root hair deformation activity of nodulation factors and their fate on *Vicia sativa*. - Plant Physiol. 105: 787–797.
Heidstra R., Yang W.C., Yalcin Y., Peck S., Emons A.M., van Kammen A., Bisseling T. 1997. Ethylene provides positional information on cortical cell division but is not involved in Nod factor-induced root hair tip growth in *Rhizobium*-legume interaction. - Development 124: 1781-1787.
Horvath B., Heidstra R., Lados M., Moerman M., Spaink H.P., Prome J.C., van Kammen A., Bisseling T. 1993. Lipo-oligosaccharides of *Rhizobium* induce infection-related early nodulin gene expression in pea root hairs. - Plant J. 4 :727-733
Hunt S., Layzell D.B. 1993. Gas exchange of legume nodules and the regulation of nitrogenase activity. – Annu. Rev. Plant Physiol. Plant. Mol. Biol. 44: 483-511.
Igarashi R.Y., Seefeldt L.C. 2003. Nitrogen fixation: the mechanism of the mo-dependent nitrogenase. - Crit. Rev. Biochem. Mol. Biol. 38: 351-384.
Jarvis, B.D.W., van Berkum P., Chen W.X., Nour S.M., Fernandez M.P., Cleyet-Mariel J.C. and Gillis M. 1997. Transfer of *Rhizobium loti*, *Rhizobium huakuii*, *Rhizobium ciceri*, *Rhizobium mediteraneum* and *Rhizobium tianshanese* to *Mesorhizobium* gen. nov. - Int. J. Syst. Bacteriol. 47: 895-898.
Jordan D.C. 1982. Transfer of *Rhizobium japonicum* Buchanan 1980 to *Bradyrhizobium* gen. Nov., a genus of slow-growing, root nodule bacteria from leguminous plants. - Int. J. Syst. Bacteriol. 32: 136-139.
Jordan D.C. 1984. *Rhizobiaceae*. - In: Kreig N.R. (Ed.) Bergey's manual of systematic bacteriology, vol. 1. Baltimore, USA, Williams and Wilkins, pp. 234-256.
Journet E.P., Pichon M., Dedieu A. de Billy F., Truchet G., Barker D.G. 1994. *Rhizobium meliloti* Nod factors elicit cell-specific transcription of the ENOD12 gene in transgenic alfalfa. - Plant J. 6: 241-249.
Kaminski P.A., Batut J., Boistard P. 1998. A survey of symbiotic nitrogen fixation by rhizobia. - In: Spaink H.P., Kondorosi A., Hooykaas J.J. (Eds.) The *Rhizobiaceae*. - Dordrecht, The Netherlands, Kluwer Academic Publishers, pp. 431-460.
Kaneko T. *et al.* 2000. Complete genome structure of the nitrogen-fixing symbiotic bacterium Mesorhizobium loti. - DNA Res. 7: 331-338.
Kijne J.W. 1992. The *Rhizobium* infection process. - In: Stacey G., Burris R.H., Evans H.J. (Eds.) Biological nitrogen fixation. - New York, USA, Chapman & Hall, pp. 349-398.
Krusell L. *et al.* 2002. Shoot control of root development and nodulation is mediated by a receptor-like kinase. - Nature 420: 422-426.

Kurkdjian A.C. 1995. Role of the differentiation of root epidermal cells in Nod Factor (from *Rhizobium meliloti*)-induced root-hair depolarization of *Medicago sativa*. - Plant Physiol. 107: 783-790.
Kosuta S., Chabaud M., Lougnon G., Gough C., Dénarié J., Barker D.G., Bécard G. 2003. A diffusible factor from arbuscular mycorrhizal fungi induces symbiosis-specific *MtENOD11* expression in root of Medicago truncatula. - Plant Physiol. 131: 952-962.
Kistner C., Parniske M. 2002. Evolution of signal transduction in intracellular symbiosis. - Trends Plant. Sci. 7: 511-518.
Kuzma M.M., Hunt S., Layzell D.B. 1993. Role of oxygen in the limitation and inhibition of nitrogenase activity and respiration rate in individual soybean nodules. - Plant Physiol. 101: 161-169.
Lanzilotta W.N., Seefeldt L.C. 1996. Changes in the midpoint potentials of the nitrogenase metal centers as a result of iron protein-molybdenum-iron protein complex formation. - Biochemistry 36: 12976-12983.
Lea P.J. , Miflin B.J. 1980. The energetics of nitrogen metabolism. 1978. – In: Gnanam A., Krishnaswamy S., Kahn J.S. (Eds.) Proceedings of the international symposium on biological applications of solar energy, pp. 97-101.
Lechevalier M.P., Lechevalier H.A. 1979. The taxonomic position of the actinomycete endophytes. – In: Gordon J.C., Wheleer C.T., Perry D.A., (Eds.) Symbiotic nitrogen fixation in the management of temperate forests. - Corvallis, USA, Oregon State University Forest Research Laboratory, pp 111-121.
Leigh J.A., Walker G.C. 1994. Exopolysaccharides of *Rhizobium*: synthesis, regulation and symbiotic function. - Trends Genet. 10: 63-67.
Lerouge P., Roche P., Faucher C., Maillet F., Truchet G., Promé J.C., Dénarié J. 1990. Symbiotic host-specificity of *Rhizobium meliloti* is determined by a sulphated and acylated glucosamine oligosaccharide signal. - Nature 344: 781–784.
Lev-Yadun S., Gopher A., Abbo S. 2000. Archaeology. The cradle of agriculture. - Science 288: 1602-1603.
Limpens E., Bisseling T. 2003. Signaling in symbiosis. - Curr. Opin. Plant Biol. 6: 343-350.
Limpens E., Franken C., Smit P., Willemse J., Bisseling T., Geurts R. 2003. LysM domain receptor kinases regulating rhizobial Nod factor-induced infection. - Science 302: 630-633.
Lodwig E.M., Hosie A.H., Bourdes A., Findlay K., Allaway D., Karunakaran R., Downie J.A., Poole P.S. 2003. Amino-acid cycling drives nitrogen fixation in the legume-*Rhizobium* symbiosis. - Nature 422: 722-726.
Long S.R. 1996. *Rhizobium* symbiosis: Nod factors in perspective. - Plant Cell 8: 1885-1898.
Madsen E.B. *et al.* 2003. A receptor kinase gene of the LysM type is involved in legume perception of rhizobial signals. - Nature. 425: 637-640.
Marketon M.M., Gronquist M.R., Eberhard A., Gonzalez J.E. 2002. Characterization of the *Sinorhizobium meliloti sinR/sinI* locus and the production of novel N-acyl homoserine lactones. - J.Bacteriol. 184: 5686-5695.
Matamoros M.A., Dalton D.A., Ramos J., Clemente M.R., Rubio M.C., Becana M. 2003. Biochemistry and molecular biology of antioxidants in the rhizobia-legume symbiosis. - Plant Physiol. 133: 499-509.
Matthysse A.G., Kijne J.W. 1998. Attachment of *Rhizobiaceae* to plant cells. - In: Spaink H.P., Kondorosi A., Hooykaas J.J. (Eds.) The *Rhizobiaceae*. - Dordrecht, The Netherlands, Kluwer Academic Publishers, pp. 235-249.
Meeks J.C., Elhai J. 2002. Regulation of cellular differentiation in filamentous cyanobacterial in free-living and plant-associated symbiotic growth states. - Microb. Mol. Biol. Rev. 66: 94-121.
Miflin B.J., Habash D.Z. 2002. The role of glutamine synthetase and glutamate dehydrogenase in nitrogen assimilation and possibilities for improvement in the nitrogen utilization of crops. - J. Exp. Bot. 53: 979-987.
Miller I.M., Baker D.D. 1985. The initiation, development and structure of root nodules in *Eleagnus angustifolia* L. (Eleagnaceae). - Protoplasma 128: 107-119.

Mylona P., Pawlowski K., Bisseling T. 1995. Symbiotic nitrogen fixation. - Plant Cell 7: 869-885.
Nap J.-P., Bisseling T. 1990. Developmental biology of a plant-prokaryote symbiosis: the legume root nodule. - Science 250: 948-954.
Noonan B., Motherway M, O'Gara F. 1992. Ammonia regulation of the *Rhizobium meliloti* nitrogenase structural and regulatory genes under free-living conditions: involvement of the *fixL* gene product? - Mol. Gen. Genet. 234: 423-428.
Oldroyd G.E.D. 2001. Dissecting symbiosis: Development in Nod factor signal transduction. - Ann. Bot. 87: 709-718.
Pathirana S.M., Vance C.P., Miller S.S., Gantt J.S. 1992. Alfalfa root nodule phosphoenolpyruvate carboxylase: characterization of the cDNA and expression in effective and plant-controlled ineffective nodules. - Plant Mol. Biol. 20: 437-450.
Paustian T.D., Shah V.K., Roberts G.P. 1989. Purification and characterization of the *nifN* and *nifE* gene products from *Azotobacter vinelandii* mutant UW45. - Proc. Natl. Acad .Sci.U S A. 86: 6082-6086.
Pawlowski K., Bisseling T. 1996. Rhizobial and actinorhizal symbioses: What are the shared features. - Plant Cell 8: 1899-1913.
Penmetsa R.V., Frugoli J.A., Smith L.S., Long S.R., Cook D.R. 2003. Dual genetic pathways controlling nodule number in *Medicago truncatula*. - Plant Physiol. 131: 998-1008.
Perotto S., Brewin N.J., Kannenberg E.L. 1994. Cytological evidence for a host defense response that reduces cell and tissue invasion in pea nodules by lipopolysaccharide-defective mutants of *Rhizobium leguminosarum* strain 3841. - Mol. Plant-Microbe Interact. 7: 99-112.
Perret X., Staehelin C., Broughton W.J. 2000. - Molecular basis of symbiotic promiscuity. - Microbiol. Mol. Biol. Rev. 64: 180-201.
Phillips R.D. 1993. Starchy legumes in human nutrition, health and culture. - Plant Foods Hum. Nutr. 44: 195-211.
Poole P., Allaway D. 2000. Carbon and nitrogen metabolism in *Rhizobium*. - Adv. Microb. Physiol. 43: 117-163.
Pueppke S.G., Broughton W.J. 1999. *Rhizobium* sp. NGR234 and *R. fredii* USDA257 share exceptionally broad, nested host-ranges. - Mol. Plant-Microbe. Interact. 12: 293-318.
Radutoiu S., Madsen L.H., Madsen E.B., Felle H.H., Umehara Y., Gronlund M., Sato S., Nakamura Y., Tabata S., Sandal N., Stougaard J. 2003. - Plant recognition of symbiotic bacteria requires two LysM receptor-like kinases. - Nature 425: 585-592.
Rae A.L., Bonfante-Fasolo P., Brewin N.J. 1992. Structure and growth of infection threads in the legume symbiosis with *Rhizobium leguminosarum*. - Plant J. 2: 385-395
Rai A.N., Södebäck E., Bergman B., 2000. Cyanobacterium- plant symbioses. - New Phytol. 147: 449-481.
Rasmussen U., Svenning M. 2001. Characterization by genotypic methods of symbiotic *Nostoc* strains isolated from five species of *Gunnera*. - Arch. Microbiol. 176: 204-210.
Ribeiro A., Akkermans A.D., van Kammen A., Bisseling T., Pawlowski K. 1995. A nodule-specific gene encoding a subtilisin-like protease is expressed in early stages of actinorhizal nodule development. - Plant Cell. 7: 785-794.
Rolfe B.G., Carlson R.W., Ridge R.W., Dazzo F.B., Mateos P.F., Pankhurst C.E. 1996. Defective infection and nodulation of clovers by exopolysaccharide mutants of *Rhizobium leguminosarum* bv. *trifolii*. - Aust. J. Plant Physiol. 23: 285-303.
Roth L.E., Stacey G. 1989. Bacterium release into host cells of nitrogen-fixing soybean nodules: The symbiosome membrane comes from three sources. - Eur. J. Cell. Biol. 49: 13-23.
Sainju U.M., Whitehead W.F., Singh B.P. 2003. Agricultural management practices to sustain crop yields and improve soil and environmental qualities. - Sci. World J. 3: 768-789.
Schauser L., Roussis A., Stiller J., Stougaard J. 1999. A plant regulator controlling development of symbiotic root nodules. - Nature 402: 191-195.
Schlaman H.R.M., Phillips A.D., Kondorosi E. 1998. Genetic organization and transcriptional regulation of rhizobial nodulation genes. - In: Spaink H.P., Kondorosi A., Hooykaas J.J. (Eds.) The *Rhizobiaceae*. - Dordrecht, The Netherlands, Kluwer Academic Publishers, pp. 361-386.

Schwencke J. 1991. Rapid, exponential growth and increased biomass yield of some *Frankia* strains in buffered and stirred mineral medium (BAP) with phosphatidyl choline. - Dev. Plant Soil Sci. 48: 615-619.

Searle I.R., Men A.E., Laniya T.S., Buzas D.M., Iturbe-Ormaetxe I., Carroll B.J., Gresshoff P.M. 2003. Long distance signalling for nodulation control in legumes requires a clavata-1 like receptor kinase. - Science 299: 109-112.

Sevilla M., Burris R.H., Gunapala N., Kennedy C. 2001. Comparison of benefit to sugarcane plant growth and $^{15}N_2$ incorporation following inoculation of sterile plants with *Acetobacter diazotrophicus* wild-type and Nif- mutants strains. - Mol. Plant-Microbe Interact. 14: 358-366.

Shaw S.L., Long S.R. 2003. Nod factor elicits two separable calcium responses in *Medicago truncatula* root hair cells. - Plant Physiol. 13: 976-984.

Shirley B.W. 1996. Flavonoid biosynthesis: new functions for an "old" pathway. - Trends Plant Sci. Bacteriol. 1: 377-381.

Skerman V.B.D., McGowan V., Sneath P.H.A. (Eds.). 1980. Approved lists of bacterial names. - Int. J. Syst. Bacteriol. 30: 225-420.

Smit, G., de Koster C.C., Schripsema J., Spaink,H.P., van Brussel A.A.N., Kijne J.W. 1995. Uridine, a cell division factor in pea roots. - Plant Mol. Biol. 29: 869-873.

Smith P.M.C., Atkins, A.C.. 2002. Purine biosynthesis. Big in cell division, even bigger in nitrogen assimilation. - Plant Physiol. 128: 793-802.

Socolow R.H. 1999. Nitrogen management and the future of food: lessons from the management of energy and carbon. - Proc. Natl. Acad. Sci. USA 96: 6001-6008.

Steenhoudt O., Vanderleyden J. 2000. *Azospirillum*, a free-living nitrogen-fixing bacterium closely associated with grasses: genetic, biochemical and ecological aspects. - FEMS Microbiol. Rev. 24: 487-506.

Stewart W.D., Rowell P., Rai A.N. 1983. Cyanobacteria-eukaryotic plant symbioses. - Ann. Microbiol. (Paris) 134B: 205-228.

Stitt M., Müller C., Matt P., Gibon Y., Carillo P., Morcuende R., Scheible W.-R., Krapp A. 2002. Steps towards an integrated view of nitrogen metabolism. - J. Exp. Bot. 53: 959–970.

Stougaard J. 2001. Genetics and genomics of root symbiosis. - Curr. Opin. Plant Biol. 4: 328-335.

Stracke S., Kistner C., Yoshida S., Mulder L., Sato S., Kaneko T., Tabata S., Sandal N., Stougaard J., Szczyglowski K., Parniske M. 2002. A plant receptor-like kinase required for both bacterial and fungal symbiosis. - Nature 417: 959-962.

Thumfort P.P., Layzell D.B., Atkins C.A. 1999. Diffusion and reaction of oxygen in the central tissue of ureide-producing legume nodules. - Plant Cell Environ. 22: 1351-1365.

Tighe S.W., de Lajudie P., Dipietro K., Lindstrom K., Nick G., Jarvis B.D. 2000. Analysis of cellular fatty acids and phenotypic relationships of *Agrobacterium*, *Bradyrhizobium*, *Mesorhizobium*, *Rhizobium* and *Sinorhizobium* species using the Sherlock Microbial identification system. - Int. J. Syst. Evol. Microbiol. 50: 787-801.

Torrey J.G., Callaham D. 1982. Structural features of the vesice of *Frankia* sp Cp11 in culture. - Can. J. Bot. 28: 749-757.

Trinick M.J. 1979. Structure of nitrogen-fixing nodules formed by *Rhizobium* on roots of *Parasponia andersonii* Planch. - Can. J. Microbiol. 25: 565-578.

Truchet G., Roche P., Lerouge P., Vasse J., Camut S., De Billy F., Prome J. C., Dénarié, J 1991. Sulphated lipo-oligosaccharide signals of *Rhizobium meliloti* elicit root nodule organogenesis in alfalfa. - Nature 351: 670-673.

Udvardi M.K. 2001. Legume models strut their stuff. – Mol. Plant-Microbe. Interact. 14: 6-9.

Udvardi M.K., Day D. 1997. Metabolite transport across symbiotic membranes of legume nodules. - Annu. Rev. Plant. Physiol. Plant Mol. Biol. 48: 493-523.

van Batenburg, F.H.D., Jonker, R., and Kijne, J.W. 1986 *Rhizobium* induces marked root hair curling by redirection of tip growth, a computer simulation. - Physiol. Plant. 66: 476-480.

van Berkum P., Eardly B.D. 1998. Molecular evolutionary systematics of the *Rhizobiaceae*. - In: Spaink H.P., Kondorosi A., Hooykaas J.J. (Eds.) The *Rhizobiaceae*. - Dordrecht, The Netherlands, Kluwer Academic Publishers, pp. 1-24.

van Brussel A.A.N., Bakhuizen R., van Spronsen P.C., Spaink H.P., Tak T., Lugtenberg B.J.J., Kijne J.W. 1992. Induction of pre-infection thread structures in the leguminous host plant by mitogenic lipo-oligosaccharides of *Rhizobium*. - Science 257: 70–71.
Vance C.P., Heichel G.H., Phillips D.A. 1988. Nodulation and symbiotic dinitrogen fixation. – In: Hanson A.A. (Ed.) Alfalfa and alfalfa improvement. - Madison, USA, American Society of Agronomy, pp. 229-257.
Vance C.P. 1996. Root-bacteria interactions: symbiotic nitrogen fixation. - In: Waisel Y., Eshel A., Kafkafi U. (Eds.) Plant roots: The hidden half. - New York, USA, Marcel Dekker, pp. 723-756.
VandenBosch K.A., Brewin N.J., Kannenberg E.L. 1989. Developmental regulation of a *Rhizobium* cell surface antigen during growth of pea root nodules. - J. Bacteriol. 171: 4537-4542.
VandenBosch K.A., Rodgers L.R. Sherrier D.J., Kishinevsky B.D. 1994. A peanut nodule lectin in infected cells and in vacuoles and the extracellular matrix of nodule parenchyma. - Plant Physiol. 104: 327-337.
VandenBosch K.A., Stacey G. 2003. Summaries of legume genomics projects from around the globe. Community resources and crop models. - Plant Physiol. 131: 840-865.
van de Wiel C., Scheres B., Franssen H., van Lierop M.J., van Lammeren A., van Kammen A., Bisseling T. 1990. The early nodulin transcript *ENOD2* is located in the nodule parenchyma (inner cortex) of pea and soybean root nodules. - EMBO J. 9: 1-7.
van Ghelue M., Løvaas E., Ringø E., Solheim B. 1997. Early interactions between *Alnus glutinosa* (L.) Gaertn. and *Frankia* strain ArI3. Production and specificity of root hair deformation factor(s). - Physiol. Plant. 99: 579-587.
Vasse J., de Billy F., Camut S., Truchet G. 1990. Correlation between ultrastructural differentiation of bacteroids and nitrogen fixation in alfalfa nodules. - J. Bacteriol. 172: 4295-4306.
Verma D.P.S. 1992. Signals in root nodule organogenesis and endocytosis of *Rhizobium*. - Plant Cell 4: 373-382.
Verma D.P., Hong Z. 1996. Biogenesis of the peribacteroid membrane in root nodules. - Trends Microbiol. 4: 364-368.
Wais R.J., Galera C., Oldroyd G., Catoira R., Penmetsa R.V., Cook D., Gough C., Dénarié J., Long S.R. 2000. Genetic analysis of calcium spiking responses in nodulation mutants of *Medicago truncatula*. - Proc. Natl. Acad. Sci USA 97: 13407-13412.
Walker S.A., Viprey V., Downie J.A. 2000. Dissection of nodulation signaling using pea mutants defective for calcium spiking induced by Nod factors and chitin oligomers. - Proc. Natl. Acad. Sci. USA 97: 13413-13418.
Wall L.G. 2000. The Actinorhizal Symbiosis. - J. Plant Growth Regul. 19: 167-182.
Wall L.G., Huss-Danell K. 1997. Regulation of nodulation in *Alnus incana* - *Frankia* "symbiosis". - Physiol. Plant. 99: 594-600.
Willems A., Collins M.D. 1993. Phylogenetic analysis of rhizobia and agrobacteria based on 16S rRNA gene sequences. - Int. J. Sys. Bacteriol. 43: 305-313.
Witty J.F., Minchin F.R., Skot L., Sheehy J.E. 1986. Nitrogen fixation and oxygen in legume root nodules. – In: Miflin B.J., Miflin H.F. (Eds.) Oxford surveys of plant molecular and cell biology. - Oxford, UK, Oxford University Press, pp. 275-314.
Whitehead L., Day D.A. 1997. The peribacteroid membrane of legume nodules. - Physiol. Plant. 100: 30-44.
Woronin M.S. 1866. Über die bei der Schwarzerle (*Alnus glutinosa*) und bei der gewöhnlichen Gartenlupine (*Lupinus mutabilis*) auftretenden Wurzelanschwellungen. - Mémoires de l'Academie Impériale des Sciences de St. Pétersbourg, VII Series, vol. X.
Yang W.C., Katinakis P., Hendriks P., Smolders A., de Vries F., Spee J., van Kammen A., Bisseling T., Franssen H. 1993. Characterization of *GmENOD40*, a gene showing novel patterns of cell-specific expression during soybean nodule development. - Plant J. 3: 573-585.

Yang W.C., de Blank C., Meskiene I., Hirt H., Bakker J., van Kammen A., Franssen H., Bisseling T. 1994. *Rhizobium* nod factors reactivate the cell cycle during infection and nodule primordium formation, but the cycle is only completed in primordium formation. - Plant Cell. 6:1415-1426.

Yang F.-L., Lin L.-P. 1998. Characteristics from *Rhizobium fredii*. - Bot. Bull. Acad. Sin. 39: 261-267.

Young J.P.W., Haukka K.E. 1996. Diversity and phylogeny of rhizobia. - New Phytol. 133: 87-94.

Zahran H.H. 1999. *Rhizobium*-legume symbiosis and nitrogen fixation under severe conditions and in an arid climate. - Microbiol. Mol. Biol. Rev. 63: 968-989.

Zehr J.P., Jenkins B.D., Short S.M., Steward G.F. 2003. Nitrogenase gene diversity and microbial community structure: a cross-system comparison. - Environ. Microbiol. 5: 539-554.

Zeng S., Tjepkema J.D. 1994. The wall of the infected cell may be the major diffusion barrier in nodules of *Myrica gale* L. - Soil Biol. Biochem. 26: 633-639.

Zuanazzi J.A.S., Clergeot P.H., Quirion J.-C., Husson H.P., Kondorosi A., Ratet P. 1998. Production of *Sinorhizobium meliloti nod* gene activator and repressor flavonoids from *Medicago sativa* roots. - Mol. Plant-Microbe Interact. 11: 784-794

Chapter 5

NITROGEN METABOLISM AND PLANT ADAPTATION TO THE ENVIRONMENT - THE SCOPE FOR PROCESS-BASED MODELLING

Marcel van Oijen and Peter Levy

INTRODUCTION

Nitrogen is the fourth most common element in plant tissues. It is an essential component of enzymes and structural proteins, nucleic acids, pigments and other secondary metabolites. If a plant's access to N is interrupted, its longevity is not directly endangered, but growth becomes impossible within days. The importance of N for plant growth has been understood well since the pioneering work of Von Liebig (Liebig 1840) on plant nutrition. Since his work, many treatises have been written on plant nutrition in general and N in particular. The present book is an example.

Every process occurring in plants can be characterised as either transport or conversion of material. Because of the long history of plant N research, we now have a large body of knowledge on the flows of N into, within and out of plants, as well as the conversions between different N compounds. Moreover, we have learned about the key role that N compounds like enzymes play in the interconversion of non-N compounds. The quantitatively most important example is the carboxylation of ribulose-bisphosphate mediated by the Rubisco enzyme, which is but one example out of many. Wall-charts representing plant biochemistry show a myriad of reactions, most of which are enzymatically regulated and many of which involve nitrogenous reactants. Much of this physiological knowledge is reviewed in depth in other chapters of this book. In the present chapter, we will focus on efforts to integrate such information in dynamic, process-based models of plant growth. Process-based models describe the behaviour of systems in terms of the underlying processes (Van Oijen 2002). For example, weather forecasters use process-based models to predict changes in air flow and cloud distribution from air pressure differences. We will discuss modelling approaches that try to explain whole-plant behaviour on the basis of the underlying processes, with emphasis on the N relations.

Our knowledge about the N pools in plant tissues, the flows between them, and interconversions of N compounds is increasingly quantified. A beautiful early example is the work of Pate and co-workers (Pate *et al.* 1979), who quantified N pools and flows in *Lupinus albus*. A very much-simplified representation of the type of data that Pate gathered is shown in Figure 1B. Measuring pool sizes is generally easier than measuring rates of transport or conversion because the latter processes are interrupted by destructive measurement. However, Pate and many others before and after him have found methods, including the use of isotopes, to quantify the rates. Once pools and flows were known to some extent, their interaction and regulation became the focus of interest. This is where models were of some use. A dynamic model of plant growth is in essence nothing more than a set of so-called state variables that represent the pools, and a set of rules that determine rates of transport and conversion between and within the pools. The state variables and rules are expressed in mathematical form - often differential equations - or, nowadays more commonly, in the form of a computer program. The only way we can know that the rate-determining rules in a model are realistic, is by having information about transports and conversions.

Various types of rules are used in dynamic plant growth models. Mechanistic rules are common. For many of the processes occurring in plants, physical or chemical mechanisms are known or proposed. Transport often involves physical mechanisms (water-potential gradients, osmosis, diffusion, *etc.*) and conversions follow chemical mechanisms (mostly enzyme kinetics). Modellers have often represented such mechanisms explicitly in their models, following the belief that the mechanism underlying a process determines its rate.

However, what we need is knowledge of how processes are controlled, rather than how they are mechanistically implemented. The rules that we want to put in our models should reflect control, which only sometimes coincides with mechanism. A non-plant biological example is riding a bicycle. The force applied by the cyclist drives the bicycle forward, so the mechanism is simple mechanics, but the cyclist controls the rate. The mechanism does impose constraints on the rate: cycling is slower than driving a car. In plants, transport by diffusion is a common mechanism in which at any time transport rate is proportional to concentration difference, but this knowledge is irrelevant if the plant has means to control the concentration gradient or to change the diffusivity of the pathway.

If the plant has no means to affect a mechanism, then the mechanism itself does constitute an appropriate rule for the model. Various examples of this distinction between mechanism and control, and the consequences for implementing rate rules in models, will be discussed in this chapter. The four steps in modelling, from observation, via measurement to identification of mechanisms and controlling factors, are shown in Figure 1.

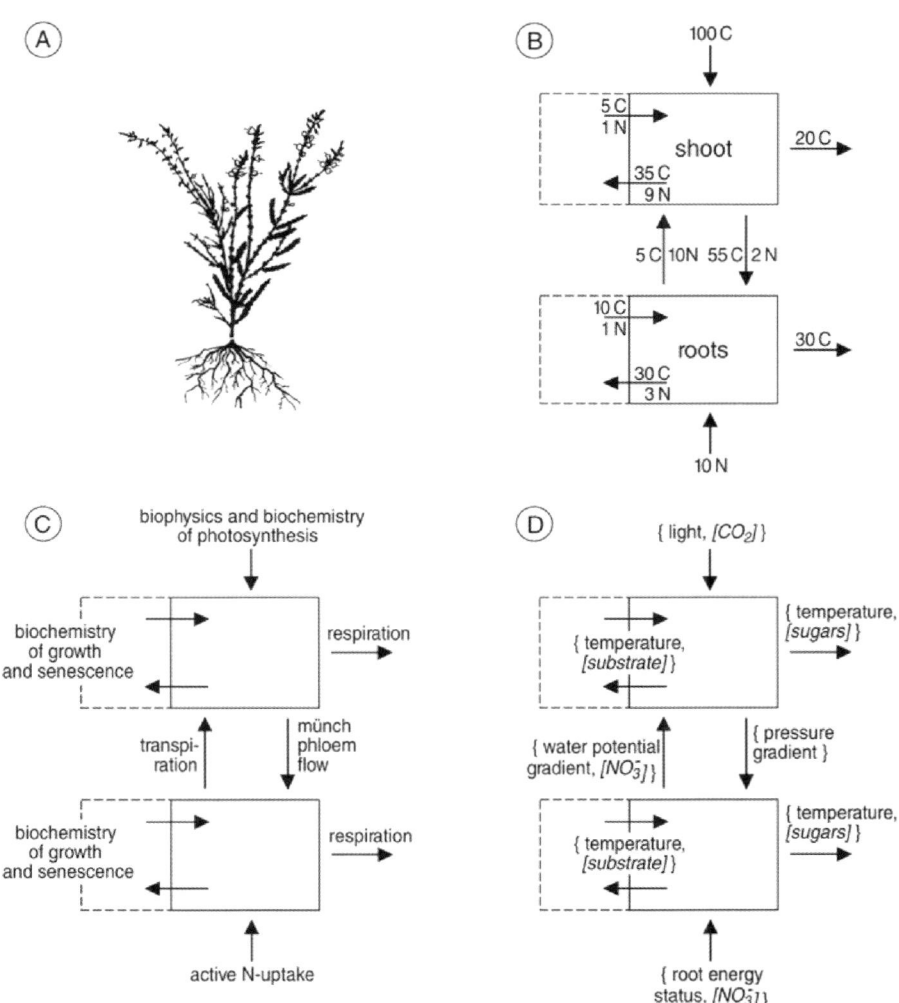

Figure 1. The four stages in the development of plant physiological models. A. Observation; B. Measurement of flows (numbers indicate fluxes of C and N in arbitrary units); C. Processes and mechanisms; D. Controlling factors.

When considering control in any system, a key question is whether the control is local or global. To what extent are processes in different cells or leaves (or in roots versus shoots) independent? How is integration of activity in the whole plant realised? Do plants have means to ensure that processes in different organs respond in a well-coordinated way to changes in the growing environment? This is of course the central problem of plant physiology, which cannot be solved by modelling. However, modelling can contribute to the science by demonstrating whether, for example, local control may be sufficient to produce adaptive behaviour of the whole plant. In this respect, adaptive behaviour is defined as a form of control that makes the plants behave "better" after a change in the growing environment. This is of key importance for plants, which are sessile and cannot, like animals, keep their growing environment constant by moving away when local resources are depleted.

Although process-based models of plant growth can be useful to explain how plants adapt to their environment, they are generally used for more applied purposes. Models predict the response of vegetation to environmental change such as pollution and climate change. However, this predictive task of modelling can only be executed well if the models incorporate adequate understanding of adaptive processes, and that is what we focus on in this chapter. We will begin by reviewing the modelling of individual N-related plant processes, before we move on to integration at the whole-plant level and the modelling of adaptation to the environment.

MODELLING N TRANSPORT AND CONVERSION

Plant physiological modelling started in the 1950s and 1960s and originally focused on C relations. Plant growth was largely equated with C assimilation in photosynthesis followed by allocation of the assimilates to different organs. Water relations started to become included from the 1960s, and N relations from the 1970s (Bouman et al. 1996). However, even today many plant models do not consider N, and much work remains to be done. In this section, we will review N-related processes that have been implemented in models, beginning with transport processes and then conversions. For each process, we will discuss modelling approaches in relation to availability of data and to knowledge of mechanisms and controls.

Uptake and loss of N from the plant

Von Liebig (Liebig 1840) stated that N is taken up by the roots, whereas C enters via the leaves. However, it is becoming increasingly clear that plants can take up N from the air as well, and we will begin with this process. The influx of NH_4^+ into leaves can be modelled as diffusion from the ambient atmosphere into the substomatal cavities at a rate proportional to stomatal conductance. Riedo et al. (2002) showed that this mechanism accounted well for observed rates of stomatal uptake and loss of ammonia in grasslands. However, as for photosynthesis, the control is shared between the stomata and the processes determining the internal concentration of the diffusing compound, *i.e.* apoplastic ammonia.

Uptake of N compounds by the roots, on the other hand, is an active process (chapters 1 and 2). A vast literature exists on the measuring and modelling of NO_3^+ and NH_4^+ uptake by plant roots (*e.g.* Nye and Tinker, 1977, Apel *et al.* 1985, Tinker and Nye 2000). Recent literature suggests that plant roots may also take up N in organic form (Nasholm *et al.* 1998), but for this only few data are available and no modelling attempt seems to have been made. Data on the uptake of mineral N by roots do exist in abundance, but precise data on short-term flow rates exist only for controlled conditions, like plants growing on solution. Experiments with well-stirred culture solutions have shown that the rate of N uptake is largely independent of the concentration in the medium until the concentration drops below very low levels. However, the relevance of such observations for understanding uptake under realistic conditions in soil is debatable. The most detailed models of mineral N uptake in soils include transport of ions to the root surface by means of mass flow and diffusion through the soil pores, followed by uptake according to asymptotic kinetics (Nye and Tinker 1977, Tinker and Nye 2000). This combined process of soil flow and root uptake has been worked out in great detail for individual roots growing in homogeneous soils (De Willigen *et al.* 2002). The role of the three-dimensional architecture of the root system has generally not been incorporated in uptake models. One exception is the modelling work by Fitter *et al.* (1991) who used a model to show that the herring bone and other root system architectures tend to be optimum solutions to the trade-off between exploration and exploitation: exploration of large volumes of soil to find new nutrient-rich patches is balanced with exploitation of the rich patches already found. However, root architecture is generally more asymmetric than the herringbone and other structures suggest. It has been shown that roots grow preferentially into nutrient-rich soil patches (Hutchings and De Kroon 1994, chapter 1). Over the lifetime of the plant, uptake rates may depend more on such adaptation of the root system to soil heterogeneity than on uptake kinetics. Models that offer a mechanistic explanation of preferential root growth exist (Van Oijen *et al.* 1986) and will be discussed below.

There are various elements missing from the uptake models described above. First, it has become clear that what we measure as N uptake is in fact the net result of simultaneous influx and efflux (Bouma and De Visser 1993, chapter 1). The relevance for whole-plant physiology is that energy expenditure is proportional to gross uptake rather than to net uptake. Bouma and de Visser (1993) used a process-based model to quantify and analyse the efflux kinetics. This work showed that the net rate of N uptake is partly controlled by the concentrations within the root, rather than solely by the root surface concentration. However, at longer timescales than minutes, control of the uptake rate is more likely to rest with the use of N within the plant after uptake. Therefore many physiological models disregard the fine details of the uptake mechanism, and model uptake as being proportional to the demand of the whole plant for N, as long as demand is less than the supply capacity of the soil. The supply function may be a simple function of bulk soil N content but can be made dependent on soil water flow and pore size distribution as well. Demand is often quantified by multiplying photosynthetic C uptake with an appropriate N/C ratio, or some similar approach dependent on tissue N concentration.

Besides losing N through efflux from the root apoplast, plants lose N by discarding senesced leaves and roots. The mechanisms underlying senescence are poorly known, as are the controlling factors. For turnover of roots, we do not even have good data on the rate of the process (Swinnen *et al.* 1994). Because of these deficiencies, senescence and redistribution are generally simulated empirically with constant or temperature dependent organ turnover rates.

A weakness of current models of N uptake is that the availability of N in the soil is assumed to be determined by abiotic processes only. The effects of root exudates on soil chemistry and weathering is poorly known and usually ignored, as is the role of mycorrhizae (chapter 3).

Transport of nitrogenous compounds within the plant

The two main transport systems within the plant are the xylem and the phloem. N compounds are transported in both, with transport in the phloem being limited to organic N. The two systems are connected in shoot and root (Van Bel 1993), so N can cycle between plant parts (Lambers *et al.* 1982). The mechanisms underlying the two transport systems are highly different. Xylem flow is driven by the water potential gradient imposed by transpiration, whereas phloem flow is driven by pressure differences between source and sink, which are maintained by phloem loading and unloading respectively as explained by Münch's hypothesis (Munch 1930). Active loading and unloading of N compounds into and out of the vessels requires energy but the transport itself occurs passively.

Although the description of plant transport systems just given is only very sketchy, many models of plant growth simplify transport even further and apply diffusion-like equations according to which N-transport rate is proportional to differences between organs in non-structural N concentration. The characteristics of the transport systems are then reduced to one single number: the diffusion coefficient. Models that disregard N transport altogether, and assume that every plant organ has access to one common N substrate pool, are also employed.

The validity of the different approaches depends on how transport is controlled. If transport of N within the plant is ultimately controlled by the rates of N uptake in roots and N use in meristems, then simplification of the transport mechanism is valid. The simplification may be invalid if we consider short-term responses of plant growth to changes in the environment, when the speed of plant response may be limited by the capacity of the transport systems. However, N availability is not likely to change abruptly except under artificial experimental conditions.

Conversions of nitrogenous compounds within the plant

Whole-plant physiological models, which we focus on in this paper, usually do not include any details of biochemical pathways. Generally, at most three types of N compound are distinguished: mineral N (nitrate, ammonium), mobile organic N (amino acids, amides) and non-mobile organic N (proteins). Greater detail is found in organ-level models, such as Hikosaka's (Hikosaka and Terashima 1995) model for the optimal allocation within leaves of N among Rubisco, other Calvin cycle enzymes, electron carriers, photosystem I and II core complexes and light-harvesting chlorophyll-protein complexes I and II. The key problem for physiological models that represent conversion of N compounds in some detail is how to maintain an appropriate balance between the rates of formation and turnover of N and non-N compounds. The chemical composition of plant tissues can vary, but within bounds. For example, if N uptake is hampered, plant-N concentration may decrease, but the decrease will be limited because C assimilation and growth will be hampered as well, and enhanced old leaf turnover followed by N redistribution may keep N concentration in young leaves high. How these processes are controlled is poorly known and models therefore maintain stable chemical composition by empirical approaches such as fixed C/N ratios in growing tissue.

INTEGRATION OF PROCESSES IN THE WHOLE PLANT

Bottom-up

In the previous section, we discussed transport and conversion processes separately. Here, we discuss how the processes are integrated in models. Transport and conversion are interlinked: conversion processes require import of substrate and export of products, and transport processes tend to be inhibited if the transported compounds accumulate at the sink without being converted. How are these couplings between transport and conversion realised in models?

The discussion of N-related processes has already highlighted various ways in which N, water and C relations are coupled. The flux of N in the soil towards the roots depends on soil water flows. Transport of N in the plant depends on the transpiration stream in the xylem, and on the carbohydrate pump maintaining pressure gradients in the phloem. The stepwise conversion of mineral N, from nitrate reduction to protein formation, requires C substrates. All this leads to nearly constant stoichiometry of elements in structural plant tissue, but possibly varying concentrations of substrate. Simple models in which organ growth rates are proportional to concentrations of both C and N substrate (Cooper and Thornley 1976) are able to explain the functional equilibrium between roots and shoots, in which the growth of roots is predominant when N is deficient, and shoot growth is favoured when photosynthetic rates are low (Brouwer 1983).

Similar models, but of slightly greater complexity, are also able to explain observations of preferential root growth into nutrient-rich soil patches. Van Oijen *et al.* (1986) implemented the following mechanism in their model: The root system was divided into two parts, one in N-rich medium, and the other in N-poor medium. Growth rate of each root part, and of the shoot, depended on the concentrations of N and C substrate in that plant part, and transport between root parts was only possible via the shoot. N uptake of the root in poor medium was obviously hampered, so growth of this root depended fully on organic N received from the shoot in the phloem. This allowed some growth, but at low rate, leading to accumulation of unused carbohydrates and subsequent inhibition of the Münch-type phloem inflow. High rates of phloem import, and growth, were maintained in the root part growing in rich medium. This model was able to account with great accuracy for observations on maize roots in split-root experiments that showed preferential growth into rich medium, but some continued root growth in medium without any N at all (Spek and Van Oijen 1988).

Top-down

The previous section described efforts to model integrated plant activity without explicitly incorporating whole-plant level control mechanisms (Cheeseman 1993, Cheeseman *et al.* 1996). The idea was that integration of activity would arise automatically from the interactions of many low-level plant processes. This is a bottom-up approach. In the present section, we will look at top-down approaches, in which integrated plant behaviour is achieved by constraining the lower-level processes in some coordinated way.

Mechanistic modelling of whole-plant level integration

Earlier we made a distinction between mechanism and control, with the first referring to how something works and the second to how something is affected by external factors. We can of course try to model control itself as a mechanism, for example by inspecting how control is effectuated in plants. One possible method through which plants can achieve whole-plant communication is through molecules that function as signals, *e.g.* hormones (Trewavas 2003). Modellers have tended to shy away from explicitly modelling hormone production, transport, sensing and signal transduction because of the lack of quantitative data. An exception is the work by Tardieu *et al.* (1993) who modelled root-shoot transport of abscisic acid and subsequent stomatal closure. We know of no examples of models on hormonal control in N relations, but see Van der Werf and Nagel (1996) for suggestions on how to extend an allocation model with signalling by sucrose and cytokinins. However, besides the lack of quantitative data, there is another reason why modelling hormonal control of plant processes has been attempted only rarely. The idea exists among modellers that hormonal control mainly affects plant behaviour in the very short-term (up to days), and that integration of plant activity over longer periods is primarily determined by the availability and within-plant transport and use of resources. Hormones thus are seen as accelerators of plant responses to changes in the growing environment, rather than as essential agents causing specific responses (De Wit and Penning de Vries 1983).

Empirical modelling of whole-plant level integration

Early models of plant growth used empirical functions of thermal time (*i.e.* the number of degree-days since the start of the growing season) to modify the allocation of assimilates and N compounds between plant organs. Annual plants, for example, were started off with high fractional allocation to root growth, with later shifts towards foliage growth and finally reproductive growth. The advantage of such methods was their simplicity, and the ease with which observations of plant growth over calendar time could be translated into the required allocation functions of thermal time. The obvious downside of such methods was their rigidity: the flexible response of plants to changes in their growing conditions could not easily be accommodated with such methods.

Phenomenological modelling of whole-plant level integration

To model the flexible adaptive responses to environmental change that we observe in plants, without attempting to model poorly known control mechanisms, phenomenological methods are increasingly being used. In particular, optimality approaches are becoming more frequent. Studies with simple models have shown how plant photosynthesis is maximised if the vertical distribution of N compounds associated with photosynthesis is the same as the vertical distribution of light (Goudriaan 1995). The same models suggest that such a distribution of N will make overall plant light-use efficiency (LUE), defined as carbon assimilated per unit intercepted light, relatively independent of light intensity itself (Dewar 1996). More recently, these N-profile models have been extended with the assumption that plants shed those bottom leaves whose predicted N level is below a minimum level required for photosynthesis (Yin *et al.* 2000). This method allowed successful prediction of leaf senescence in rice and barley. The predicted exponential distribution of N with height, and the consequent robustness of the LUE have also been confirmed empirically, but not always completely (Dreccer et al. 2000). Often the vertical N distribution has been found to be slightly more uniform than the assumption of maximum photosynthesis would suggest. Such a state-of-affairs, with good but not perfect simulations, poses particular problems for phenomenological modellers. Faced with an imperfect model, a mechanistic modeller would say: let's add more biological detail about how the plant works, and sooner or later we will get it right. The empirical modeller would simply incorporate the observed N profiles, and possibly some statistical relationships between profile and environmental conditions, into the model. The phenomenological modeller has no such clear way forward. His optimality approach might have failed because the plant may in reality not have any means to optimise N distribution. Furthermore, even if an assumption of optimality is made, the problem may be that the plant does not maximise overall photosynthetic rate but something else, such as overall C economy.

MODELLING PLANT ADAPTATION TO THE ENVIRONMENT

In the previous two sections, we described the different methods that modellers have employed to simulate integration of activity in higher plants. In the present section, we discuss how the models can be used to analyse how plants adapt their N relations and overall behaviour to their growing environment.

Adaptation at different scales

For simplicity, we may distinguish adaptation at three different scales: the whole-plant level, the level of the foliage or the root system, and the level of individual leaves or roots.

At the whole-plant level, the best known example of plant adaptation to its environment is the functional equilibrium between root and shoot, in which changes in availability of nutrients, water, CO_2 or light cause allocation to shift towards the organ system that has to absorb the increasingly limiting resource. We have seen above how the functional equilibrium can be explained by mechanistic models. Phenomenological models for partitioning, in which the optimal root/shoot ratio is calculated directly as a function of resource availabilities, can be and have been made as well. However, the phenomenological models cannot predict how fast partitioning adapts - they therefore commonly assume that adaptation is instantaneous and root/shoot ratio thus always optimal - so they are less useful than the mechanistic models for the analysis of the dynamics of adaptation.

At the level of the foliage or the root system, examples of adaptation are redistribution of photosynthetic N between leaf layers when light conditions change, and preferential root growth into nutrient-rich patches. Both processes have been modelled mechanistically and by means of optimisation approaches.

We know of no models that attempt to simulate adaptation of individual roots to their growing environment, even though it is known that along the length of individual roots, anatomical differences may exist that result from adaptation to local soil conditions, *e.g.* extra suberisation in dry patches (North and Nobel 2000). Models do exist that account for the adaptation of individual leaves to their environment. We have mentioned the model of Hikosaka and Terashima (1995) that accounts for the optimal allocation within leaves of N among different compounds. Another example is the coordination theory of Reynolds and Chen (1996), according to which N is distributed to keep the capacities for photosynthetic electron transport and carboxylation in balance. However, both models are phenomenological; we lack the physiological understanding to attempt mechanistic modelling of leaf adaptation.

Quantifying adaptation processes

In the previous section, we have shown by examples that plants adapt to their environment at various scales. However, we have also seen that our understanding of adaptation processes is fragmentary. Much of our knowledge derives from experiments that compare plants growing under different, but constant, conditions from seedling to harvest. Such experiments do not reflect the ever-changing environments in which plants live naturally. We need to study the dynamics of adaptation.

It may be that the study of adaptation has been hampered by the absence of a conceptual framework, which allows to compare different adaptation processes. Mechanistic modelling might provide this framework. Mechanistic modelling of adaptation requires quantifying the following four characteristics of adaptation processes: 1) speed, 2) metabolic costs, 3) reversibility, and 4) physiological and morphological constraints.

1) Speed. Is the adaptation process fast or slow? Does its speed match the rate at which the environment changes? There may be a general rule here that the speed of reaction decreases when going from leaf to canopy to whole-plant processes. However, within one scale level, different adaptation processes may also have different time constants. Sage and Reid (1994) showed how Rubisco activation status of *Pueraria lobata* leaves decreased within one day of initiation of shading, whereas Rubisco concentration decreased more slowly over a period of weeks. Adaptation of chlorophyll content may be slower still.

2) Metabolic costs. Few adaptation processes are completely passive, without energy demand. Energy demand may be low for reactivating previously deactivated enzyme, but much higher for synthesising new enzyme. Even more

expensive will be processes like N redistribution which involve local breakdown of compounds, followed by transport across membranes via energy-requiring carriers to other organs and reassembly. Is there a general rule here that slow adaptation processes are generally expensive? We need to have more quantitative information about the economics of adaptation.

3) Reversibility. Few plant processes are truly irreversible - anabolic processes usually come paired with catabolic ones - but we can define reversibility in an economic sense as low cost of undoing and redoing a previous action. We thus need to quantify not only the metabolic costs of adaptation processes themselves (the preceding item) but also the costs involved in reversing any adaptation. Rubisco formation is strictly reversible (as observable in senescing tissues) but plants do not break down all Rubisco every night because it would have to be reassembled the next morning incurring the high costs of protein synthesis. An economic definition of reversibility may help us understand the different strategies plants employ to adapt to unstable environments.

4) Physiological and morphological constraints. All plant activity, including adaptation, is subject to constraints. Changes in the environment may cause changes in allocation patterns, but allocation to stems, branches and petioles must continue to satisfy the demands of biomechanics. Similarly, shade-leaves tend to have higher specific leaf area than sun-leaves, but all leaves must be thin enough to permit efficient within-leaf CO_2 diffusion, and thick enough to allow leaf blade angles that are appropriate for light interception. Adaptation involving N redistribution cannot proceed faster than what the capacity of the transport system allows. It is in the constraints to adaptation that differences between genotypes are likely to be most clear.

If we quantify the above characteristics for adaptation processes, it will not only be possible to construct mechanistic models of adaptation, but it will also be possible to compare the advantages and disadvantages of different plant adaptation strategies in a structured manner.

DISCUSSION AND OUTLOOK

In this chapter, we have tried to show how process-based modelling of plant-N relations may benefit our understanding of plant adaptation to the environment. We have given examples that show how models can explain observed plant responses at various scales, from leaves and roots to the whole plant. The modelling work has shown that integration of activity at the whole-plant level need not necessarily depend on explicit integrating mechanisms, like plant hormone systems, although such systems play at least a facilitating role. However, we should not overstate the modelling state-of-the-art. We have given examples of plant responses to changes in the environment that can be understood, and modelled, using ideas from optimal control theory, but for which we do not have mechanistic models. Without mechanistic underpinning such phenomenological explanations remain tentative at best. In other words, the modelling work has not yet led to a full biological theory of adaptation; it has only contributed some building blocks for such a theory.

We have discussed plant adaptation to the environment, but omitted discussion of the equally important adaptation of the environment to the plants. In particular, soil structure and composition is affected by root activity and litterfall. For the effects of plants on soils, and the impact of air pollution, see chapter 3.

The focus of this chapter has been on modelling N relations. However, it must be emphasised that physiological models of plant-N relations as such do not and cannot exist. Nitrogen relations are intertwined with C metabolism and water relations at various scales within and around the plants. The uptake of N from the soil, the functional equilibrium between root and shoot, the turnover of photosynthetic enzymes – in fact all of N physiology is tightly linked to C and water metabolism. We hope to have shown how modelling can help demonstrate the importance of these linkages, and how they are key to whole-plant integrated behaviour. However, there is a major gap concerning the modelling of the coupling between plant N relations and other nutrients. Models of uptake of P, K and other nutrients exist, but very little quantitative work has been done on the role of non-N nutrients in whole-plant physiology. We may well have overemphasised the importance of N relative to other elements.

Models can only operate if they are internally complete and consistent, with all their cause-effect relations properly quantified, and they can only be tested against quantitative data. Because of this dependency on an adequate knowledge- and database, plant models often are useful in identifying gaps in our physiological knowledge. In particular, whenever processes are integrated in a physiological model, there is a need to quantify the rate at which each process proceeds, the consequences of the process for the C and energy budget of the plant, and the constraints that affect the process. So, to explain adaptation we need to quantify the speed, metabolic costs, reversibility and physiological and morphological constraints of adaptive processes. To some extent this task will be made easier by the rapid advances that are being made in molecular biology. There is an emerging discipline which calls itself "systems biology" (Kitano 2002), which aims to measure the simultaneous activity of different genes in cellular systems and to model the activity of a set of genes/proteins as a functional network (Raikhel and Coruzzi 2003). Systems biology may be misnamed - what are plant physiology and ecology if not systems sciences? - but it may help us understand and quantify the genetic constraints to plant adaptation.

REFERENCES

Apel P., Bauwe H., Frank R., Peisker M. 1985. Beziehungen zwischen Kohlenstoff und Stickstoffmetabolismus bei Pflanzen - Eine Ubersicht uber ausgewahlte Aspekte. - Kulturpflanze. 33: 41-72.
Bouma T.J., De Visser R. 1993. Energy requirements for maintenance of ion concentrations in roots. - Physiol. Plant. 89: 133-142.
Bouman B.A.M., Van Keulen H., Van Laar H.H., Rabbinge R. 1996. The 'School of de Wit' crop growth simulation models: A pedigree and historical overview. - Agric. Sys. 52: 171-198.
Brouwer R. 1983. Functional equilibrum: sense or nonsense? - Neth. J. Agric. Sci. 31: 335-348.
Cheeseman J.M. 1993. Plant growth modelling without integrating mechanisms. - Plant Cell Environ. 16: 137-147.
Cheeseman J.M., Barreiro R., Lexa M. 1996. Plant growth modelling and the integration of shoot and root activities without communicating messengers: Opinion. - Plant Soil 185: 51-64.
Cooper A., Thornley J.H.M. 1976. Response of dry matter partitioning, growth, and carbon and nitrogen levels in the tomato plant to changes in root temperature: Experiment and theory. - Ann. Bot. 40: 1139-1152.
De Willigen P., Heinen M., Mollier A., Van Noordwijk M. 2002. Two-dimensional growth of a root system modelled as a diffusion process. I. Analytical solutions. - Plant Soil 240: 225-234.
De Wit C.T., Penning de Vries F.W.T. 1983. Crop growth models without hormones. - Neth. J. Agric. Sci. 31: 313-323.
Dewar R.C. 1996. The correlation between plant growth and intercepted radiation: An interpretation in terms of optimal plant nitrogen content. - Ann. Bot. 78: 125-136.
Dreccer M.F., Van Oijen M., Schapendonk A.H.C.M., Pot C.S., Rabbinge R. 2000. Dynamics of vertical leaf nitrogen distribution in a vegetative wheat canopy. Impact on photosynthesis. - Ann. Bot. 86: 821-831.
Fitter A.H., Strickland T.R., Harvey M.L., Wilson G.W. 1991. Architectural analysis of plant root systems. 1. Architectural correlates of exploitation efficiency. - New Phytol. 118: 375-382.
Goudriaan J. 1995. Optimization of nitrogen distribution and of leaf area index for maximum canopy assimilation rate. - In: Thiyagarajan T.M., Ten Berge H.F.M., Wopereis M.C.S. (Eds.) Nitrogen management studies in irrigated rice. - Los Banos, Philippines, SARP Res. Proc., pp. 85-97.
Hikosaka K., Terashima I. 1995. A model of the acclimation of photosynthesis in the leaves of C3 plants to sun and shade with respect to nitrogen use. - Plant Cell Environ. 18: 605-618.
Hutchings M.J., De Kroon H. 1994. Foraging in plants - The role of morphological plasticity in resource acquisition. - Adv. Ecol. Res. 25: 159-238.
Kitano H. 2002. Systems biology: A brief overview. - Science 295: 1662-1664.
Lambers H., Simpson R.J., Beilharz V.C., Dalling M.J. 1982. Growth and translocation of C and N in wheat (*Triticum aestivum*) grown with a split root system. - Physiol. Plant. 56: 421-429.
Liebig J.v. 1840. Die Chemie in ihrer Anwendung auf Agrikultur und Physiologie. - Braunschweig, Germany, Vieweg.
Munch E. 1930. Die Stoffbewegungen in der Pflanze. - Jena, Germany, Gustav Fischer.
Nasholm T., Ekblad A., Nordin A., Giesler R., Hogberg M., Hogberg P. 1998. Boreal forest plants take up organic nitrogen. - Nature 392: 914-916.
North, G.B., Nobel, P.S. 2000. Heterogeniety in water availability alters cellular development and hydraulic conductivity along roots of a desert succulent. Ann. Bot. 85: 247-255.
Nye P.H., Tinker P.B. 1977. Solute movement in the soil-root system. - Oxford, U.K., Blackwell.
Pate J.S., Layzell D.B., McNeil D.L. 1979. Modeling the transport and utilization of carbon and nitrogen in a nodulated legume. - Plant Physiol. 63: 730-737.
Raikhel N.V., Coruzzi G.M. 2003. Plant systems biology. - Plant Physiol. 132: 403.
Reynolds J.F., Chen J.L. 1996. Modelling whole-plant allocation in relation to carbon and nitrogen supply: Coordination versus optimization. - Plant Soil 185: 65-74.

Riedo M., Milford C., Schmid M., Sutton M.A. 2002. Coupling soil-plant-atmosphere exchange of ammonia with ecosystem functioning in grasslands. - Ecol. Modell. 158: 83-110.

Sage R.F., Reid C.D. 1994. Photosynthetic response mechanisms to environmental change in C_3 plants. - In: Wilkinson R.E. (Ed.) Plant-environment interactions. - New York, Marcel Dekker, pp. 413-499.

Spek L.Y., Van Oijen M. 1988. A simulation model of root and shoot growth at different levels of nitrogen availability. - Plant Soil 111: 191-197.

Swinnen J., Van Veen J.A., Merckx R. 1994. Rhizosphere carbon fluxes in field-grown spring wheat - Model calculations based on C-14 partitioning after pulse-labelling. - Soil Biol. Biochem. 26: 171-182.

Tardieu F., Zhang J., Gowing D.J.G.1993. Stomatal control by both [ABA] in the xylem sap and leaf water status - A test of a model for droughted or ABA-fed field-grown maize. - Plant Cell Environ. 16: 413-420.

Tinker P.B., Nye P.H. 2000. Solute transport in the rhizosphere. - Oxford, U.K, Oxford University Press.

Trewavas A. 2003. Aspects of plant intelligence. - Ann. Bot. 92:1-20.

Van Bel A.J.E. 1993. Strategies of phloem loading. - Annu. Rev. Plant Physiol. Plant Mol. Biol. 44: 53-281.

Van der Werf, A., Nagel O.W. 1996. Carbon allocation to shoots and roots in relation to nitrogen supply is mediated by cytokinins and sucrose. - Plant Soil. 185: 21-32.

Van Oijen, M. 2002. On the use of specific publication criteria for papers on process-based modelling in plant science. - Field Crops Res. 74: 197-205.

Van Oijen, M., Spek L.Y., Brouwer R. 1986. A simulation model of growth and C and N metabolism in young maize plants. - In: Lambers, H., Neeteson, J.J., Stulen, I. (Eds.) Fundamental, ecological and agricultural aspects of nitrogen metabolism in higher plants. - Dordrecht, The Netherlands, Martinus Nijhoff Publishers, pp. 323-327.

Yin, X.Y., Schapendonk A., Kropff M.J., van Oijen M., Bindraban P.S. 2000. A generic equation for nitrogen-limited leaf area index and its application in crop growth models for predicting leaf senescence. - Ann. Bot. 85: 579-585.

Chapter 6

LIGHT REGULATION OF NITRATE UPTAKE, ASSIMILATION AND METABOLISM

Cathrine Lillo

INTRODUCTION

Evolving from a germinating seedling, dependent on a stored restricted energy supply, into an active photosynthetic plant capturing the energy from sunlight is obviously a vulnerable time period in the life of a plant. Expression of genes involved in nitrogen reduction and assimilation during this time span is strongly influenced by low intensity red and blue light acting through special light receptors, phytochromes and cryptochromes. After greening and establishing of the photosynthetic apparatus, light acts on nitrogen metabolism through photosynthesis and the products thereof, sugars and certain carbon compounds. The ratio between products of nitrogen and carbon metabolism regulates transcription of genes in the nitrogen pathway and acts post-transcriptionally on these gene products. Later in a plant's life, day length, measured with the help of the red and blue-light absorbing receptors and circadian rhythms, are important for transition into different developmental stages, like reproduction or dormancy. Abundant nitrogen supply is known to sometimes counteract transition into flowering or dormancy, while on the other hand scarce nitrogen supply may induce flowering or dormancy. These interactions between nitrogen metabolism and development in the mature plant are well known but poorly understood and will not be covered in this review. The subject of this review is the influence of light on transcriptional and post-transcriptional regulation of genes involved in nitrate uptake, reduction and incorporation into organic compounds during greening and the vegetative stage.

Photomorphogenesis

There are two general means by which light modifies plant growth. Long-term growth requires a sufficient supply of energy via photosynthesis, which provides essential raw material, *i.e.* organic carbon, as well as energy for all biochemical reactions and growth. Light also influences growth via special photomorphogenetic pigments. These include the red/far-red light absorbing phytochromes and blue/UV light absorbing pigments. These pigments detect one or more conditions in the environment such as quality, quantity, direction and duration of light (Cosgrove 1993).

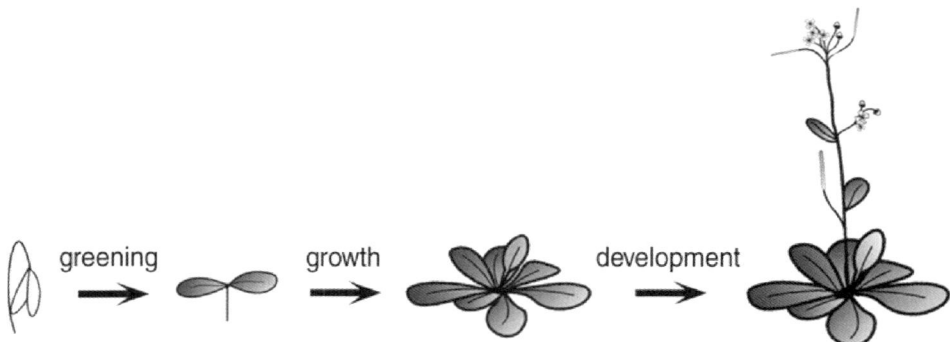

Figure 1. Deetiolation, growth and development. Greening: Low intensity red and blue light, in combination with nitrate induces expression of genes involved in nitrogen assimilation during deetiolation and greening. Growth: Photosynthetic active light sustains further growth. Nitrate and products of both photosynthesis and nitrogen assimilation regulate expression of genes involved in nitrogen metabolism. Development: Transformation into different developmental stages like flowering or dormancy can be induced by day length perceived by red and blue-light receptors. Nitrogen limitation may promote flowering and dormancy.

In the dark, plants develop in a special manner, stems become long and thin, leaves remain folded and small and the apical part of the stem often forms a hook (in dicotyledons). Furthermore, chloroplast development is suppressed and pigmentation reduced. Such appearance is referred to as etiolated and is a useful form of development for bringing a seedling or other propagules buried in the soil to the surface as rapidly as possible and without damage (Cosgrove 1993). On reaching the surface and light exposure, the developmental pattern changes, with leaves becoming green and photosynthetically active.

Efficiency in the greening process is obviously important for survival since the plant needs to capture energy and carbon through photosynthesis in order to replace the supply rapidly consumed from stored resources in seeds and propagules. This critical transformation in the plant's life is much studied and reveals changes in metabolism preparing the plant not only for photosynthesis, but also for assimilation of nitrogen. The red and blue-light absorbing pigments are crucial for this transformation from etiolated into green, photosynthetically active plants.

The same groups of pigments are important for induction of flowering and dormancy, shading response etc. Promotion of flowering by nitrogen limitation in otherwise day-length-influenced plants is well known (Tanaka 1986, Ishioka *et al.* 1991). However, the interaction with nitrogen metabolism in processes like flowering induction or dormancy is not well understood at the molecular level.

Plants react differently to light during development. For example exposure to light often alters the sensitivity to further treatments of light exposure. The concentration of different light receptors changes as a result of illumination. There are certainly also species differences as well as tissue differences in response to light treatments. Action spectra related to a special process, for instance hypocotyl elongation, show some variations concerning effects of red and blue light among species (Cosgrove 1993). Similarly, effects from red and blue light on expression of enzymes in the nitrogen pathway also vary among species.

Photoreceptors

Three sensory photoreceptor families are known: the phytochromes, cryptochromes, and phototropins (Quail 2002). In *Arabidopsis* 5 different phytochrome genes *PHYA-PHYE* have been identified, and the overall picture is that PHYB has a role at all stages of the life cycle, whereas PHYA, PHYD and PHYE exert their principal functions at selected stages (Smith 2000). Phytochromes have a tetrapyrrole chromophore, absorb light in the red and far-red spectrum, and undergo light-induced interconversion from a biologically inactive form into an active form. The phytochromes are cytosolically localised, but translocate into the nucleus upon irradiation where they may directly interact with transcription factors (Smith 2000). PHYB is activated by red light and inactivated by far-red light and is the form present in green tissue. PHYA is the abundant form in dark-grown tissue and responds to very low-fluence red light and high irradiance far-red light. Phytochromes also absorb in the blue-light region, which therefore results in overlapping action spectra of phytochromes and blue-light receptors (Casal 2002). The second class of photoreceptors, the cryptochromes, are blue-light receptors carrying two different chromophores, a flavine type and a pterin type. Two cryptochrome genes are found in *Arabidopsis*, *CRY1* and *CRY2*. Both cryptochromes are localised to the nucleus. The photochemical signal capture of CRY is not known, but is likely to involve a redox reaction (Quail 2002). The third class of photoreceptors, phototropins, are also flavoproteins and hence blue-light receptors. Phytochromes and cryptochromes are generally found to be involved in morphogenetic processes, for example germination, de-etiolation and flowering, whereas the phototropins are involved in movements, like bending of the stem towards light or movements of chloroplasts in the cell and opening/closing of stomata (Briggs and Christie 2002). Phytochromes and cryptochromes are generally found to be involved in regulation of the nitrogen pathway in plants. The picture may still not be complete concerning photoreceptors, another UV-B receptor with spectral properties different from cryptochromes and phototropins appears to be present in plants as well (Briggs and Christie 2002).

The signal transduction chains connecting light perception by the various photoreceptors to regulation of gene expression is far from clear. Phosphorylation steps have been suggested to be involved, and a handful of proteins interacting with phytochromes or cryptochromes have been identified. COP1 is a protein found to interact directly with blue-light receptors (CRY and PHY) by protein-protein contact. COP1 has a role in mediating the regulated proteolysis of another protein, HY5, which is a bZIP transcription factor that promotes the expression of light-induced genes (Wang et al. 2001). PHY and CRY interact with COP1 and direct the migration of COP1 to the cytoplasm thereby reducing the activity of COP1. Mutations in *COP1* and *HY5* influence expression of nitrate reductase (see below) as well as genes involved in photosynthesis. Another example of a transcription factor interacting with light is PIF3 (phytochrome interacting factor), which is constitutively expressed and member of the basic helix-loop-helix class of transcription factors. PIF3 binds to a G-box motif present in promoters of various light-regulated genes. It has been suggested that the active form of PHY translocates into the nucleus and bind to already promoter-bound PIF3 and thereby facilitate transcription (Quail 2002).

Light responsive elements (LRE)

When studying photosynthesis-associated nuclear genes a number of cis-acting elements involved in the control of transcription by light were identified (Argüello-Astorga and Herrera-Estrella 1998). These elements were classified into different types, according to their nucleotide sequence and binding of transcription factors. Put simply, certain core sequence units are recognised in the DNA motifs, although not strictly identical in different plants or promoters. Examples of such motifs are: G-box (core unit ACGT), GTI-box (core unit GGTTAA), GATA-box (core unit GATA) and a special GATA motif, I-box (core unit GATAAGA/G). A single motif can not function as a light responsive element (LRE), at least two different motifs have been found to be involved (Martínez-Hernández et al. 2002). A minimal LRE containing an I-box (GATAAGA) and G-box (CACGTGGC) linked to a reporter gene was studied in *Arabidopsis* (Martinez-Hernández et al. 2002). Such an LRE resulted in reporter gene activation by both PHY and CRY, and red, blue or white light all were active in inducing a response. The minimal LRE also confined tissue specificity in adult plants, and expression of the reporter gene was only found in cells containing chloroplasts, like mesophyll cells of leaves, stems and petioles as well as stomata, but not in epidermis, vascular tissue or roots. The reporter gene was strongly expressed in *cop1* mutants, as expected, since COP1 mediates proteolysis of a factor (HY5) involved in transcription of light activated genes. Decreased expression was found in *hy5* mutants as predicted. However, 50% activity was retained in the *hy5* background showing that some other factor could partially replace HY5 (Martínez-Hernández *et al.* 2002). Although putative light-responsive motifs are present upstream of many of the genes involved in the nitrogen pathway, very few elements have been examined for their function and shown to be involved in light-activation of the nitrogen pathway. Much, almost everything remains to be clarified concerning light reactive elements in promoters and enhancer regions of genes in the nitrogen pathway of plants.

Light activation by thioredoxins

Thioredoxins are small, ubiquitous, proteins found in both prokaryotes and eukaryotes. The highly conserved active site, Trp-Cys-Gly-Pro-Cys, contains a disulphide group that undergoes reversible redox changes between S-S and 2SH (Ruelland and Miginiac-Maslow 1999, Schürmann and Jacquot 2000). Thioredoxins serve as electron donors in a variety of cellular redox reactions. Plants contain multiple forms, two in the chloroplasts (m and f), one in the cytosol and one in the nucleus. In chloroplasts and cyanobacteria, reduction of thioredoxins are coupled to light-driven photosynthetic electron transport and mediated through reduced ferredoxin. It is known for long that several Calvin cycle enzymes are activated by the ferredoxin-thioredoxin system. In addition to these enzymes related to CO_2-fixation, also other chloroplastic enzymes are regulated by thioredoxins, such as the coupling factor CF1 providing ATP for biosynthetic reactions, and NADPH-malate dehydrogenase which is important for the export of reducing equivalents (malate) to the cytosol. Also some chloroplast-located enzymes involved in nitrogen metabolism are activated by thioredoxins, *i.e.* glutamine synthetase, ferredoxin-dependent glutamate synthase, and the first enzyme in the pathway of aromatic amino acids (Ruelland and Miginiac-Maslow 1999, Schürmann and Jacquot 2000).

PII

PII signal transduction proteins are widespread, being present in eukaryotes, bacteria and archaea and are essential for regulation of nitrogen metabolism. In bacteria and archaea, PII proteins serve as the central processing unit for the integration of signals of carbon and nitrogen status (Ninfa and Atkinson 2000). A homologous gene has been identified in plants, and there is evidence that the same metabolites as in bacteria, α-ketoglutarate and glutamate, are crucial for regulation of nitrogen metabolism in plants. This points to the possibility that PII may have a similar function in plants as in prokaryotes.

In *Escherichia coli* the PII protein is covalently modified by uridylylation of a conserved tyrosyl residue (gives PII-UMP). High concentration of glutamine triggers this uridylylation of PII. The carbon-status indicator molecule α-ketoglutarate binds directly to PII. Overall, in *E. coli* PII is converted into different forms in response to the concentration of nitrogen and carbon compounds, and acts post-translationally upon the activation and inactivation of an enzyme in the nitrogen assimilation pathways, *i.e.* glutamine synthetase (Ninfa and Atkinson 2000). PII also affects regulation on the level of transcription, and does so by influencing the phosphorylation status of a protein necessary for transcription of the glutamine synthetase gene in *E.coli*.

In cyanobacteria (*Synechococcus*), PII regulates uptake of nitrate and nitrite, possibly by direct interaction with a component of the uptake system. Modification of PII in cyanobacteria is via phosphorylation of a conserved seryl residue. When PII is unphosphorylated the uptake processes are inhibited. This seryl residue is conserved also in plant PII proteins, whereas the tyrosyl residue, which is uridylylated in *E. coli*, is replaced by phenylalanine in plants (Hsieh *et al.* 1998). PII like proteins have so far been identified in *Arabidopsis thaliana* and *Ricinus communis* (Hsieh *et al.* 1998). The plant PII is a nuclear-encoded chloroplast protein. Intriguingly, expression of PII is influenced by the same factors that are supposed to be sensed by the PII protein itself. PII mRNA was induced by light in dark-adapted green *Arabidopsis* leaves, and sucrose stimulated mRNA accumulation in both darkness and light, whereas asparagine, glutamine and glutamate partly reversed the positive effect of light or sucrose. Red light did not lead to the accumulation of PII mRNA, hence phytochrome did not appear to be involved in PII expression. Ferrario-Méry *et al.* (2000) presented clear evidence with transgenic tobacco plants that the balance between the important modulators of PII proteins, glutamine and α-ketoglutarate, are essential for regulation of nitrogen metabolism also in plants. The *Arabidopsis* PII has recently been cloned and overexpressed in *E.coli*, and polyclonal antibodies have been produced towards the *Arabidopsis* PII (Smith *et al.* 2002). Expression patterns, post-translational modifications and a role of PII in control of nitrogen metabolism are expected to be revealed soon also for plants.

Fungi

Although PII is found in bacteria, archaea, algae and plants and may have a crucial role in regulation of nitrogen metabolism in these organisms, there are no reports or Blast hits found for fungi or yeast when searching for this protein. However, another common principle concerning regulation of nitrogen metabolism is widespread among fungi and yeast. In these organisms compounds like ammonia, glutamine, glutamate, and asparagine are preferentially used as nitrogen source. As long as these compounds are present, expression of genes for nitrate uptake and reduction are repressed. However, when these favoured nitrogen sources are not available, other sources, for instance nitrate can be used. The necessary genes are then transcribed (Marzluf 1997). Positive acting, global regulatory genes are found in yeast and fungi, *Nit-2* in *Neurospora*, *AreA* in *Aspergillus*, and *Gln-3* in *Saccharomyces*. These genes all code for GATA-type zinc finger transcription factors. These global factors bind to promoters of various genes regulated by nitrogen. For example the promoter region of the *Neurospora* nitrate reductase gene has three binding sites for NIT2 all containing GATA elements. When glutamine and other preferable nitrogen sources are scarce NIT2 becomes active, binds to the promoter region of nitrate reductase and, together with a pathway specific transcription factor, NIT4, activate transcription. Glutamine is a critical metabolite, which exerts catabolite repression in *Neurospora*; however, a still unknown feature is the identity of the element, or signal pathway that senses the presence of repressing levels of glutamine (Marzluf 1997). The repression of gene expression by glutamine and other reduced nitrogen compounds has a parallel in plants since glutamine and NH_4^+ are known to repress transcription of several genes in the nitrogen pathway also in plants. This raised the question of whether the regulatory mechanism is conserved between fungi and plants, and attempts have been made to identify a similar system in plants as in fungi. The NIT2 protein of *Neurospora* was found to bind upstream of the NR structural gene from tomato *in vitro* (Jarai *et al.* 1992), and a gene homologous to the *NIT2* gene was identified in tobacco. This gene was called *NTL1* (for nit-2-like) and the amino acid sequence of the Zn-finger domain showed 60% identity to the NIT2 protein. It still remains to be shown if the NTL1 protein is involved in regulation of NR or other genes in plants (Daniel-Vedele and Caboche 1993). GATA motifs are present in several of the genes involved in nitrogen uptake and metabolism in plants, hence putative targets for a NIT2 like transcription factor are present.

A complementation approach in yeast was undertaken, using a *Saccharomyces* mutant deficient in the global regulatory protein GLN3. Three *Arabidopis* cDNAs were found that restored growth of the yeast mutant on certain nitrogen sources, and led to the isolation of *RGA* and *GAI* cDNA. Surprisingly, these were the same genes already known to be involved in the response to gibberellins in plants. Loss of function (*rga2*, *gai-t6*) and gain of function (*gai-1*) mutants were also studied. These studies did not, however, identify genes involved in nitrogen metabolism or regulation thereof in plants (Bouton *et al.* 2002).

NITRATE TRANSPORTERS

Recently genes from two different families of nitrate transporters named *NRT1* and *NRT2*, were cloned from a wide range of higher plants (Forde 2000). For details on nitrate transporters see chapter 1. Physiologically different types of nitrate transporters have been identified. Some transporters have low affinity for nitrate (Low Affinity Nitrate Transporters, LATS) while others have high affinity for nitrate (High Affinity Nitrate Transporters, HATS). Generally, LATS were associated with the *NRT1* family, and HATS were related to the *NRT2* family. However, the picture is more complex. For instance the *Arabidopsis NRT1.1* transporter contributes to high as well as low affinity transport (Touraine *et al.* 2001, chapter 1).

Nitrate transporters and phytochrome

There is some evidence that red light may influence uptake of nitrate in algae. Red and blue light pulses for 5 min stimulated nitrate uptake in the red alga *Corallina elongata*, and the light effects were reversed by far-red light in accordance with phytochrome being involved (Figueroa 1993). Also in a green alga, *Ulva rigida*, positive effects of red and blue light pulses on nitrate uptake reversed by far-red light were found (Lopez-Figueroa and Ruediger 1991). *Spirodela polyrhiza* (*Lemnaceae*), a monocot water-plant, has been much used for studying nitrogen metabolism in relation to phytochrome; however, no influence on nitrate uptake by the phytochrome system was found (Appenroth *et al.* 1992). The literature gives very little information on phytochrome effects on nitrate uptake in higher plants, and when examined generally no or very small effects on nitrate uptake as influenced by the phytochrome system have been found (Sasakawa and Yamamoto 1979).

Interestingly, oligonucleotide arrays revealed a nitrate transporter gene regulated by PHYA in *Arabidopsis*. One hour of far-red light induced the expression of this nitrate transporter gene *NTP3* (CAB38706) (Tepperman *et al.* 2001). The *NTP3* gene codes for a protein predicted to contain the 12 putative trans-membrane helices found in other transporters of the *NRT1* family. However, there is as yet no information on the transport function of the NTP3 protein. Four other nitrate transporters present on the Affymetrix oligonucleotide arrays were not identified as targets of PHYA signalling. It should be noted that plants were grown on half-strength MS-medium, and the high concentration of NH_4^+ and NO_3^- experienced by the plants for several weeks may inhibit expression of nitrate uptake genes. In addition to nitrate transporters providing uptake of NO_3^- into the roots, other nitrate transporters are involved in allocation of NO_3^- to the vacuole, or loading of NO_3^- into the xylem. Since phytochrome effects on nitrate uptake into the roots is not generally found, possibly the NTP3 transporter may be involved in NO_3^- translocation inside the cell, or loading of NO_3^- into the xylem to provide NO_3^- to other parts of the plant. The effect of far-red light on a nitrate transporter gene actualises results obtained several years ago showing that non-photosynthetic light reactions were important for intracellular NO_3^- movement in etiolated pea and barley leaves (Jones and Sheard 1979).

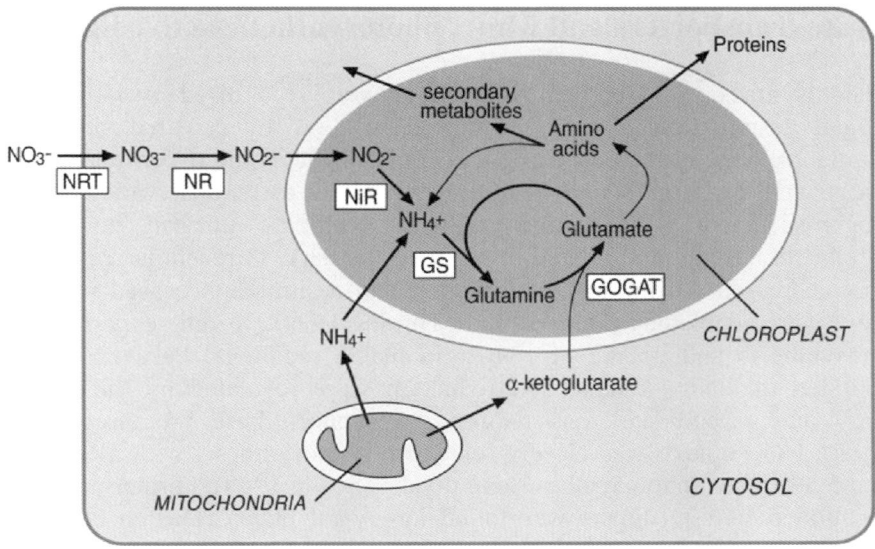

Figure 2. Overview of the nitrogen assimilation pathway in (green) cells. Nitrate is transported into the cell by the help of special nitrate transporters (NRT). Nitrate is then reduced in the cytosol by help of NAD(P)H and nitrate reductase (NR) to nitrite, which is further reduced in the chloroplast by reduced ferredoxin and nitrite reductase (NiR). Ammonium is then incorporated into amino acids by glutamine synthetase (GS) which converts glutamate and NH_4^+ to glutamine, and this reaction requires ATP. Glutamine and α-ketoglutarate are converted to two molecules of glutamate by glutamate synthase (GOGAT). Several amino acids are formed from glutamate and keto acids by transaminases, and form the basis also for further synthesis of other amino acids. Amino acids are used for building proteins, and different nitrogen compounds as well as being the start point for synthesis of secondary compounds. All the enzymes shown in brackets are known to be up-regulated by light in most plants tested.

Nitrate transporters and blue light

In various algae, blue light was often found to have a positive effect on nitrate uptake. Low irradiance blue light has been found to stimulate nitrate uptake in the green algae *Monoraphidium braunii* (Aparicio and Quinones 1991, Ullrich *et al.* 1998), and *Ulva rigida* (Lopez-Figueroa and Ruediger 1991). Also in *Chlorella* blue light stimulated nitrate uptake independent of photosynthesis (Maetschke *et al.* 1997). The blue-light effect appears to act through a signal transduction chain involving phosphorylation since the blue-light effect was sensitive to protein kinase inhibitors (Tischner 2000). In higher plants, *i.e.* rice seedling, different light qualities, including blue light, were tested, but neither 5 min nor 6 h illumination with blue light stimulated nitrate uptake (Sasakawa and Yamamoto 1979).

Nitrate transporters and white, photosynthetic active light

Photosynthesis and carbon metabolism were recognised as important for nitrogen assimilation already in the early days of research on nitrogen assimilation back in the 50's using unicellular algae as the experimental system as reviewed by Huppe and Turpin (1994). An early observation was that nitrogen-sufficient algae required light to assimilate nitrogen, but nitrogen-limited cells rapidly assimilated nitrogen in the dark. Intracellular carbohydrate stores decreased during dark assimilation and assimilation ceased when these stores were depleted (Huppe and Turpin 1994). From experiments on photosynthesis and uptake of $^{15}NO_3^-$ in maize seedlings, Pace *et al.* (1990) concluded that NO_3^- uptake and reduction were regulated by the supply of energy and carbon skeletons required to support these processes. Also in ryegrass NO_3^- uptake was closely related to photosynthesis with a time lag of about 5 h between photosynthesis and uptake (Scaife 1989). Furthermore diurnal regulation of nitrate uptake was found in several plants (Delhon *et al.* 1995, Cardenas-Navarro *et al.* 1998). In *Arabidopsis* roots, the transcript levels of both *AtNRT1.1* and *AtNRT2.1* were found to undergo diurnal variations, with transcript levels in the light reaching more than 5-fold of those in the dark (Lejay *et al.* 1999). Onset of darkness was accompanied by a rapid decrease in mRNA transcript levels as well as a decrease in $^{15}NO_3^-$ influx. Supply of 1% sucrose at the time of light to dark transition hindered the decrease in transcript levels and nitrate influx, supporting the view that the positive effect of light is due to products of photosynthesis being translocated to the roots. It is well known that nitrogen uptake genes are regulated by the N status in the plant, a demand for N will enhance expression of these genes and increase uptake of nitrate. The effect exerted through products of photosynthesis or N deficiency appears to be differently mediated because *AtNRT1.1* and *AtNRT1.2* are affected in the same way by light and sucrose supply, but are differently affected by nitrogen deficiency (Lejay *et al.* 1999). The *Arabidopsis* genome project reveals a range of potential *cis*-regulating elements upstream of nitrate transporter genes, like GATA, ACGT (G-box core), AG/CTCA (nitrate element) and for some also GATAAGA/G (I-box). However, the significance for regulation of transcription by these elements remains to be examined. The mechanism of signal transduction from carbohydrates and nitrogen compounds leading to changes in expression are still unknown, as are the transcription factors and *cis*-elements actively involved.

Tissue specific expression

Since the nitrate transporter genes are differentially expressed in leaves and roots, specific tissue factors, as well as interaction with light are expected to be involved in achieving the specific expression of these genes. One of the *AtNRT2* genes, i.e. *AtNRT2.7* was expressed in leaves but not in roots (Orsel *et al.* 2002). A minimal promoter with an I-box and a G-box which confined light responsiveness to a reporter gene, and at the same time restricted expression to cells containing chloroplasts had previously been described (see above). However, the mechanism for tissue specificity of *AtNRT2.7* is not known, and the I-box is not present in the promoter area of this gene. Interestingly *AtNRT2.7* is the only nitrate transporter (-like) gene known to code for a putative chloroplast targeted peptide, and therefore this transporter is a candidate for being located in the chloroplast membranes. *NRT2.3* is the other *NRT2* gene strongly expressed in leaves. The gene was expressed in the shoots of young plants, but at the reproductive stage stronger expression was found in the roots (Orsel *et al.* 2002). The function of all the *NRT* genes is not known, and some may be involved in transportation of other anions than NO_3^-, for instance ions that need to be transported into chloroplasts. Most nitrate transporters are preferably expressed in roots, and generally, not much is known about root-specific expression. A member of the *NRT1* family, *NRT1.1*, was expressed in young leaves as well as in roots, and auxin, was found to be important for targeting *NRT1.1* expression to nascent organs (Guo *et al.* 2002).

NITRATE REDUCTASE (NR)

$$NO_3^- + NAD(P)H + H^+ \rightarrow NO_2^- + NAD(P)^+ + H_2O$$

NR catalyses the reaction where electrons from NAD(P)H reduce NO_3^- to NO_2^-. In *Arabidopsis* two different genes coding for NADH:NR have been identified, *NIA1* and *NIA2*. Most higher plants examined have been shown to possess two or more genes coding for NR; however, *Solanaceae* species such as *Nicotiana* spp. and tomato have only one *NIA* gene per haploid genome (Rouzé and Caboche 1992). NR using NADH as an electron donor (EC 1.6.1.1) is the dominant form in higher plants. A NADH/NADPH bispecific form (EC 1.6.1.2) is found in addition to the NADH specific form in N_2 fixing plants like soybean, and also in monocot species including maize, rice and barley (Miyazaki *et al.* 1991). In birch only a bispecific form is present (Friemann *et al.* 1991). Monospecific NADPH:NR (EC 1.6.1.3) occurs in mosses and fungi but is not found in higher plants. Only small changes in the structure of NR are required to change the enzyme from a NADH specific to a NADPH (bi)specific form. In birch, altering only one amino acid changed the enzyme from a bispecific to a NADH specific enzyme (Schondorf and Hachtel 1995). The reason why plants possess NRs with different specificities for NADH and NADPH is not clear. Those plants that have both forms tend to express the bispecific form in the roots. For details on the modulation of NR activity see chapter 7.

NR and phytochrome

Light has been known to influence nitrate reductase expression for several decades. NR induction mediated by phytochrome was first reported in etiolated peas in 1972 (Jones and Sheard). Since then the involvement of phytochrome in NR induction was established for a large number of dicot and monocot plants (Sopory and Sharma 1990, Lillo and Appenroth 2001). Light absorbed by phytochrome generally leads to increased levels of NR mRNA, protein and enzyme activity in etiolated plants. Often, the light effect was found to be dependent on availability of nitrate (Appenroth and Oelmüller 1995, Li and Oaks 1995, Migge *et al.* 1997), although this was not always the case (Sopory and Sharma 1990).

Interestingly, different *NIA* genes within a species may show different responses to irradiation as demonstrated for *Arabidospis NIA1* and *NIA2* in the wild type (Cheng *et al.* 1991), and the *cr88* mutant (Lin and Cheng 1997). *Cr88* is a chlorate resistant mutant of *Arabidopsis* that exhibits the etiolated phenotype in darkness but also in continuous red light; with long hypocotyls and closed, or partially expanded cotyledons, and delayed greening. At the seedling stage NR activity was low in *cr88*, and *NIA2* was not induced by light as in wild type seedlings. In mature *cr88* plants light effects were as for wild type, thus demonstrating that light effects in mature plants are mediated very differently from light effects in etiolated seedlings. Stimulation of NR expression by sucrose was similar in *cr88* and wild type plants, and is consistent with light effect in mature plants being mediated through products of photosynthesis such as sucrose or other carbohydrates. Although light absorbed by phytochrome increased *NIA* expression in a vast number of species, *NIA* expression and NR activity generally continued to increase to a much higher level when plants were placed under conditions allowing photosynthesis, or when given sucrose. For instance in etiolated barley and wheat, NR activity increased approximately 2-fold in response to red light pulses, although, the activity was still only 5 and 35%, respectively, of the activity found in plants grown in photosynthetic active light (Lillo and Henriksen 1984, Melzer *et al.* 1989). In tomato cotyledons, NR protein after red or far-red light treatments was 20 to 30% of NR protein in white light grown seedlings. The NR mRNA level was, however, nearly the same in response to red light or photosynthetic active light. White light apparently acts post-transcriptionaly by stabilising, or increasing synthesis of the NR protein in tomato (Migge *et al.* 1997).

NR and photosynthesis

Products of photosynthesis, *i.e.* various carbohydrates are well known to stimulate *NIA* expression and NR activation. The products of nitrogen assimilation on the other hand, and especially glutamine, are known to exert a negative feedback on *NIA* expression (Sivasankar and Oaks 1996, Coruzzi and Bush 2001). In leaf discs of wild type tobacco plants the NR mRNA level was low when discs were fed with glutamine, whereas feeding sucrose or α-ketoglutarate resulted in a high NR mRNA level. In transgenic plants expressing antisense GOGAT, and thereby possessing only 10% of wild type GOGAT activity, the NR mRNA level was high in spite of high *in situ* concentrations of glutamine (Ferrario-Méry *et al.* 2001). However, in these transgenic plants the concentration of α-ketoglutarate was also high. It appears that the expression of NR depends mostly on the ratio between glutamine and α-ketoglutarate, and the inhibiting effect of glutamine on transcription is overcome by high concentration of α-ketoglutarate. The experiments of Ferrario-Méry and co-workers are strongly supportive of PII (see above) being involved in regulation of nitrogen metabolism in plants, and PII may be a "missing link" between photosynthesis and regulation of NR expression. However, sucrose, which increased the NR mRNA level, did not lead to increase α-ketoglutarate concentration *in situ*, as would have been expected if α-ketoglutarate were a component in the regulatory system. It still remains to be established that PII really is a sensor for products of photosynthesis and reduced nitrogen in plants. In *E. coli*, PII influences the phosphorylation status of a protein that interacts with transcription factors. In plants this would necessarily be more complicated since PII is a plastidic protein and hence separated from active transcription factors in the nucleus. Depending on C/N status PII may influence the translocation of certain factors across chloroplast membranes and subsequently influence the components in the nucleus, which then influence transcription. Such a signal transduction pathway from PII to the transcription level remains to be revealed.

NR promoters and light regulation

Deletion analysis of the 1.5-kb 5'-flanking regions of the *Arabidopsis NIA1* and *NIA2* genes showed that 238 and 188 bp, respectively, were important for nitrate induction of a reporter gene (*CAT*) in transgenic tobacco (Lin *et al.* 1994). A 12 bp conserved sequence with a core consensus AG/CTCA, a "nitrate element", was necessary for nitrate dependent transcription (Hwang *et al.* 1997). Sequences involved in light regulation of *Arabidopsis* NR have not been reported as investigated so far.

In bean (*Phaseolus vulgaris*) two NADH:NR genes were identified and their promoters studied in transgenic tobacco with the help of a GUS reporter gene. The one *NIA* promoter studied in more details showed that a 900 bp region proximal to the transcription initiation site was necessary for high expression in leaves and roots. Surprisingly it was not possible to find any significant effect of nitrate on the expression of GUS. Circadian rhythms of the GUS mRNA were retained, and this may indicate that light responsive elements were present within the construct since light responsive and circadian elements may be inseparable (Lillo *et al.* 2001). Many light and circadian regulated genes contain GATA motifs in their promoters (Teakle and Kay 1995). Five GATA motifs were found within the 700 bp upstream of the transcription start site, but their role in regulation of transcription of the bean *NIA* genes has not been clarified (Jensen *et al.* 1996).

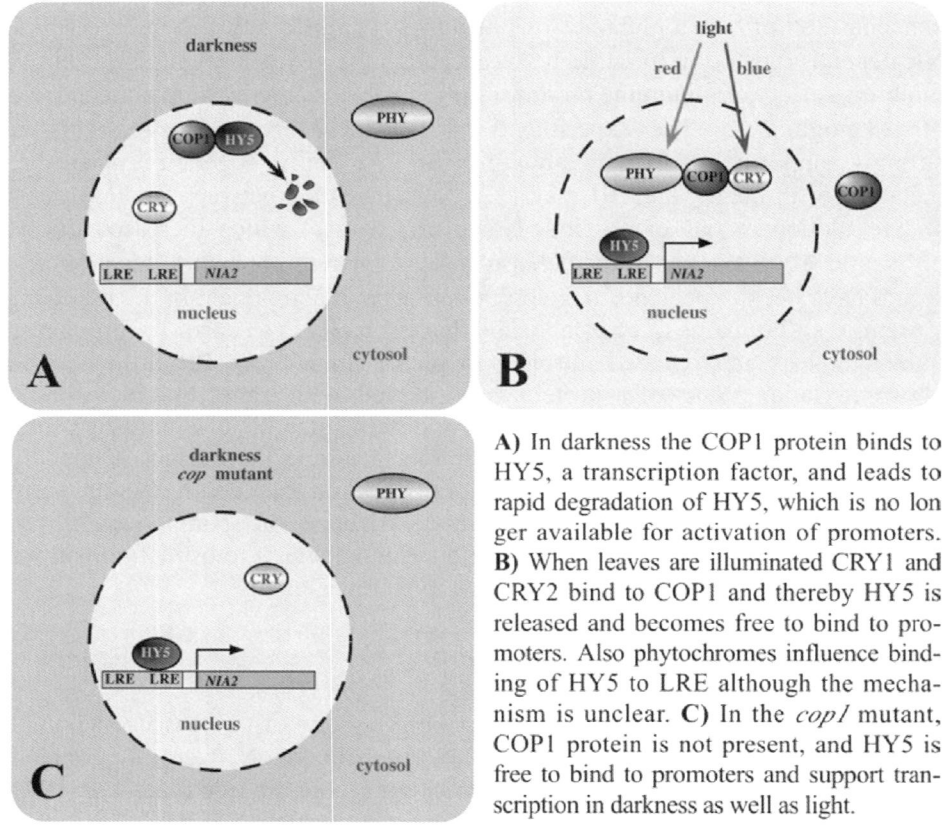

Figure 3. Working model for blue and red light effects on *NIA* transcription.

A) In darkness the COP1 protein binds to HY5, a transcription factor, and leads to rapid degradation of HY5, which is no longer available for activation of promoters. B) When leaves are illuminated CRY1 and CRY2 bind to COP1 and thereby HY5 is released and becomes free to bind to promoters. Also phytochromes influence binding of HY5 to LRE although the mechanism is unclear. C) In the *cop1* mutant, COP1 protein is not present, and HY5 is free to bind to promoters and support transcription in darkness as well as light.

Deletion analysis of the 1.6-kb 5'-flanking sequence of the birch *NIA* gene were performed by fusing promoter sequences with the GUS reporter gene and introducing them into *Nicotiana plumbaginifolia* (Strater and Hachtel 2000). Interestingly, the promoter length conferring maximal light stimulation was different in roots and leaves. Light responsiveness was retained within 237 bp and 320 bp proximal of the transcription start site for roots and leaves, respectively. Only high intensity, white light was tested. Nitrate responsiveness was found for the 643 bp promoter fragment, but not for the 535 bp, indicating that important sequences for nitrate responsiveness were between 535 bp and 643 bp. As for bean, a leaf specific, negative *cis*-element was found around 1000 bp upstream from the transcription start site. Both the nitrate motif AG/CTCA and GATA motifs are present in the birch NR promoter, but their function has not been further evaluated. When *NIA* promoters from birch, bean or *Nicotiana* linked to a reporter gene or *NIA* structural gene were introduced into the *N. plumbaginifolia* genome only a small number of transgenic plants were obtained expressing that gene. Furthermore, expression was often very low (Strater and Hachtel 2000, Warning and Hachtel 2000). This points to the importance of location of the *NIA* gene in the genome for expression, a factor not yet fully understood.

Specific transcription factors or proteins mediating light effects on transcription of *NIA* genes are largely unknown. The HY5 protein appears to be the only example so far of an identified transcription factor likely to be involved in regulation of *NIA* expression. The *HY5* locus encodes a bZIP transcription factor known to be important for deetiolation and expression of several genes related to photosynthesis. In *Arabidopsis* the HY5 protein binds to Light Responsive Elements (LRE) in promoters, for instance a G-box (Chattopadhyay *et al.* 1998). HY5 degradation is stimulated by another protein, COP1. In darkness COP1 binds to HY5 and this leads to rapid degradation of HY5, hence HY5 is not available for activation of promoters (Figure 2A). When leaves are illuminated, cryptochromes bind to COP1 and thereby releasing HY5, which becomes free to bind to promoters. Phytochromes influence binding of HY5 to LRE as well although the mechanism is still unclear (Quail 2002) (Figure 2B). In the *cop1* mutant, seedlings deetiolate in total darkness as if they had perceived a light signal. Genes involved in photosynthesis are induced in darkness, as is also *NIA2* in this mutant (Deng *et al.* 1991). Since COP1 protein is not present, HY5 is free to bind to the promoters of these genes and support initiation of transcription (Figure 2C). Other, still unknown transcription factors and components are likely also to be involved in *NIA* expression.

Light and post-transcriptional regulation of NR

Post-translational regulation of NR is also strongly influenced by light (Lillo 1994, Lillo and Appenroth 2001). Most of what is known about post-translational NR regulation, with few exceptions (Appenroth *et al.* 2000), has been deduced from experiments using green leaves in which mainly effects of photosynthetic active light have been studied. NR in higher plants is phosphorylated at a special serine residue in the hinge between the molybdenum co-factor binding domain and the heme-binding domain (serine 534 in *Arabidopspis*). Generally NR is inactivated in darkness by phosphorylation, and activated by dephosphorylation in the light (Kaiser and Spill 1991, MacKintosh 1992). However, the system is more complex than simply phosphorylation and dephosphorylation because members of the 14-3-3 protein family also bind to phosphorylated NR (reviewed by Kaiser and Huber 2001, Lillo and Appenroth 2001, MacKintosh and Meek 2001). 14-3-3 proteins belong to a highly conserved protein family with regulatory roles in plant, fungal and mammalian cells (MacKintosh and Meek 2001). It is after the binding of these 14-3-3 proteins that phosphorylated NR is actually inhibited, and inhibition is only observed in the presence of cations. The most important cations for this inhibition are Mg^{2+} and polyamines (Provan et al. 2000).

In etiolated barley leaves the inactive (phosphorylated) form of NR was dominant, and red light absorbed by phytochrome did not influence the activation state of NR. The potential for activation of NR was present in etiolated leaves since acid loading, a well known treatment to activate NR, is efficient for NR activation also in etiolated leaves (Appenroth et al. 2000). Post-translational activation of NR is triggered by photosynthesis, and mediated through the balance of kinases and phosphatases acting on NR. Photosynthesis may lead to a decrease in Ca^{2+} in the cytosol and an increase in phosphorylated sugar compounds, with both changes leading to decreased NR kinase activity and hence activation of NR. The significance of these factors for regulating NR *in situ* is not clarified. Phosphatases are involved directly by dephosphorylating NR, and also indirectly by dephosphorylating and thereby inactivating, NR kinases. Still, little is known about regulation of phosphatases in plants. In prokaryotes PII is involved in transcriptional as well as post-translational regulation of enzymes in nitrogen metabolism (see above). In plants this has not yet been deeply studied, but so far no indications of involvement of PII in post-translational regulation were found. In fact, high concentration *in situ* of the signal component, α-ketoglutarate, had no effect on the phosphorylation state of NR (Ferrario-Méry et al. 2001).

Light activates transcription of NR, and light activates NR post-translationaly. Light acts on yet another level, which became clear when examining expression of the NR gene linked to the 35S CaMV promoter (Vincentz et al. 1993). As expected, NR was constitutively expressed and NR mRNA levels were high in darkness (after 56 h) as well as in the light. Accumulation of NR protein and activity were, surprisingly, still promoted by light. This could be explained by increased rate of NR synthesis or decreased rate of degradation in the light, or both. Since light favours dephosphorylation of NR it is tempting to assume that the non-phosphorylated form, which dominates in the light is more stable than the phosphorylated form. Post-translational modification of proteins has often been found to be a signal for degradation (Callis 1995). Experiments with spinach did indeed support this as phosphorylated NR was shown to be more rapidly degraded than non-phosphorylated NR (Kaiser and Huber 2001). However this did not appear to be the case in *Arabidopsis* cell cultures (Cotelle et al. 2000) or in *Nicotiana* species

(Lillo *et al.* 2003). The findings for *Nicotiana* species are in agreement with light having a positive effect on the translation process.

Toxic products and by-products of NR: nitrite, peroxynitrite and nitric oxide

The very complex means of regulation of NR may have evolved not only to optimise NR activity in response to nitrate and energy (light) supply, but evolution may also have been driven by the importance of down-regulating NR activity under certain conditions. For instance in darkness, further assimilation of the product from NR, NO_2^-, is slow, because this step is usually closely linked to photosynthesis. Unless NR activity is rapidly down-regulated, NO_2^- may accumulate in darkness. Nitrite is mutagenic and accumulation of NO_2^- may be detrimental to the plant in the long run. Furthermore, NR reduces not only NO_3^- to NO_2^- (Ruoff and Lillo 1990), but under certain conditions reduces also O_2 to form O_2^- (superoxide anion) and NO_2^- to form NO (nitric oxide) (Yamasaki 2000, Rockel *et al.* 2002). Nitric oxide may have both detrimental and beneficial effects in plants, and the subject is far from fully understood. Peroxynitrite (ONOO$^-$), a highly reactive and very toxic compound, is made non-enzymatically from NO and O_2^-. A strict regulation of NR may, therefore, be necessary to avoid by-products causing oxidative damage and mutations in the plants (chapter 7).

Nitric oxide also plays a key role in atmospheric chemistry and is important for atmospheric radical balance and for generation of photooxidants (Wildt *et al.* 1997). Previously, vegetation was not taken into account as a source of NO in the atmosphere, probably because NO emission from plants was firstly only measured under certain conditions. More recently, NO emission from a variety of nitrate nourished plant species was observed under normal growth conditions. During daytime the NO emission was closely correlated to CO_2 uptake. Furthermore, when nitrate content of the nutrient solution was enhanced, NO emission was observed also in darkness. On a global basis NO emission from plants was calculated to be 1-5%, only, compared to the NO evolved from soil due to microbial activity. However, over areas with dense plant cover the emission from plants is not negligible (Wildt *et al.* 1997).

NITRITE REDUCTASE (NiR)

$$NO_2^- + 6\ Fd_{red} + 8H^+ \rightarrow NH_4^+ + 6\ Fd_{ox} + 2\ H_2O$$

After reduction of nitrate to nitrite in the cytosol, nitrite is translocated into the chloroplasts/plastids where further reduction takes place with the help of nitrite reductase (NiR, EC 1.7.7.1). Plant NiR uses reduced ferredoxin formed in photosynthesis as electron source. In darkness or in non-green tissue, ferredoxin or a ferredoxin-like protein can be reduced by NADPH from the oxidative pentose phosphate pathway. NiR is a nuclear-encoded protein with an N-terminal signal peptide that directs it to the chloroplasts (Meyer and Stitt 2001). Some higher plants contain only a single *NII* gene (gene encoding NiR) per haploid genome, whereas other plant species contain two or more copies (Wray 1993). The amphidiploid *Nicotiana tabacum* contains four genes, two from each ancestor. Because nitrite is toxic, cells must possess enough NiR to reduce all the nitrite produced by NR. As for *NIA* genes, nitrate and light are the two basic factors necessary for strong expression, and *NII* and *NIA* genes are generally regulated in co-ordination. In etiolated plants with no nitrate source, expression of *NII* genes is often very low, or not detectable. Variations among species and different organs and tissues are, however, clearly seen.

NiR and light effects in various plants

There are not many reports on light effects on NiR in *Arabidopsis*, but *NII* expression was examined in the chlorate-resistant *cr88* mutant. At the seedling stage *cr88* was altered in the regulation of *NIA2* expression compared to wild type plants. *NIA2* expression was not induced following 8 h (white) light exposure, nor were the photosynthetic genes coding for RBCS and CAB, but the seedlings still expressed *NII* as in the wild type. The work with *cr88* therefore showed that the signal transduction chains leading to activation of *NIA2* and *NII* expression are different in *Arabidopsis* (Lin and Cheng 1997).

A phytochrome deficient tomato mutant, the *aurea* mutant, was used to study expression of both *NIA* and *NII* genes. This mutant has less than 5% of spectrophotometrically active phytochrome compared with wild type tomato. Both *NIA* and *NII* expression was impaired in the mutant when testing etiolated seedlings (Becker *et al.* 1992). This clearly showed that phytochrome was important for inducing both *NIA* and *NII* expression in young etiolated seedlings. Analysis of wild type tomato plants revealed the presence of two *NII* genes. One gene being mainly expressed in etiolated leaves, whereas the other gene was expressed in green leaves (Migge *et al.* 1998). Red, blue or UV-A irradiation induced *NII* expression, consistent with the involvement of phytochrome and possibly a blue-light receptor (Goud and Sharma 1994, Migge *et al.* 1998). Blue light especially stimulated accumulation of one of the NiR proteins (NiR2). As pointed out by Migge *et al.* (1998), a UV-A effect does not necessarily imply involvement of a specific blue/UV-light receptor because phytochrome may also be converted, to some extent, into its active form by UV-A light. In fact Goud and Sharma (1994) found that the positive effect of blue light on NiR in wild type tomato disappeared in the *aurea* mutant, indicating that phytochrome was indeed the important photoreceptor. In mature, green leaves of the *aurea* mutant activities of NR and NiR were identical with activities in wild type on a per mg chlorophyll basis. Levels of NR and NiR mRNA and their diurnal variations were also very similar in leaves of the mature green *aurea* mutant and wild type plants (Becker *et al.* 1992). This indicates that different signal transduction chains from light to expression of *NIA* and *NII* are operating in seedlings contra mature plants. An interesting difference observed for tomato in comparison with other species was a strong, positive light effect on expression also when nitrate is removed from the growth medium and replaced with an alternative nitrogen source, such as glutamine.

Effects of different light sources on *NII* expression have also been studied in *Nicotiana tabacum* and *N. plumbaginifolia*. Both red and blue light enhanced *NII* expression in etiolated tobacco cotyledons, and light effects were only seen when nitrate was present in the growth medium. Light intensity was also important, and increasing light intensity had a positive effect on expression (Neininger *et al.* 1992). Four *NII* genes are present per haploid *Nicotiana* genome. The two genes more thoroughly tested, were found to be differently expressed in roots and leaves; one gene mainly in leaves and the other mainly in roots. Expression of both genes was induced by nitrate, and diurnal variation followed the same pattern in leaves and roots (Kronenberger *et al.* 1993). *NIA* and *NII* genes were similarly regulated in *Nicotiana* with respect to inhibition by glutamine and glutamate. An interesting difference was that in green leaves glucose apparently could replace light for the enhancement of *NIA* expression, while a positive effect on *NII* expression was much less pronounced (Vincentz *et al.* 1993).

In barley, where only one *NII* gene is present, nitrate was strictly necessary for detection of NiR activity and protein. In leaves, induction also depended on light, whereas in the roots expression was just as strongly induced in darkness (by nitrate). This showed that the gene was derepressed in the roots by nitrate only, whereas some factor in the leaves apparently hindered its expression (Duncanson *et al.* 1992). In soybean at least three different *NII* genes were present, and one of these genes were constitutively expressed in the cotyledons, the others were induced by nitrate as in other plants (Kim *et al.* 2001).

NII promoters

As for *NIA* genes, nitrate and light are the most important factors known to enhance transcription of *NII* genes. Promoter analysis of bean (Sander *et al.* 1995), spinach (Rastogi 1993, Neininger *et al.* 1994, Sivasankar *et al.* 1998) tobacco (Dorbe *et al.* 1998) and birch (Warning and Hachtel 2000) has revealed *cis*-acting elements involved in nitrate induction of *NII* transcription. Promoter analysis was performed by linking the promoters, or promoter fragments, to a reporter gene and introducing this into tobacco, or in *Arabidopsis* (Dorbe *et al.* 1998). For the *NIA* genes, reporter genes linked to the promoter very often showed only low expression in transgenic plants. For the *NII* promoter no such problems with gene extinction were encountered (Stitt and Meyer 2001). For all promoters tested, *cis*-acting elements providing nitrate induction of *NII* transcription were found to be present within 0.7 kb upstream of the transcription start site. In the fungus *Neurospora crassa* special GATA elements in the promoter have been shown to bind a regulatory protein, NIT2, resulting in induction of transcription by nitrate (Marzluf 1997). GATA elements were identified in the spinach, tobacco and birch *NII* promoters. However, apparently the GATA elements were not strictly necessary for nitrate induced transcription, because transcription was induced by nitrate also when these elements were deleted.

Neininger *et al.* (1994) found that the -200 to +131 bp (relative to transcription start) provided phytochrome mediated induction of transcription of the spinach *NII* gene. Warning and Hachtel (2000) identified light sensitive regions in the birch promoter between -155 and -267. Analysis of fragments of the birch *NII* promoter indicated that an inhibitor would bind to this region in the dark and thereby preventing transcription. When testing various promoter fragments linked to the GUS reporter gene it was also clear that a specific promoter fragment resulted in unlike activities in roots and leaves, possibly due to different concentrations of transcription factors present in the tissues.

Argüello-Astorga and Herrera-Estrella (1998) found that a single motif can not function as an LRE (see above). Since the presence of nitrate was often a prerequisite for obtaining light effects on *NII* expression, possibly also one motif involved in nitrate activation could act in co-operation with one motif for light activation. Certainly both *cis* and *trans*-acting elements still need to be explored for fully understand light regulation of *NII* genes.

GLUTAMINE SYNTHETASE (GS)

$$\text{glutamate} + NH_4^+ + ATP \rightarrow \text{glutamine} + ADP + P_i + H_2O$$

In primary nitrogen metabolism the product of the nitrite reductase reaction, ammonium, is further assimilated into amino acids by glutamine synthetase (GS, EC 6.3.1.2). Importantly, the nitrite reductase reaction is certainly not the only source of ammonium in plants (chapter 2). Ammonium is released from amino acids by deamination during nitrogen remobilisation. Ammonium is also released in special metabolic pathways as in the link between aromatic amino acid synthesis and secondary metabolism catalysed by phenylalanine ammonia lyase. Ammonium is formed in the glycine decarboxylase step during photorespiration, a process that may exceed primary nitrogen assimilation 10-fold. Thus over the life span of a plant, nitrogen is released as ammonium and refixed several times (Miflin and Habash, 2002). For details on the characterization and regulation of GS expression and activity see chapter 2.

Red and blue-light effects on GS gene expression

Regulation of GS mediated by phytochrome has been found in a wide range of higher plants: Scots pine (Elmlinger and Mohr 1991), *Spirodela* (Teller and Appenroth 1994), lettuce (Sakamoto 1990), mustard (Weber *et al.* 1990), pea (Edwards and Coruzzi 1989), tomato (Migge *et al.* 1998) and *Arabidopsis* (Oliveira and Coruzzi 1999). Generally GS2 is strongly induced by light while GS1 is more developmentally regulated. Following illumination of etiolated leaves, an increase in both GS2 transcript and GS protein have been observed in a majority of plant species, and both red and blue-light receptors were involved (Cren and Hirel 1999).

Effects of sugars and amino acids on GS gene expression

As for other enzymes in the nitrogen assimilation pathways strong positive effects of carbohydrates are found. Light-induced changes not fully accounted for by phytochrome, were observed in pea (Edwards and Coruzzi 1989). A study performed with *Arabidopsis* showed that in addition to phytochrome-mediated red light effects, which enhance expression of the GS2 gene, different sugars stimulated accumulation of GS2 mRNA. Sucrose, fructose and glucose all mimicked the effect of light and enhanced GS2 gene expression when added to the growth medium in the dark. A moderate stimulation by sugars on GS1 gene expression was evident. Several amino acids, aspartate, asparagine, glutamate, glutamine added to the growth medium showed an antagonistic effect to sugar (Oliveira and Coruzzi 1999). The effects of sugars and amino acids on *Arabidopsis* GS are thus reminiscent of the regulatory mechanism found in *E. coli* where the PII protein is known to be a key component in regulation of GS (see above).

Promoters

Analysis of the GS gene promoters of soybean, *Phaseolus*, rice and tobacco revealed promoter fragments responsible for organ specific and developmental expression (Hirel and Lea 2001, Morey *et al.* 2002). Light responsive GS2 promoters were demonstrated for *Phaseolus vulgaris* (Cock *et al.* 1992) and pea (Tjaden *et al.* 1995). The GS2 pea promoter contains an AT-rich 33 bp region at 807 bp upstream of the transcription start site, and when this AT-rich region was deleted, GS2 expression was reduced 10-fold Similarity to AT-rich elements in light-regulated photosynthetic genes in various species has been pointed out. A *trans*-acting factor binding to this AT-rich sequence was identified, and found to be similar to AT-binding factors for other plant promoters (Tjaden and Coruzzi 1994). Interestingly, a phosphorylation step inhibited binding of the *trans*-acting factor to the AT-rich region. Important elements for light responsiveness were confined to a short promoter fragment 323 bp upstream from the transcriptional start site and sequences already known to be important for light responsiveness in other genes, like I-box and GT-1 were identified.

The upstream region of the *Arabidopsis* GS2 (At5G35630) also has I-boxes, but these are about 2000 and 2400 bp upstream from the transcription start site. Two nitrate elements within an AT-rich region are also present between -1645 and -1607 bp upstream of the transcription start site in *Arabidopsis* GS2. But no detailed promoter analysis or interacting *trans*-acting factors have been reported for *Arabidopsis*.

Post-translational regulation of GS

Cytosolic and chloroplastic GS from cauliflower were found to bind 14-3-3 proteins. Phosphorylation of GS may be important for its stability as found in *Brassica napus* (Finnemann and Schjoerring 2000). However, 14-3-3 binding did not influence the activity of GS (Moorhead *et al.* 1999, Riedel *et al.* 2001). As for NR, 14-3-3 proteins may be important for the regulation of GS; however, this point is still not clarified.

A redox change was found to modulate GS2 activity in *Canavalia lineata* and two cystein residues necessary for this activation was identified. In various plants these cysteine residues were present in GS2, but not in GS1. GS2 was activated by the sulfhydryl-reducing agent DTT like enzymes of the photosynthetic electron transport were activated by the ferredoxin-thioredoxin system (Choi *et al.* 1999). Regulation of GS2 by a redox change was further confirmed by identification of *Arabidopsis* GS as a binding target for thioredoxin (Motohashi *et al.* 2001).

GLUTAMATE SYNTHASE (GOGAT)

glutamine + α-ketoglutarate + 2 Fd $_{red}$ (NADH + H$^+$)
$$\rightarrow 2 \text{ glutamate} + 2 \text{ Fd}_{ox} (NAD^+)$$

Glutamate synthase (GOGAT) catalyses the conversion of glutamine and α-ketoglutarate to two molecules of glutamate. For the characteristics, localisation and regulation of enzyme expression and activity see chapter 2.

Light and sugar effects on GOGAT expression

Light acting through phytochrome has been shown to enhance expression of Fd-GOGAT in several plants whereas the NADH-GOGAT activity generally was not influenced by light. *Arabidopis* has two genes coding for Fd-GOGAT

(*GLU1* and *GLU2*). *GLU1* was expressed at the higher level in leaves, and its mRNA level was specifically enhanced by light (only high intensity white light was tested). Sucrose could partly replace light in inducing expression of *GLU1* when light-grown seedlings were dark-adapted for 3 days before testing effects of light and sucrose (Coschigano *et al.* 1998). In contrast, *GLU2* was expressed at a lower, constitutive, level and accumulated primarily in roots (Coschigano *et al.* 1998).

Fd-GOGAT was expressed in etiolated tomato cotyledons, but expression increased by exposure to light. Phytochrome was involved in light perception. UV-A light also increased Fd-GOGAT expression, and the effect of UV-A was mediated through a special blue-light receptor or phytochrome (Becker *et al.* 1993, Migge *et al.* 1998). As for *Arabidopsis*, no light effects were seen on NADH-GOGAT expression.

When etiolated barley seedlings were exposed to light (white, high intensity) for 48 h a 3-fold increase in Fd-GOGAT was observed. Although light did have a positive effect on Fd-GOGAT expression in barley a substantial level of Fd-GOGAT activity was present in the absence of light and also in the absence of nitrate. Early work on Fd-GOGAT activity in barley showed no effects of daily day/night shifts on Fd-GOGAT activity, whereas NR activity varied during the same period (Lillo 1983, 1984). In contrast to the situation in *Arabidopis*, light could not be replaced by sucrose or glucose added to the nutrient solution. No positive effect of sugar was found for Fd-GOGAT, although NR activity increased 2-fold in the same experiment. Hence in barley, Fd-GOGAT and NR did not react in parallel in response to the sugar treatment or circadian control (Lillo 1984, Pajuelo *et al.* 1997).

In maize, exposure of etiolated seedlings to white light led to a 3-4-fold increase in Fd-GOGAT activity and mRNA levels. Red light also stimulated Fd-GOGAT expression, and far-red reversibility confirmed that phytochrome was involved. NADH-GOGAT activity was not affected in these experiments. Again a relatively high Fd-GOGAT level was found also in etiolated seedlings confirming that light was not strictly necessary for Fd-GOGAT expression (Suzuki *et al.* 1987, 2001).

Phytochrome was important for appearance of Fd-GOGAT activity in etiolated pine seedlings. Apparently also a blue-light/UV-A receptor like cryptochrome was involved. Blue or red light had no effect on NADH-GOGAT (Elmlinger and Mohr 1991). Fd-GOGAT was induced by red light and blue light in etiolated turions of *Spirodela*, and apparently both phytochrome and a blue-light receptor were involved. NADH-GOGAT was not influenced by red or blue light. Again a relatively high background activity of Fd-GOGAT was seen also in darkness (Teller *et al.* 1996).

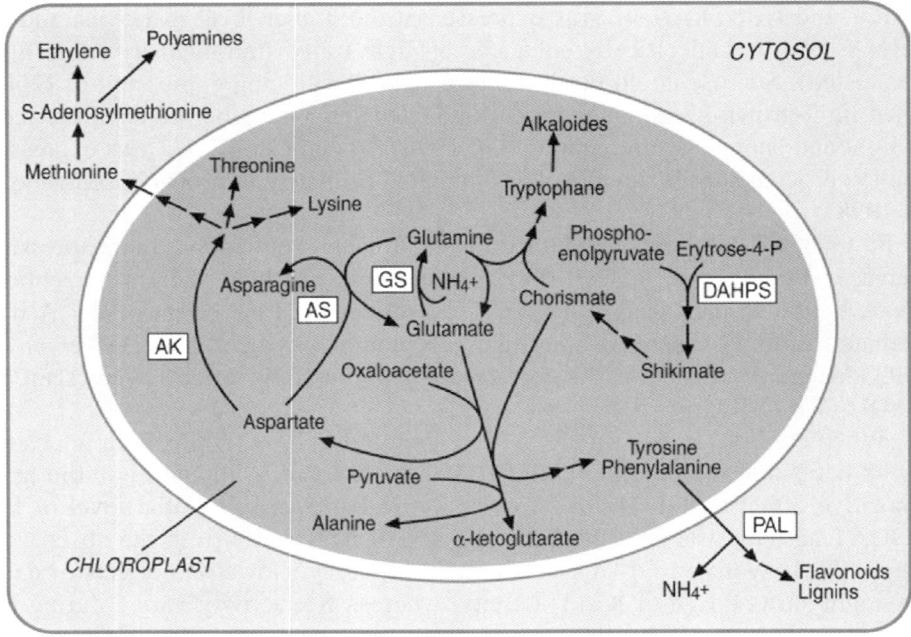

Figure 4. Overview of metabolism from the amino acid glutamate, and pathways leading to secondary metabolites. Glutamate and glutamine serve as nitrogen donors in important metabolic pathways. Glutamate is an amino donor in transaminase reactions generating for instance alanine, and aspartate from keto acids. Glutamate or glutamine also donate an amino group to chorismate in the pathways leading to tyrosine/phenylalanine or tryptophane, respectively. The first enzyme in the pathway for synthesis of amino acids in the aspartate family, aspartate kinase (AK) is stimulated by light. Asparagine synthetase (AS) activity is generally decreased in light, and is the main route for synthesis of asparagine, using aspartate and glutamine as substrates. Phosphoenolpyruvate and erythrose-4-phosphate are condensed by the 3-deoxy-arabinoheptulosonate-7-phosphate synthase (DAHPS) enzyme which is stimulated by light in many plants, and is the first step in the shikimate pathway leading to chorismate, and then aromatic amino acids. The first enzyme in the pathway converting aromatic amino acids to flavonoids and lignin is phenylalanine ammonia-lyase (PAL). PAL is influenced by many environmental factors, also light. The PAL reaction is a main point for release of NH_4^+.

Promoter analysis has been carried out for alfalfa NADH-GOGAT indicating four regulatory elements involved in the expression within the nodules (Trepp *et al.* 1999). Promoter analysis is otherwise scarce for NADH and Fd-GOGAT. The *GLU1* promoter of *Arabidopsis* has two AGTCA within AT-rich stretches (nitrate elements) in the 1200 bp region upstream of the transcription site but involvement of these elements in regulation remains to be analysed.

LIGHT INFLUENCED AMINO ACID SYNTHESIS

The pathways proposed for amino acids synthesis in plants are largely derived from those known in *Escherichia coli* and yeast, but show more complexity and additional ways of regulation. Amino acids are usually divided into families related to a common precursor compound, and amino acids of one family share a common pathway in their scheme of synthesis.

The majority of the amino acids can be synthesised within the chloroplasts; hence synthesis may be positively influenced by photosynthetic light. It has been known for a long time that the concentration of free amino acids in plants show diurnal variations and varies in response to light and darkness (Matt 1998, Coruzzi and Last 2000).

Asparagine

Asparagine is often present in very high concentrations in both xylem and phloem sap, and is used to carry nitrogen away from source tissue. Asparagine is more soluble, less reactive and has a nitrogen to carbon ratio higher than glutamine, all of which make it a better transport and storage compound. Asparagine synthetase (AS, EC 6.3.5.4) is considered as the major route for asparagine biosynthesis in plants. In an ATP-dependent reaction AS catalyses the transfer of the amino group from glutamine to aspartate generating glutamate and asparagine (Coruzzi and Last 2000):

$$\text{glutamine} + \text{aspartate} + \text{ATP} \rightarrow \text{glutamate} + \text{asparagine} + \text{AMP} + \text{PP}_i$$

Three AS genes (*ASN*) have been identified in *Arabidopis* (Lam *et al.* 1996, 1998). The expression of *ASN1* was repressed in leaves by light or presence of sucrose. The repression was reversed by asparagine, glutamine and glutamate. The two additional *ASN* genes were regulated in a reciprocal manner to *ASN1*, *i.e.*, their mRNA levels were increased by light or sucrose. These latter genes were, however, expressed at much lower levels, and belonged to a different class of *ASN* genes. These genes may be involved in NH_4^+-detoxification (Lam *et al.* 1998). In pea leaves, two *ASN* genes were identified and the expression of both was stimulated by darkness. *ASN* genes from various plants have been cloned and in the majority of cases light and /or carbohydrates were shown to repress gene expression (Hirel and Lea 2001). Light up-regulates genes which are involved in assimilation of ammonium into glutamine and glutamate, and phytochrome is involved in this process. In contrast, light generally down-regulates *ASN* genes. Interestingly, this negative effect of light is also mediated, at least partially, by phytochrome. The negative effect of sucrose on *ASN* expression can be antagonised by amino acids. A model has been proposed where GS2 is enhanced by light, or when carbon skeletons are abundant. Thus in the light, nitrogen is assimilated into and transported as glutamine, which is a substrate for numerous reactions. In darkness, when no photosynthesis can take place, *ASN* expression increases, and nitrogen is directed into the ideal storage and transport compound, asparagine.

ASN promoter and light repression

Using complementary loss and gain of function experiments, a 17 bp *cis*-element with a core unit TGGG (reverse) was identified at -43 upstream of the transcriptional start site in the *ASN1* pea promoter. This element was both necessary and sufficient for down-regulation by light. Other light depressed enes, like *PHYA* from several species, and *NPR* genes from Lemna also possess ≥ TGGG element (Ngai *et al.* 1997).

The tomato *aurea* mutant, that has only 20% of wild-type levels of PHYA, used to study the signal transduction chain leading from light to change in ription. The results were in agreement with PHYA being the light receptor, *ASN1* gene was light repressed, rather than dark activated (Neuhaus *et*

≥-derived amino acids

acids lysine, methionine, threonine and S-adenosylmethionine spartate family of amino acids. Important regulatory compounds nd polyamines are derived from S-adenosylmethionine. S-
ne itself is the donor of virtually all the methyl groups in ns. The aspartate-derived amino acids are of special interest, has been much studied, because these amino acids may be ncentrations in food to support optimal growth and health humans confined to imbalanced diets.
C-labelled aspartate have shown that chloroplasts are sine, threonine, isoleucine and homocysteine in light-he enzymes required are present in the chloroplast. quences also confirms that plastid-target sequences rminal extensions of these enzymes. The final he conversion of homocysteine to methionine has ytosol. It has long been known that the pathway nisms on the protein level. The different end branch point and beginning of the pathway nce, aspartate kinase (AK), the committing sis of aspartate-derived amino acids, is adenosylmethionine (Aarnes and Rognes K expression has revealed that this key-transcriptional level. The threonine-

sensitive AK isoenzyme is a bifunctional protein that also contains homoserine dehydrogenase activity (HSD). In *Arabidopis*, a single AK/HSD gene was detected, and the promoter was linked to the GUS reporter gene and introduced into tobacco. Light stimulated expression of this reporter gene in cotyledons upon germination (Zhu-Shimoni *et al.* 1997). Red light acting through the phytochrome system was found to increase the AK mRNA level in etiolated chickpea (Dey and Guha-Mukherjee 1999) and barley (Rao *et al.* 1999). Expression of another key-enzyme, cystathionine γ-synthase, at the branchpoint leading to methionine, was also stimulated by light (Hughes *et al.* 1999).

Aromatic amino acids and phenylpropanoids

The shikimate pathway is the first part of aromatic amino acid synthesis and comprises 7 enzymatic reactions whose endproduct is chorismate. After chorismate the pathway divides for synthesis of tryptophane, or tyrosine and phenylalanine (Figure 4). At this branch-point, feedback inhibition by the aromatic amino acids regulates the flow into the different branches. The shikimate pathway is found only in microorganism and plants, not in animals. The pathway is, therefore, an important target for herbicides and antibiotics. An well known example is the herbicide glyphosate, which inhibits the second last enzyme of the pathway. In bacteria, the pathway is almost entirely used for making aromatic amino acids for protein synthesis. In higher plants these amino acids are not only important constituents of proteins, but are also precursors of many important secondary compounds like: lignin, pigments, defensive phytoalexins, and alkaloids. Certainly this additional function of the pathway demands special regulation and capacity. Twenty per cent of the carbon fixed by plants may flow through the aromatic amino acid pathway (Coruzzi and Last 2000).

The shikimate pathway is strongly influenced by environmental stimuli such as light, pathogens and wounding. The first reaction of the pathway is the condensation between phophoenolpyruvate and erythrose-4-phosphate catalysed by the 3-deoxy-arabinoheptulosonate-7-phosphate synthase (DAHPS) enzyme. Activity of DAHPS is increased following wounding and correlates with an increase in secondary metabolites. In bacteria DAHPS is feedback inhibited by the aromatic amino acids, but this is not the case in plants. Early work revealed that aromatic amino acid biosynthesis can occur in isolated spinach chloroplasts. Molecular analysis also shows that all enzymes involved have N-terminal extensions characteristic of chloroplast transit sequences. The localisation to chloroplasts makes light and photosynthesis plausible candidates for regulation of the pathway, and in plants light may be the main regulator of the DAHPS. Light has been shown to upregulate DAHPS at the transcriptional level in parsley cell cultures (Henstrand 1992). However, although the shikimate pathway is located in the plastids, non-green tissue, like roots and flowers, contain the greatest amount of mRNA for the inducible isoenzymes of the pathway (Weaver and Herrmann 1997). Recently *Arabidopsis* DAHPS has been shown to be activated by the ferredoxin/thioredoxin system (Entus *et al.* 2002). The activity of *Arabidopsis* DAHPS expressed in *E. coli* and purified needs reduced thioredoxin, which could be replaced to some extent by DDT. This indicates a likely way of linking carbon fixation and aromatic amino acid synthesis, since DAHPS and enzymes of the Calvin cycle share reduced ferredoxin as an activator (Entus *et al.* 2002).

Phenylalanine is the startpoint for synthesis of phenylpropanoids, which include, lignins, anthocyanins, flavonoids, coumarines, and small phenolic molecules like salicylic acid. These compounds have multiple functions in structural support, pigmentation, defence, and signalling. Taking into account the function of these compounds, it is likely that the evolution of the phenylpropanoid pathways played a key role in the ability of plants to colonise land (Douglas 1996). Manipulation of the phenylpropanoid pathway has important implications for instance with respect to pulping and forage digestibility. Since these factors are of great economical importance the subject continues to be an active area of research. All phenylpropanoids are derived from cinnamic acid, which is made from phenylalanine by the action of the key-enzyme phenylalanine ammonia-lyase (PAL). PAL represents the branchpoint between primary (shikimate pathway) and secondary (phenylpropanoid) metabolism (Dixon and Palva 1995) (Figure 4). PAL is a cytosolic enzyme, however, it may be attached to membranes in response to phosphorylation. Many of the subsequent enzymes in phenylpropanoid synthesis are localised to he endoplasmic reticulum. The amino group cleaved from phenylalanine by PAL is released as ammonium and recaptured by glutamine synthetase. PAL is encoded by a single gene or a multigene family, depending on the species, and is subjected to a number of control mechanisms. For example, transcription of the PAL gene is enhanced by light, plant growth regulators and different types of stress. In tomato leaves, PAL activity is induced by wounding (excision) of the leaves, and reaches maximum activity 24 h after excision. In response to stressful light conditions, for example high UV irradiation, PAL expression and concentrations of phenylpropanoids like anthocyanins and flavonoids increase (Dixon and Palva 1995). These UV- absorbing compounds then act as a sunscreen and protect the plant against UV damage (Hirner *et al.* 2001).

Glycine and serine

Glycine and serine are two interconvertible amino acids present in many different compartments of the cell. There are two major pathways leading to serine and glycine synthesis. The first pathway is linked to glycolysis and leads to serine formation from 3-phosphoglycerate. The second route is linked to the C_2 cycle (photorespiratory cycle). As part of the C_2 cycle glycolate produced in the chloroplast by the oxygenase activity of Rubisco is exported to the peroxisomes where it is converted to glycine by transamination. Glycine is further transferred to the mitochondria where GDC (glycine decarboxylase) catalyses the oxidative decarboxylation of glycine to produce CO_2, NH_3, NADH and methylenetetrahydrofolate. Serine hydroxymethyltransferase then synthesises serine from the methylenetetrahydrofolate and a second molecule of glycine. It is assumed that (part of) the photorespiratory pathway is the major route for synthesis of glycine and serine in photosynthetic active tissue. It is still not clear if this pathway is active in all tissue or during non-photosynthetic conditions (Morot-Gaudry et al. 2001).

GDC is a complex enzyme that may constitute as much as 40% of the protein in the mitochondria, hence this is one of the most abundant proteins in green leaves, and a major metabolic route is directed through GDC. The cycle liberates large amounts of NH_3 that needs to be refixed. Leaves in the dark have low activity of the GDC enzyme complex, but expression is stimulated upon exposure to light and development of the photosynthetic apparatus (Douce et al. 2001).

CONCLUSIONS

Light acts on nitrogen uptake and metabolism through special red- and blue-light receptors. This has been well documented at the seedling stage and for dark-grown tissue. The red and blue-light receptors are important also in mature plants. Low light irradiance absorbed through the special receptors may trigger for instance flowering, but not much is known concerning the interaction between these signals and nitrogen metabolism in the mature plant. Generally, many of the genes involved in nitrogen uptake, and metabolism are activated by light. A certain enzyme or transporter protein is often coded for by several genes, which are differently regulated by light. This diversity assures rational expression in different tissues, cell compartments or developmental stages. In the green plant light acting through photosynthesis stimulates nitrogen uptake and metabolism. There has been considerable progress in defining components in the signal transduction chains leading from light to regulation of transcription or post-translational modifications. However, complete comprehension and identification of all components in any such chain is still lacking. The *Arabidopsis* genome project gives a powerful tool to identify putative regulatory elements, but characterisation of specific transcription factors and interacting *cis*-acting elements involved in nitrogen metabolism is still scarce.

ACKNOWLEDGEMENTS

The author is indebted to Dr Fioana Provan and Ms Melinda Christensen for helpful comments on the manuscript, and to Prof. Peter Ruoff for drawing the figures.

REFERENCES

Aarnes H., Rognes S.E. 1974. Threonine-sensitive aspartate kinase and homoserine dehydrogenase from *Pisum sativum*. - Phytochemistry 13: 2717-2724.
Aparicio P.J., Quinones M.A. 1991. Blue light a positive switch signal for nitrate and nitrite uptake by the green alga *Monoraphidium braunii*. - Plant Physiol. 95: 374-378.
Appenroth K.-J., Meco R., Jourdan V., Lillo C. 2000. Phytochrome and post-translational regulation of nitrate reductase in higher plants. - Plant Sci. 159: 51-56.
Appenroth K.J., Augsten H., Mohr H. 1992. Photophysiology of turion germination in *Spirodela polyrhiza* L. Schleiden X. Role of nitrate in the phytochrome-mediated response.- Plant Cell Environ. 15: 743-748.
Appenroth K.J., Oelmüller R. 1995. Regulation of transcript level and nitrate reductase activity by phytochrome and nitrate in turions of *Spirodela polyrhiza*. - Physiol. Plant. 93: 272-278.
Argüello-Astorga G., Herrera-Estrella L. 1998. Evolution of light-regulated plant promoters. - Annu. Rev. Plant Physiol. Plant Mol. Biol. 49: 525-55.
Becker T.W., Foyer C., Caboche M. 1992. Light-regulated expression of the nitrate-reductase and nitrite-reductase genes in tomato and in the phytochrome-deficient *aurea* mutant of tomato. - Planta 188: 39-47.
Becker T.W., Nef-Campa C., Zehnacker C., Hirel B. 1993. Implication of the phytochrome in light regulation of the tomato gene(s) encoding ferredoxin-dependent glutamate synthase. - Plant Physiol. Biochem. 31: 725-729.
Bouton S., Leydecker M.-T., Meyer C., Truong H.-N. 2002. Role of gibberellins and of the *RGA* and *GAI* genes in controlling nitrate assimilation in *Arabidopsis thaliana*. - Plant Physiol. Biochem. 40: 939-947.
Briggs W.R., Christie J.M. 2002. Phototropins 1 and 2: Versatile plant blue-light receptors. - Trends Plant Sci. 7: 204-210.
Cardenas-Navarro R., Adamowicz S., Robin P. 1998. Diurnal nitrate uptake in young tomato (*Lycopersicon esculentum* Mill.) plants: test of a feedback-based model. - J. Exp. Bot. 49: 721-730.
Callis J. 1995. Regulation of protein degradation. - Plant Cell 7: 845-857.
Casal J.J. 2002. Environmental cues affecting development. - Curr. Opin. Plant Biol. 5: 37-42.
Chattopadhyay S., Ang L.-H., Puente P., Deng X.-W., Wei N. 1998. Arabidopsis bZIP protein HY5 directly interacts with light-respovise promoters in mediating light control of gene expression. - Plant Cell 10: 673-683.
Cheng C.L., Acedo G.N., Dewdney J., Goodman H.M., Conkling M.A. 1991. Differential expression of the 2 Arabidopsis nitrate reductase genes. - Plant Physiol. 96: 275-279.
Choi Y.A., Kim S.G., Kwon Y.M. 1999. The plastidic glutamine synthetase activity is directly modulated by means of redox change at two unique cysteine residues. - Plant Sci. 149: 175-182.
Cock J.M., Hemon P., Cullimore J.V. 1992. Characterization of the gene encoding the plastid-located glutamine synthetase of *Phaseolus vulgaris*: regulation of beta-glucuronidase gene fusions in transgenic tobacco. - Plant Mol. Biol. 18: 1141-1149.
Coruzzi G., Bush D.R. 2001. Nitrogen and carbon nutrient and metabolite signaling in plants. - Plant Physiol. 125: 61-64.

Coruzzi G., Last R. 2000. Amino acids. - In: Buchanan B.B., Gruissem W., Jones R.L. (Eds.) Biochemistry and molecular biology of plants. - Maryland, USA, American Society Plant Physiologists, pp 358-410.
Coschigano K.T., Melo-Oliveira R., Lim J., Coruzzi G.M. 1998. Arabidopsis *gls* mutants and distinct Fd-GOGAT genes: implications for photorespiration and primary nitrogen assimilation. - Plant Cell 10: 741-752.
Cosgrove D.J. 1993. Photomodulation of growth. - In: Kendrick R.E., Kronenberg. G.H.M. (Eds.) Photomorphogenesis in plants, 2nd edition. - Dordrecht, The Netherlands, Kluwer Academic Publishers, pp 631-658.
Cotelle V., Meek S.E.M., Provan F., Milne F.C., Morrice N., MacKintosh C. 2000. 14-3-3s regulate global cleavage of their diverse binding partners in sugar-starved *Arabidopsis* cells. - EMBO J. 19: 2869-2876.
Cren M., Hirel B. 1999. Glutamine synthetase in higher plants: Regulation of gene and protein expression from the organ to the cell. - Plant Cell Physiol. 40: 1187-1193.
Daniel-Vedele F., Caboche M. 1993. A tobacco cDNA clone encoding a GATA-1 zinc finger protein homologous to regulators of nitrogen metabolism in fungi. - Mol. Gen. Genet. 240: 365-373.
Delhon P., Gojon A., Tillard P., Passama L. 1996. Diurnal regulation of NO_3^- uptake in soybean plants. 4. Dependence on current photosynthesis and sugar availability to the roots. - J. Exp. Bot. 47: 893-900.
Deng X.-W., Caspar T., Quail P.H. 1991. *cop1*: a regulatory locus involved in light-controlled development and gene expression in *Arabidopsis*. - Genes Dev. 5: 1172-1182.
Dey H., Guha-Mukherjee S. 1999. Phytochrome activation of aspartate kinase in etiolated chickpea (*Ciecer arietinum*) seedling. - J. Plant Physiol. 154: 454-458.
Dixon R.A., Palva N.L. 1995. Stress-induced phenylpropanoid metabolism. - Plant Cell 7: 1085-1097.
Dorbe M.-F., Truong H.-N., Crété P., Daniel-Vedele F. 1998. Deletion analysis of the tobacco *Nii1* promoter in *Arabidopsis thaliana*. - Plant Sci. 139: 71-82.
Douce R., Bourguignon J., Neuburger M., Rébeillé F. 2001. The glycine decarboxylase system: a fascinating complex. - Trends Plant Sci. 6: 167-176.
Douglas C.J. 1996. Phenylpropanoid metabolism and lignin biosynthesis: from weeds to trees. - Trends Plant Sci. 1: 171-178.
Duncanson E., Ip S.-M., Sherman A., Kirk D.W., Wray J.L. 1992. Synthesis of nitrite reductase is regulated differently in leaf and root of barley (*Hordeum vulgare* L.) - Plant Sci. 87: 151-160.
Edwards J.W, Coruzzi G.M. 1989. Photorespiration and light act in concert to regulate the expression of the nuclear gene for chloroplast glutamine synthetase. - Plant Cell 1: 241-248.
Elmlinger M.W., Mohr H. 1991. Coaction of blue/ultraviolet-A light and light absorbed by phytochrrome in controlling the appearance of ferredoxin-dependent glutamate synthase in Scots pine (*Pinus sylvestris* L) seedling. - Planta 183: 374-380.
Entus R., Poling M., Herrmann K.M. 2002. Redox regulation of Arabidopsis 3-deoxy-D-arabino-heptulosonate 7-phosphate synthase. - Plant Physiol. 129: 1866-1871.
Ferrario-Méry S., Suzuki A., Kunz C., Valadier M.-H., Roux Y., Hirel B., Foyer C.H. 2000. Modulation of amino acid metabolism in transformed tobacco plants deficient in Fd-GOGAT. - Plant Soil 221: 67-79.
Ferrario-Méry S., Masclaux C., Suzuki A., Vladier M.-H., Hirel B., Foyer C.H. 2001. Glutamine and α-ketoglutarate are metabolic signals involved in nitrate reductase gene transcription in untransformed and transformed tobacco plants deficient in ferredoxin-glutamine-α-ketoglutarate aminotransferase. - Planta 231: 265-271.
Figueroa F.L. 1993. Photoregulation of nitrogen metabolism and protein accumulation in the red alga *Corallina elongata* Ellis et Soland. – Z. Naturforsch., (Biosciences) 48: 788-794.
Finnemann J., Schjoerring J.K. 2000. Post-translational regulation of cytosolic glutamine synthetase by reversible phosphorylation and 14-3-3 protein interaction. - Plant J. 24: 171-181.

Forde B.G. 2000. Nitrate transporters in plants: structure, function and regulation. - Biochim. Biophys. Acta 1465: 219-235.

Friemann A., Brinkmann K., Hachtel W. 1991. Sequence of a cDNA encoding bi-specific NAD(P)H-nitrate reductase from the tree *Betula pendula* and identification of conserved protein regions. - Mol. Gen. Genet. 227: 97-105.

Goud K.V., Sharma R. 1994. Retention of photoinduction of cytosolic enzymes in *aurea* mutant of tomato (*Lycopersicon esculentum*). - Plant Physiol. 105: 643-650.

Guo F.Q., Wang R., Crawford N.M. 2002. The *Arabidopsis* dual-affinity nitrate transporter gene *AtNRT1.1 (CHL1)* is regulated by auxin in both shoots and roots. - J. Exp. Bot. 53: 835-844.

Henstrand J.M., McCue K.F., Brink K., Handa A.K., Herrmann K.M., Conn E.E. 1992. Light and fungal elicitor induce 3-deoxy-D-*arabino*-heptulosonate 7-phosphate synthase mRNA in suspension cultured cells of parsley (*Petroselinum crispum* L.). - Plant Physiol. 98: 761-763.

Hirel B., Lea P.J. 2001. Ammonia assimilation. - In: Lea P.J., Morot-Gaudry J.F. (Eds.) Plant nitrogen. - Berlin, Germany, Springer Verlag, pp. 79-99.

Hirner A.A., Stefan V.H., Seitz U. 2001. Regulation of anthocyanin biosynthesis in UV-A irradiated cell culture of carrot and in organs of intact carrot plants. - Plant Sci. 161: 315-322.

Hsieh M.H., Lam H.M., Loo F.J. von, Coruzzi G. 1998. A PII like protein in *Arabidopsis*: putative role in nitrogen sensing. – Proc. Natl. Acad. Sci. USA 95: 13965-13970.

Hughes C.A., Gebhardt J.S., Reuss A., Matthews B.F. 1999. Identification and expression of a cDNA encoding cystathionine gamma-synthase in soybean. - Plant Sci. 146: 69-79.

Huppe H.C., Turpin D.H. 1994. Integration of carbon and nitrogen metabolism in plant and algal cells. - Annu. Rev. Plant Physiol. 45: 577-607.

Hwang C.-F., Lin Y., D'Souza T., Cheng C.-L. 1997. Sequences necessary for nitrate-dependent transcription of Arabidopsis nitrate reductase genes. - Plant Physiol. 113: 853-862.

Ishioka N., Tanimoto S., Harada H. 1991. Roles of nitrogen and carbohydrate in floral-bud formation in *Pharbitis* apex cultures. - J. Plant Physiol. 138: 573-576.

Jarai G., Truong H.N., Daniel-Vedele F., Marzluf G. 1992. NIT2, the nitrogen regulatory protein of *Neurospora crassa*, binds upstream of *nia*, the tomato nitrate reductase gene, in vitro. - Curr. Genet. 21: 37-41.

Jensen P.E., Hoff T., Stummann B.M., Henningsen K.W. 1996 Functional analysis of two bean nitrate reductase promoters in transgenic tobacco. - Physiol. Plant. 96: 351-358.

Jones R.W., Sheard R.W. 1972. Nitrate reductase activity: phytochrome mediation of induction in etiolated peas. - Nature 238: 221-222.

Jones R.W., Sheard R.W. 1979. Light factors in nitrogen assmilation. In: Hewitt E.J., Cutting C.V. (Eds.) Nitrogen assimilation of plants. - London, UK, Academic Press, pp. 521-539.

Kaiser W.M., Huber S.C. 2001. Post-translational regulation of nitrate reductase: mechanism, physiological relevance and environmental triggers. – J. Exp. Bot. 52: 1981-1989.

Kaiser W.M., Spill D. 1991. Rapid modulation of spinach leaf nitrate reductase activity by photosynthetis. II. In vitro modulation by ATP and AMP. - Plant Physiol. 96: 368-375.

Kronenberger J., Lepingle A., Caboche M. 1993. Cloning and expression of distinct nitrite reductases in tobacco leaves and roots. - Mol. Gen. Genet. 236: 203-208.

Kim C.-H., Jun S.-S., Hong Y.-N. 2001. *GmNiR-1*. A soybean nitrite reductase gene that is regulated by nitrite and light. - Aust. J. Plant Physiol. 28: 1031-1039.

Kronenberger J., Lepingle A., Caboche M., Vaucheret H. 1993. Cloning and expression of distinct nitrite reductases in tobacco leaves and roots. - Mol. Gen. Genet. 236: 203-208.

Lam H.-M., Hsieh M.-H., Coruzzi G. 1998. Reciprocal regulation of distinct asparagine synthetase genes by light and metabolites in Arabidopis thaliana. - Plant J. 16: 345- 353.

Lam H.-M., Coschigano K.T., Oliveira I.C., Melo-Oliveira R., Coruzzi G.M. 1996. The moleculargenetics of nitrogen assimilation into amino acids in higher plants. - Annu. Rev. Plant Physiol. Mol. Biol. 47: 569-5693.

Lejay L., Tillard P., Lepetit M., Olive F.D., Filleur S., Daniel-Vedele F., Gojon A. 1999. Molecular and functional regulation of two NO_3^- uptake systems by N-and C-status of *Arabidopsis* plants. - Plant J. 18: 509-519.

Li X.-Z. Oaks A. 1995. The effect of light on the nitrate and nitrite reductases in *Zea mays*. - Plant Sci. 109: 115-118.
Lillo C. 1983. Studies of diurnal variations of nitrate reductase activity in barley leaves using various assay methods. - Physiol. Plant. 57: 357-362.
Lillo C. 1984. Diurnal variations of nitrite reductase, glutamine synthetase, glutamate synthase, alanine aminotransferase and aspartate aminotransferase in barley leaves. - Physiol. Plant. 61: 214-218.
Lillo C. 1994. Light regulation of nitrate reductase in green leaves of higher plants. - Physiol. Plant. 62: 89-94.
Lillo C. Appenroth K.-J. 2001. Light regulation of nitrate reductase in higher plants: Which photoreceptors are involved? - Plant Biol. 3: 455-465.
Lillo C. Henriksen A. 1984. Comparative studies of diurnal variations of nitrate reductase activity in wheat, oat and barley. - Physiol. Plant. 62: 89-94.
Lillo C., Lea U.S., Leydecker M.-T., Meyer C. 2003. Mutation of the regulatory phosphorylation site of tobacco nitrate reductase results in constitutive activation of the enzyme in vivo and nitrite accumulation. - Plant J. 35: 566-573.
Lillo C., Meyer C., Ruoff P. 2001. The nitrate reductase circadian system. The central clock dogma contra multiple oscillatory feedback loops. - Plant Physiol 125: 1554-1557.
Lin Y., Cheng C.-L. 1997. A chlorate-resistant mutant defective in the regulation of nitrate reductase gene expression in Arabidopsis defines a new *HY* locus. - Plant Cell 9: 21-35.
Lin Y., Hwang C.-F., Brown J.B., Cheng C.-L. 1994. 5' proximal regions of *Arabidopsis* nitrate reductase genes direct nitrate-induced transcription in transgenic tobacco. - Plant Physiol. 106: 477-484.
Lopez-Figueroa F., Ruediger W. 1991. Stimulation of nitrate net uptake and reduction by red and blue light and reversion by far-red light in the green alga *Ulva rigida*. - J. Phycology 27: 389-394.
MacKintosh C. 1992. Regulation of spinach-leaf nitrate reductase by reversible phosphorylation. - Biochim. Biophys. Acta 1137: 121-126.
MacKintosh C., Meek S.E.M. 2001. Regulation of plant NR activity by reversible phosphorylation, 14-3-3 proteins and proteolysis. - Cell Mol. Life Sci. 58: 205-214.
Maetschke M., Riedel J., Tischner R. 1997. Evidence for signal transduction in the stimulation of nitrate uptake by blue light in *Chlorella saccharophila*. - Photochem. Photobiol. 66: 128-132.
Martinez-Hernández A., López-Ochoa L., Argüello-Astorga G., Herrera-Estrella L. 2002. Functional properties and regulatory complexity of a minimal *RBCS* light-responsive unit activated by phytochrome, cryptochrome, and plastid signals. - Plant Physiol. 128: 1223-1233.
Marzluf G.A. 1997. Genetic regulation of nitrogen metabolism in the fungi. - Microb. Mol. Biol. Rev. 61: 17-32.
Matt P., Geiger M., Walch L.P., Engels C., Krapp A., Stitt M. 2001. Elevated carbon dioxide increases nitrite uptake and nitrate reductase activity when tobacco is growing on nitrate, but increases ammonium uptake and inhibits nitrate reductase activity when tobacco is growing on ammonium nitrate. - Plant Cell Envioron. 24: 1119-1137.
Melzer J.M., Kleinhofs A., Warner R.L. 1989. Nitrate reductase regulation. Effects of nitrate and light on nitrate reductase mRNA accumulation. Mol. Gen. Genet. 217: 341-346.
Meyer C., Stitt M. 2001. Nitrate reduction and signalling. - In: Lea P.J., Morot-Gaudry J.F. (Eds.) Plant nitrogen. - Berlin, Germany, Springer Verlag, pp 37-59.
Miflin B.J., Habash D.Z. 2002. The role of glutamine synthetase and glutamate dehydrogenase in nitrogen assimilation and possibilities for improvement in the nitrogen utilization of crops. - J. Exp. Bot. 53: 979-987.
Migge A., Carrayol E., Hirel B., Lohmann M., Meya G., Becker T.W. 1998. Influence of UV-A or UV-B light and of the nitrogen source on the induction of ferredoxin-dependent glutamate synthase in etiolated tomato cotelydons. - Plant Physiol. Biochem. 36: 789-797.

Migge A., Meya G., Carryol E., Hirel B., Becker T.W. 1997. Coaction of light and the nitrogen substrate in controlling the expression of the tomato genes encoding nitrite reductase and nitrate reductase. - J. Plant Physiol. 151: 151-158.

Miyazaki J., Juricek M., Angelis K., Schnorr K.M., Kleinhofs A., Warner R.L. 1991. Characterization and sequence of a novel nitrate reductase from barley. - Mol. Gen. Genet. 228: 329-334.

Moorhead G., Douglas P., Cotelle V., Harthill J., Morrice N., Meek S., Deiting U., Stitt M., Scarabel M., Aitken A., MacKintosh C. 1999. Phosphorylation-dependent interactions between enzymes of plant metabolism and 14-3-3 proteins. - Plant J. 18: 1-12.

Morey K.J., Ortega J.L., Sengupta-Gopalan C. 2002. Cytosolic glutamine synthetase in soybean is encoded by a multigene family, and the members are regulated in an organ-spesific and developmental manner. - Plant Physiol. 128: 182-193.

Morot-Gaudry J.F., Job D., Lea P.J. 2001. Amino acid metabolism. - In: Lea P.J., Morot-Gaudry J.F. (Eds.) Plant nitrogen. - Berlin, Germany, Springer Verlag, pp. 167-211.

Motohashi K., Kondoh A., Stumpp M.T., Hisabori T. 2001. Comprehensive survey of proteins targeted by chloroplast thioredoxin. - Proc. Natl. Acad. Sci. USA 25: 11224-11229.

Neininger A., Back E., Bichler J., Schneidenbauer A., Mohr H. 1994. Deletion analysis of nitrite reductase promoter from spinach in transgenic tobacco. - Planta 194: 186-192.

Neininger A., Kronenberger J., Mohr H. 1992. Coaction of light, nitrate and plastidic factor in controlling nitrite-reductase gene expression in tobacco. - Planta 187: 381-387.

Neuhaus G., Bowler C., Hiratsuka K., Yamagata H., Chua N.-H. 1997. Phytochrome-regulated repression of gene expression requires calcium and cGMP. - EMBO J. 16: 2554-2564.

Ngai N., Tsai F.-Y., Coruzzi G. 1997. Light-induced transcriptional repression of the pea *AS1* gene: identification of *cis*-elements and transfactors. - Plant J. 12: 1021-1034.

Ninfa A.J., Atkinson M.R. 2000. PII signal transduction proteins. - Trends Microbiol. 8: 172-179.

Oliveira I.C., Coruzzi G.M. 1999. Carbon and amino acids reciprocally modulate the expression of glutamine synthetase in Arabidopsis. - Plant Physiol. 121: 301-309.

Orsel M., Krapp A., Daniel-Vedele F. 2002 Analysis of the NRT2 nitrate transporter family in Arabidopsis. Structure and gene expression. - Plant Physiol. 129: 886-896.

Pace G.M., Volk R.J., Jackson W.A. 1990. Nitrate reduction in repsonse to carbon dioxide-limited photosynthesis relationship to carbohydrate supply and nitrate reductase activity in maize seedlings. - Plant Physiol. 92: 286-292.

Pajuelo P., Pajuelo E., Forde B.G., Marquez A.J. 1997. Regulation of the expression of ferredoxin glutamate synthase in barley. - Planta 203: 517-525.

Provan F., Aksland L.-M., Meyer C., Lillo C. 2000. Deletion of the nitrate reductase N-terminal domain still allows binding of 14-3-3 proteins but affects their inhibitory properties. - Plant Physiol. 123: 757-764.

Quail P.H. 2002. Photosensory perception and signalling in plant cells: new paradigms? - Curr. Opin. Cell Biol. 14: 180-188.

Rao L.V.M., Rajasekhar V.K., Soporo S.K., Guha-Mukherjee S. 1981. Phytochrome regulation of nitrite reductase - a chloroplastic enzyme in etiolated maize leaves. - Plant Cell Physiol. 22: 577-582.

Rao S.S., Kochhar S., Kochhar V.K. 1999. Analysis of photocontrol of aspartate kinase in barley (*Hordeum vulgare* L.) seedlings. - Biochem. Mol. Biol. Int. 47: 347-360.

Rastogi R., Back E., Schneiderbauer A., Bowsher C.G., Moffatt B., Rothstein S.J. 1993. A 330 bp region of the spinach nitrite reductase gene promoter directs nitrate-inducible tissue-specific expression in transgenic tobacco. - Plant J. 4: 317-326.

Riedel J., Tischner R., Mack G. 2001. The chloroplastic glutamine synthetase (GS-2) of tobacco is phosphorylated and associated with 14-3-3 proteins inside the chloroplast. - Planta 213: 396-401.

Rockel P., Strube F., Rockel A., Wildt J., Kaiser W.M. 2002. Regulation of nitric oxide (NO) production by plant nitrate reductase *in vivo* and *in vitro*. - J. Exp. Bot. 53: 103-110.

Rouzé P., Caboche M. 1992. Nitrate reduction in higher plants: Molecular approach to function and regulation. In: Wray J.L. (Ed.) Inducible plant proteins, Society for experimental biology seminar series 49. - Cambridge, U.K., Cambridge University Press, pp. 45-77.

Ruelland E., Miginiac-Maslow M. 1999. Regulation of chloroplast enzyme activities by thioredoxins: activation or relief from inhibition. - Trends Plant Sci. 4: 136-141.

Ruoff, P. Lillo, C. 1990. Molecular oxygen as electron acceptor in the NADH-nitrate reductase system. - Biochem. Biophys. Res. Commun. 172: 1000-1005.

Sakamoto A., Takeba G., Shibata D., Tanaka K. 1990. Phytochrome-mediated activation of the gene for cytosolic glutamine synthetase GS-1 during imbibition of photosensitive lettuce seeds. - Plant Mol. Biol. 15: 317-324.

Sander L., Jensen P.E. 1995. Structure and expression of a nitrite reductase gene from bean (*Phaseolus vulgaris*) and promoter analysis in transgenic tobacco. - Plant Mol. Biol. 27: 165-177.

Sasakawa H., Yamamoto Y. 1979. Effects of red, far-red and blue light on enhancement of nitrate reductase EC-1.6.6.1 activity and on nitrate uptake in etiolated rice seedlings. - Plant Physiol. 63: 1098-1101.

Scaife A. 1989. A pump/leak//buffer model for plant nitrate uptake. - Plant Soil 114: 139-141.

Schondorf T., Hachtel W. 1995. The choice of reducing substrate is altered by replacement of an alanine by proline in the FAD domain of a bispecific NADH(P)H-nitrate reductase from birch. - Plant Physiol. 108: 203-210.

Schürmann P., Jacquot J.-P. 2000. Plant thioredoxin systems revisited. - Annu. Rev. Plant Physiol. Plant Biol. 51: 371-400.

Sivasankar S., Oaks A. 1996. Nitrate assimilation in higher plants: The effect of metabolites and light. - Plant Physiol. Biochem. 34: 609-620.

Sivasankar S., Rastogi R., Jackman L., Oaks A., Rothstein S. 1998. Analysis of cis-acting DNA elements mediating induction and repression of the spinach nitrite reductase gene. - Planta 206: 66-71.

Smith H. 2000. Phytochromes and light signal perception by plants - emerging synthesis. - Nature 407: 585-591.

Smith C.S., Zaplachinski S.T., Muench D.G., Moorhead G.B. 2002. Expression and purification of the chloroplast putative nitrogen sensor, PII, of *Arabidopsis thaliana*. - Protein Expr. Purif. 25: 342-347.

Sopory S.K., Sharma A.K. 1990. Spectral quality of light, hormones and nitrate assimilation. - In: Abrol Y.P. (Ed.) Nitrogen in higher plants. - Somerset, U.K., Research Studies Press Ltd., pp. 129-157.

Strater T., Hachtel W. 2000. Identification of light- and nitrate-responsive regions of the nitrate reductase promoter from birch. - Plant Sci. 150: 153-161.

Suzuki A., Audet C., Oaks A. 1987. Influence of light on the ferredoxin-dependent glutamate synthase in maize leaves. - Plant Physiol. 84: 578-581.

Suzuki A., Rioual S., Lemarchand S., Godfroy N., Roux Y., Boutin J.-P., Rothstein S. 2001. Regulation by light and metabolites of ferredoxin-dependent glutamate synthase in maize. - Physiol. Plant. 112: 524-530.

Tanaka O. 1986. Flowering induction by nitrogen deficiency in *Lemna paucicosta* 6746. - Plant Cell Physiol. 27: 875-880.

Teakle G.R., Kay S.A. 1995. The GATA-binding protein CGF-1 is closely related to GT-1. - Plant Mol. Biol. 29: 1253-1266.

Teller S., Appenroth K..J 1994. The appearance of glutamine synthetase in turions of *Spirodela polyrhiza* (L.) Schleiden as regulated by blue and red light, nitrate and ammonium. - J. Exp. Bot. 45: 1219-1226.

Teller S., Schmidt K.-H., Appenroth K.-J. 1996. Ferredoxin-dependent but not NADH-dependent glutamate synthase is regulated by phytochrome and blue/UV-A light receptor in turions of *Spirodela polyrhiza*. - Plant Physiol. Biochem. 34: 713-719.

Temple S.J., Vance C.V., Gantt J.S. 1998. Glutamate synthase and nitrogen assimilation. - Trends Plant Sci. 3: 51-56.

Tepperman J.M., Zhu T., Chang H.-S., Wang X., Quail P.H. 2001. Multiple transcription-factor genes are early targets of phytochrome A signalling. - Proc. Natl. Acad. Sci. USA 98: 9437-9442.

Tischner R. 2000. Nitrate uptake and reduction in higher and lower plants. - Plant Cell Environ. 23: 1005-1024.

Tjaden G., Edwards J.W., Coruzzi G.M. 1995. *cis* Elements and *trans*-acting factors affecting regulation of a nonphotosynthetic light-regulated gene for chloroplast glutamine synthetase. - Plant Physiol. 108: 1109-1117.

Tjaden G., Coruzzi G.M. 1994. A novel AT-rich DNA binding protein that combines an HMG I-like DNA binding domain with a putative transcription domain. - Plant Cell 6: 107-118.

Touraine B., Daniel-Vedele F., Forde B.G. 2001. Nitrate uptake and its regulation. - In: Lea P.J., Morot-Gaudry J.-F. (Eds.) Plant nitrogen. - Berlin, Germany, Springer Verlag, pp. 1-36.

Trepp G.B., van der Mortel M., Yoshioka H., Miller S.S., Samac D.A., Gantt J.S., Vance C.P. 1999. NADH-glutamate synthetase in alfalfa root nodules. Genetic regulation and cellular expression. - Plant Physiol. 119: 817-828.

Ullrich W.R., Lazarova J., Ullrich C.L., Witt F.G., Aparicio P.J. 1998. Nitrate uptake and extracellular alkalinization by the green alga *Hydrodictyon reticulum* in blue and red light. - J. Exp. Bot. 49: 1157-1162.

Vincentz M., Moureaux T., Leydecker M.-T., Vaucheret H., Caboche M. 1993. Regulation of nitrate and nitrite reductase expression in *Nicotiana plumbaginifolia* leaves by nitrogen and carbon metabolites. - Plant J. 3: 315-324.

Wang H., Ma L.-G., Zhao H.-Y., Li J.-M., Deng X.W. 2001. Direct interaction of *Arabidopsis* cryptochromes with COP1 in light control development. - Science 294: 154-158.

Warning H.O., Hachtel W. 2000. Functional analysis of a nitrite reductase promoter from birch in transgenic tobacco. - Plant Sci. 155: 141-151.

Weaver L., Herrmann K.M. 1997. Dynamics of the shikimate pathway in plants. - Trends Plant Sci. 2: 346-351.

Weber M., Schmidt S., Schuster C., Mohr H. 1990. Factors involved in the coordinate appearance of nitrite reductase and glutamine synthetase in the mustard *Sinapsis alba* L seedlings. - Planta 180: 429-434.

Wildt J., Kley D., Rockel A., Rockel P., Segschneider H.J. 1997. Emission of NO from several higher plant species. - J Geophys. Res. 102: 5919-5927.

Wray J.L. 1993. Molecular biology, genetics and regulation of nitrite reduction in higher plants. - Physiol. Plant. 89: 607-612.

Zhu-Shimoni J.X., Lev-Yadun S., Matthews B., Galili G. 1997. Expression of an aspartate kinase homoserine dehydrogenase gene is subject to specific spatial and temporal regulation in vegetative tissues, flowers, and developing seeds. - Plant Physiol. 113: 695-706.

Yamasaki H., Sakihama, Y. 2000. Simultaneous production of nitric oxide and peroxynitrite by plant nitrate reductase: *in vitro* evidence for the NR-depndent formation of active nitrogen species. - FEBS Lett: 468, 89-92.

Chapter 7

MODULATION OF NITRATE REDUCTION - ENVIRONMENTAL AND INTERNAL FACTORS INVOLVED

Werner M. Kaiser, Elisabeth Planchet, Maria Stoimenova and Masatoshi Sonoda

INTRODUCTION

The major purpose of the cytosolic enzyme nitrate reductase (NR, EC 1.6.6.1) is to catalyse the reduction of nitrate to nitrite at the expense of NAD(P)H. In addition to that, however, NR also catalyses, though with much lower capacity, two side reactions: 1) the reduction of nitrite to nitric oxide (NO) and 2) the reduction of molecular oxygen to superoxide (Ruoff and Lillo, 1990, Barber and Kay 1996), which can both react chemically to give the highly toxic compound peroxynitrite (Figure 1, Yamasaki and Sakihama 2000). The production of such potentially toxic products (of which NO may also act as a signalling molecule in plants) is probably one reason why NR expression and activity are so tightly controlled.

For regulation of NR expression and activity by light see chapter 6, Lillo (1994) and Lillo and Appenroth (2001). In leaves, NR becomes more active upon illumination within minutes, and is as rapidly inactivated in the dark. Based on these short response times it was suggested that NR activity would not be modulated via changes in transcription or translation, but via modulation of the catalytic activity of the existing NR protein. More recently, in plants expressing NR under the control of the constitutive 35S-promoter, rapid light/dark changes in extractable NR activity from leaves were still observed, although mRNA levels expectedly did not vary in dark/light (Vincentz *et al.* 1993). Such observations strongly confirm that the existing NR molecule undergoes a reversible modulation in catalytic activity. In the past decade, the mechanism of the post-translational modulation has been investigated in detail.

Below we will summarise the present knowledge on the modulation mechanism, and on various external (environmental) and internal factors involved. Further, we will consider implications of NR modulation on the production of nitric oxide by NR. This long-known side reaction of NR is presently gaining much attention, motivated by the hope that - in analogy to animals - plants would use the NO radical as a short-lived signal for controlling growth, development and interactions with other organisms.

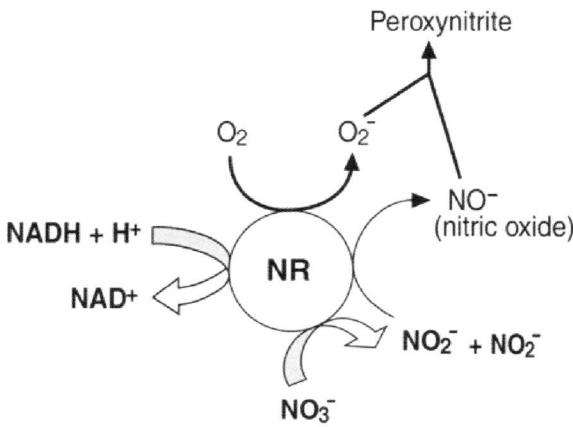

Figure 1. Reactions catalysed by NR. Broad arrows indicated the main reaction pathway, which is the two electron transfer from NAD(P)H to nitrate to yield nitrite. Narrow arrows mark two side reactions, which are both one- electron transfer reactions to reduce molecular oxygen to superoxide, and nitrite to nitric oxide (NO). Both compounds probably react rapidly to form peroxynitrite. In air and at substrate saturation, the capacity for NO formation is only 1 to 2% of the main reaction (Rockel *et al.* 2002).

MODULATION OF NR ACTIVITY BY REVERSIBLE PHOSPHORYLATION AND 14-3-3-BINDING

Evidence that NR inactivation requires NR-phosphorylation and 14-3-3-binding

When extracts were rapidly prepared from illuminated or darkened leaves and NR activity was measured in the presence of excess divalent cations, activity was usually twice as high in extracts from illuminated leaves than from darkened leaves. These changes in NR activity were not affected by desalting of extracts; thus they were not due to some direct allosteric effects of metabolites on NR (Kaiser and Brendle-Behnisch 1991). Importantly, these activity changes were observed only in buffers containing divalent cations in excess, but were absent in the presence of excess EDTA. Apparently divalent cations were required to keep NR in its "inactivated state". NR activity from "light extracts" could be rapidly decreased *in vitro* by incubation with MgATP. Based on such observations, we suggested already in 1991 that NR was probably inactivated by protein phosphorylation (Kaiser and Spill 1991). Subsequently, Huber *et al.* (1992) and MacKintosh (1992) presented direct evidence that NR was indeed more phosphorylated in dark than in light, consistent with the activity measurements. Some years later, a Ser-543 residue in the hinge 1 region was identified as the site of regulatory phosphorylation (Douglas *et al.* 1995, Bachmann *et al.* 1996a, Su *et al.* 1996).

On a first attempt to fractionate a leaf extract into a protein kinase free NR fraction and a kinase fraction, we realised that just mixing a kinase fraction with the NR fraction was sufficient to phosphorylate NR upon addition of γ-^{32}P-ATP, but without inactivating it (Glaab and Kaiser 1995). Inactivation of phosphorylated NR (P-NR) occurred only after addition of yet another protein fraction which lacked protein kinase activity. Similar observations were independently reported by MacKintosh's group, and the unknown protein was subsequently termed "IP" or "NIP", which stand for "inhibitor protein". Characterisation of that "IP" showed that it belonged to a family of binding proteins known as 14-3-3, for which 14 isoforms have been found in *Arabidopsis* (Bachmann *et al.* 1996b, Moorhead *et al.* 1996, Bachmann *et al.* 1998, MacKintosh 1998) (Figure 2).

Role of 14-3-3 proteins and cations

The obvious requirement for divalent cations in the NR inactivation cycle was originally interpreted to indicate that they were mediating the formation of a complex between 14-3-3 and P-NR. Later, the metal ion was shown to bind to 14-3-3 and activate it to bind to P-NR (Athwal *et al.* 1998, 2000). Metal cations could be replaced by polyamines, and spermidine was the most effective polyamine (Provan *et al.* 2000). As polyamines are usually accumulated in plants in response to excess-N nutrition (but also to mineral cation deficiency or to salt stress), this polyamine effect may represent a regulatory link "telling" plants to decrease nitrate assimilation in times of low demand or excess supply.

14-3-3-proteins are homodimers with two binding grooves which can be both filled by relatively small molecules (Petosa *et al.* 1998). A 3D-model (Campbell 2003) suggests that only one of the two 14-3-3 binding sites can bind to NR at once, but that two independent 14-3-3- molecules could bind to dimeric NR. It was not ruled out, however, that the docked 14-3-3 binds to another NR-dimer or another protein with a 14-3-3 recognition sequence. It is also interesting that replacement of the regulatory Ser by Asp, which also introduces a negative charge, did not result in 14-3-3-binding and in NR inactivation. Yet additional substitutions of *Arabidopsis* NR at second sites resulted in 14-3-3-binding and

NR inhibition independent of Ser-543 (Kanamaru *et al.* 1999). Interestingly, an estimation of the ratios of NR, P-NR and 14-3-3-P-NR in illuminated spinach leaves by site-specific antibodies revealed that about one third of the total NR may exist as free P-NR not bound to 14-3-3. Upon darkening, the phosphorylation state of NR increased, and much more 14-3-3s appeared bound (Table 1, Weiner and Kaiser 2001). Thus, it seems possible that not only the relative NR phosphorylation state, but also binding affinity of 14-3-3s to pre-existing P-NR may vary in light/dark. Whether that is related to a light/dark modification of 14-3-3s themselves is not known yet.

Table 1. Relative amounts of different NR-forms in extracts from illuminated (5 h) and subsequently darkened (30 min) spinach leaves, as determined with an immunological approach. NR activation state is given for comparison. Total NR was determined by immunoprecipitation with antibodies against the C- and against the N-terminus. In both cases, total NR protein did not change (not shown). NR and P-NR were immunoprecipitated with antibodies raised against synthetic peptides containing the motif around ser 543, either with ser (S-peptide) or with an asp instead (D-peptide) in order to insert a stable negative charge mimicking P-ser. The amount of 14-3-3 was obtained by coimmunoprecipitation with NR, gel fractionation and Western blotting with 14-3-3-antibodies. Methods and experimental details in Weiner and Kaiser (2001). Note that the relative amount of 14-3-3-P-NR in the light is much lower than expected from the activation state.

	% NR form			% Activation state
	Free NR	P-NR	14-3-3-P NR	
Light	65 ± 5	33 ± 4	≤ 5	50 ± 3
Dark	10 ± 2	19 ± 3	69 ± 3	20 ± 2

NR degradation

It was thought that inactivation of NR made the enzyme more accessible to proteolytic degradation (Kaiser and Huber 1997), and 14-3-3-binding to NR has been suggested to be the trigger for specific proteolysis (Weiner and Kaiser 1999, Figure 2). In the light of more recent results, this appears now less clear, however. As mentioned, the replacement of the regulatory Ser by Asp left the NR completely active (see above), which indicates the absence of 14-3-3 binding; yet NR degradation in the dark was not affected (C. Lillo, personal communication). This suggests (Figure 2) that NR degradation is eventually not initiated by 14-3-3-binding, but just by inserting a negative charge at the regulatory Ser (either via Ser phosphorylation or by replacement with Asp). Certainly, more work is required to fully understand the way in which NR is prone to specific degradation and how this is connected to the activation/inactivation cycle.

Consequences of NR inactivation on NR kinetics

Binding of 14-3-3 to P-NR has been shown to interrupt whole-chain electron transfer from NADH to nitrate, and the partial activity from the electron donor methylviologen (MV) to nitrate was also blocked (Bachmann *et al.* 1996a). Most probably, 14-3-3-binding causes a conformational change which inhibits electron transfer from the heme-Fe of the Cyt b moiety to MoMPT. That conformation change may prevent the cationic MV electron donor to interact. Interestingly, the Km values for NADH (1 to 7 µM) and nitrate (20 to 40 µM) are not changed by the above inactivation mechanism (Kaiser and Spill 1991). Thus, phosphorylation and 14-3-3 binding do not change kinetic properties of the enzyme, but only the ratios of free NR (active) to P-NR-14-3-3 complexes (inactive). Neither in leaves nor in roots was NR activity ever completely abolished or fully activated by the above modulation. In fact, as a general rule, a very low activation state in darkened leaves is about 20% (of total NR activity measured with an excess of EDTA), and a maximum activation state in illuminated leaves supplied with high CO_2 is 80%.

Protein kinases and phosphatases involved in NR modulation

The regulation of NR by phosphorylation requires, at the very end, a regulation of the participating protein kinases and/or protein phosphatases. Definite answers on how these enzymes are regulated are still lacking. Spinach leaf extracts contained several protein kinase fractions that can phosphorylate NR or synthetic peptides based on the Ser-543 sequence. Two of them are calcium dependent kinases belonging to the large family of CDPKs (Bachmann *et al.* 1996a, Douglas *et al.* 1998). One calcium independent kinase has been shown to belong to the SNF1-related kinase family (McMichael *et al.* 1995, Douglas *et al.* 1998), which are called SnRK1 (Sudgen *et al.* 1999). SnRK1 protein kinases are related to animal and yeast protein kinases which are activated by AMP. These kinases appear to be also involved in the regulation of enzymes of carbon metabolism (sucrose phosphate synthase, HMG-CoA reductase and sucrose synthase) (Sudgen *et al.* 1999), and may thus present a link between C- and N-metabolism. It is still not known which of these kinases contribute to NR phosphorylation *in situ*. Calcium-dependent kinases play a significant role as suggested by the observation that treatment of darkened leaves with EGTA + the ionophore A23187, which causes an efflux of calcium and magnesium, resulted in NR activation (Kaiser and Huber 1994). Little is known to date about potential signals that control PK activities (see below).

P-NR is dephosphorylated by a type 2A protein phosphatase, PP2A, (MacKintosh 1992), as *e.g.* indicated by the fact that PP2A-specific okadaic acid blocked NR light activation *in situ*. PP2A's are heteromeric proteins, usually composed of a catalytic C subunit and two regulatory subunits (Deruere *et al.* 1999, Shenolikar 1994). Little is known about the structure and regulation of plant PP2A's, but inorganic phosphate appears to be a PP2A inhibitor (Weiner *et al.* 1993).

INTERNAL SIGNALS CONTROLLING NR ACTIVATION/INACTIVATION

Whatever the external (environmental) triggers are that modulate nitrate assimilation in plants (see below), they have to be transformed into internal signals that act directly or indirectly (*e.g.* via phosphorylation cascades) on the protein kinases or protein phosphatases which, in turn, control the NR activation state. Thus, before discussing such environmental effectors, we will briefly summarise our (poor) knowledge on potential internal signals.

AMP

It is not known yet whether the abovementioned plant SnRK1 type protein kinases are activated by 5'-AMP (like in animals and yeasts). Actually, this seems not very probable. Activation of NR-kinases by AMP should render NR inactive. However, inactive NR in darkened leaves has been shown to be rapidly activated by feeding the cell-permeable AICAR (5-aminoimidazole-4-carboxyamid ribonucleoside) (Huber and Kaiser 1996). After entering cells, AICAR is phosphorylated inside to the 5'-AMP analogue ZMP (the monophosphate derivative of AICAR). Thus, in plants 5'-AMP either turns the NR kinases off, or the PP2As on, but it is not known whether ZMP may affect the NR activation state indirectly via changing other metabolite levels.

Sugar phosphates

The SnRK1 involved in regulation of sucrose phosphate synthase is inhibited by glucose-6-phopshate (Toroser *et al.* 2000). Thus, high cytosolic sugar-P levels should decrease NR phosphorylation, leading to a higher activation state. Indeed, in darkened leaves, levels of free hexoses, of hexose phosphates and the NR-activation state dropped rapidly, and sucrose feeding partly activated NR, and increased concentrations of sucrose and of free hexoses, but also of hexose-monophosphates (Kaiser, unpublished results). Thus, the positive effect of sugar feeding on NR activity may be in part due to increased sugar-P-levels and to a concomitant inhibition of protein kinases (but compare results on anoxia, below). As sugar feeding also increases NR transcription, it is obvious that carbon metabolism exerts a tight control over nitrate assimilation at different levels of regulation.

Cytosolic ion concentrations

Cytosolic ions may be further factors involved in the control of NR activity. Artificial acidification of darkened leaf or root pieces with permeating weak acids and low external pH activated NR, and alkalinisation with permeating bases and high external pH reversed this effect (Kaiser and Brendle-Behnisch 1995). Further, treatment of darkened leaf discs with EDTA and the ionophore A23187, also activated NR (Kaiser and Huber 1994). Unfortunately, this does not necessarily prove direct Ca^{2+} effects *e.g.* on protein kinases (CDPKs). Efflux of divalent cations may be connected to a proton influx, and NR activation may thus also be related to cytosolic acidification.

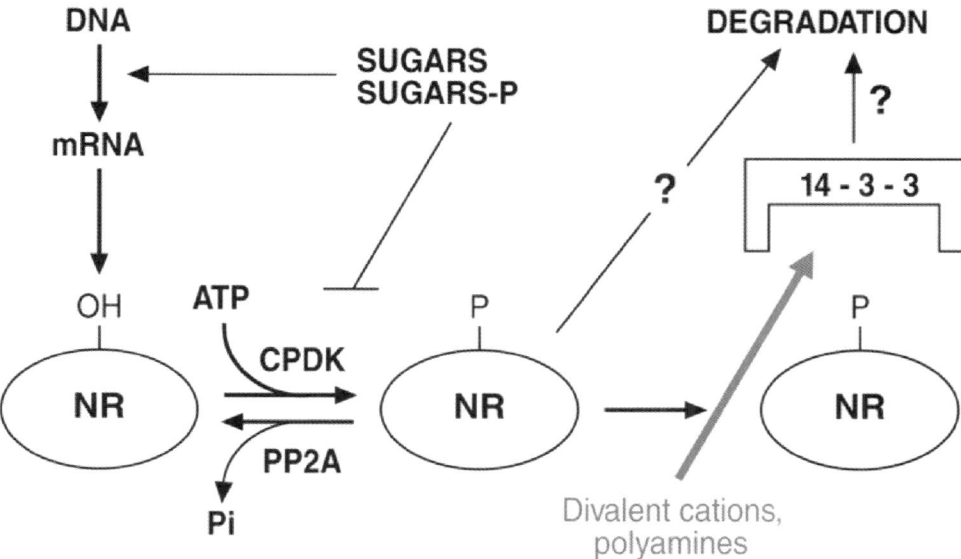

Figure 2. Scheme depicting the modulation of NR by reversible Ser phosphorylation and 14-3-3- binding. Only the 14-3-3-P-NR complex is completely inactive in the presence of divalent cations or polyamines. Also indicated are sugars and sugar-phosphates as suggested signals for control of NR transcription and NR phosphorylation, respectively. It is also shown that either the insertion of a negative charge by Ser phosphorylation, or the binding of 14-3-3 to P-NR might be the signal for NR proteolysis. For further explanations see the text.

ECOLOGICAL IMPLICATIONS - NR ACTIVATION STATE AND NITRATE REDUCTION RATES

Nitrate

NR transcription and synthesis are strongly affected by NO_3^- (chapter 6). Today, there is very little evidence that NO_3^- exerts any direct effect on the phosphorylation state of NR. Only in rare cases when plants had only a very low NR expression level, as *e.g.* under low NO_3^- supply, was the NR activation state increased in such a way that the low NR protein content was partly compensated by increased NR activation (Man *et al.* 1999, Scheible *et al.* 1997). However, even in those cases the signal for NR activation was most probably not nitrate itself, but rather the carbon metabolite levels, which are usually higher in N-deficient than in N-sufficient plants.

Besides effects on NR induction, NO_3^- in the cytosol might become a factor limiting nitrate reduction. The Km of NR for nitrate is low (40 µM), yet cytosolic NO_3^- concentrations in roots have been reported to be in the mM range, and some reports indicate that cytosolic NO_3^- concentrations are homeostatically controlled (Miller and Smith 1996, Van der Leij *et al.* 1998). This may indeed hold as long as vacuolar NO_3^- can be delivered to the cytosol. But basically it is obvious that cytosolic NO_3^- has to respond to NO_3^- supply. Otherwise it would be extremely difficult to understand how nitrate deficiency could decrease NR expression, and how nitrate supply could lead to a rapid increase in NR transcription and translation. In fact, compartmental analysis by tracer efflux (CATE) has contrasted electrode measurements, showing that the cytosolic NO_3^- pool can vary up to 80-fold, depending on plant N status (Britto and Kronzucker 2001, 2003).

Photosynthesis

The control by light of N-metabolism at the transcriptional and post-transcriptional level has been described in chapter 6. Rapid activation of NR in leaves as by light, and inactivation as in the dark is also achieved in continuous light by removing or adding CO_2 to the atmosphere around leaves (Kaiser and Förster 1989, Kaiser and Brendle-Behnisch 1991), and thus it is obvious that this modulation depends mostly on photosynthetic products rather than on light itself. As mentioned above, possible candidates for transducing the photosynthesis signal to NR are sugar phosphates. In the light, sugar phosphate concentrations increase and that may partially inhibit NR kinase, leading to an activation of NR (Figure 2).

However, the activation of NR by photosynthesis may be a transient, effect. Bloom and colleagues have shown that elevated atmospheric CO_2 over periods of hours or days decreased nitrate assimilation of leaves and led to lower NR activity in leaves and less shoot protein than with plants grown under ambient CO_2 (Bloom *et al.* 2002). One conclusion out of this work was that photorespiration would be required to maintain high nitrate assimilation rates (Bloom *et al.* 2002), and that this should be somehow related to the export of reducing power out of the chloroplast. Indeed it has been suggested that under conditions of high NR activation and sufficient nitrate supply, rates of nitrate reduction are strongly limited by NADH, even under anoxic conditions where the $NADH/NAD^+$ ratio should be high (Kaiser *et al.* 2002) and where nitrite accumulates.

Drought and salinity

As NR induction and NR activation are both affected by photosynthesis, it is obvious that nitrate assimilation will respond to any condition causing stomatal closure. Indeed, field studies have shown that plants accumulate NO_3^- under drought conditions and consume it after rain periods (Leclerc 1985), which may be caused at least in part by decreased rates of nitrate assimilation. It has long been known that NR activity in leaves also decreases under drought (Morilla *et al.* 1973, Heuer *et al.* 1979). In detached spinach leaves, water deficit prevented nitrate reduction in the light, and this inhibition was overcome by very high external CO_2 concentrations, indicating that it was probably due to stomatal closure (Kaiser and Förster 1989).

At least in non-halophytes, salinity also decreases stomatal aperture and may thereby cause similar effects as drought. In addition, salinity may interact with nitrate metabolism at various other stages. Nitrate uptake and transport within the plant appear especially sensitive to high chloride concentrations, and loading of NO_3^- into the root xylem has been defined as a very salt sensitive step (Speer *et al.* 1994, Peuke *et al.* 1996), leading to a strong decrease in shoot nitrate concentrations and an inverse chloride accumulation under salinity (Abd-El Baki *et al.* 2000). The consequences of decreased NO_3^- contents for NR induction have been outlined above. Surprisingly, in salt-stressed maize seedlings, NR activation state remained as high as in control plants, only the total NR activity and NR mRNA levels decreased, mainly during the night phase (Abd-El Baki *et al.* 2000). As a consequence of all these changes, the contribution of roots to the total nitrate reduction of the plants appeared higher than under non-saline conditions.

Water logging

While the consequences of "not enough water" for plant nitrate assimilation have been briefly discussed above, "too much water" is often as detrimental for plant growth, except for some specialists among higher plants. Water logging drastically decreases gas diffusion through soil pores and makes below ground organs (roots or shoots) hypoxic or anoxic. As a consequence of O_2 deficiency, respiration rates decrease and the energy state of cells is impaired, as indicated by low ATP levels. Cellular acidification occurs almost inevitably, where the cytosolic pH may drop from about 7.4 to 6.3, in an extreme case. Acidosis may have several reasons, like an ATP-limitation of proton pumps or a switch to a more acidifying metabolism, *e.g.* fermentative formation of lactic acid. For recent reviews on the anoxia problem consult Drew (1997) and De Sousa and Sodek (2002).

Early observations suggested that high nitrate supply has beneficial effects on waterlogged plants and helps them to survive hypoxic or anoxic phases (Drew 1997). Much effort has been invested to understand the physiological basis of this positive nitrate effect. In a recent approach, we have compared the response to anoxia of WT tobacco roots and roots of a transformant that has no NR in the roots, yet almost normal NR expression and activity in leaves. Under short term anoxia, roots of nitrate-fertilised WT plants (with NR in their roots) produced only low amounts of ethanol and lactate, whereas the NR-deficient transformant roots had much higher rates of fermentation, accumulated and excreted dramatically more ethanol and lactate, and showed a slightly stronger cytosolic acidification than WT roots (Figure 3, Stoimenova *et al.* 2003a,b). Since in WT roots fermentation rates were much higher than extractable NR activity, one conclusion was that NAD(P)H-consumption by nitrate reduction was actually too low to account for the decreased fermentation as compared to the transformant roots. Although the mechanisms behind that response are still obscure, it seems that simple substrate (NADH) competition between nitrate reduction and fermentation only insufficiently explain the phenomena.

In that context it appears interesting that in all plant organs examined so far, NR was rapidly activated under anoxia (Glaab and Kaiser 1993, Botrel *et al.* 1996, Botrel and Kaiser 1997). As pointed out above, anoxia causes cytosolic acidosis, and artificial acidification of cells also led to an activation of NR. Thus, it is tempting to suggest that the anoxic activation of NR is directly related to acidification. Whether that NR activation is due to dephosphorylation of NR or to pH effects on 14-3-3-binding has not yet been examined. However, when NR activity was determined in desalted crude extracts at different pH, activity at pH 6.5 was usually higher than at pH 7.5 (Kandlbinder *et al.* 2000), and that pH effect was reversible, indicating that the phosphorylation state had not changed (unpublished results).

Anoxic cells with activated NR accumulate NO_2^-. While this may be partly due to NR activation, NO_2^- accumulates mainly because nitrite reduction in non-green plastids (roots) is almost completely blocked under anoxia (Lee *et al.* 1979, Botrel and Kaiser 1997). The reason is not known with certainty. In non-green plastids, reductants are mainly produced by the oxidative pentose phosphate cycle (OPPC). As under anoxia hexose monophosphate (HMP) levels, just like ATP, are drastically decreased, this might be the decisive factor for the inhibition of the OPPC and of nitrite reduction.

According to what has been pointed out above on the regulation of NR-kinase(s) by HMP, one should actually expect that under anoxia, NR kinase is very active and therefore, NR should be largely inactivated. In reality, the opposite is true, as mentioned. We assume that 1) either the modulation of NR-kinase(s) by HMP is not as important as thought, or 2) PP2A's are strongly activated under these conditions, thus counterbalancing the high protein kinase activity. The question is still open.

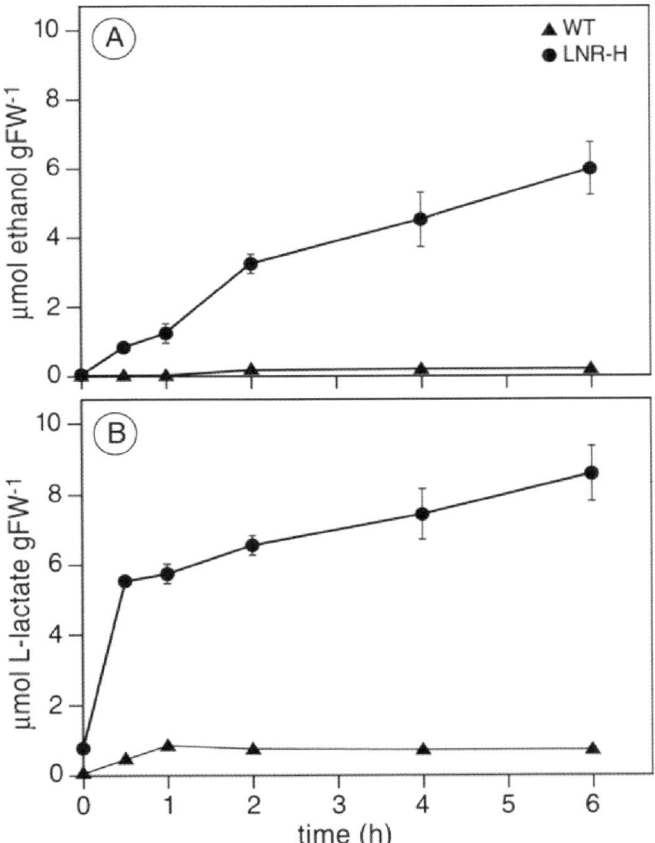

Figure 3. Ethanol and lactate production of submerged root segments from tobacco WT and from a transformant that has no NR in roots but almost normal NR activity in leaves (LNR-H). Anoxia was started at time 0 by flushing roots for 10 min with nitrogen. Subsequently the vessels were sealed with parafilm. Aerobic controls were continuously flushed with air. Loss of ethanol (15% after 2 h) by flushing with air was corrected. For experimental details compare Stoimenova *et al.* 2003a,b. Bars indicate SD (n= 6-8).

NITRATE REDUCTASE AND THE PRODUCTION AND EMISSION OF NITRIC OXIDE (NO)

Nitric oxide (NO) is a gaseous free radical that in vertebrates acts as a molecule with multiple biological functions. In animals it is mainly produced by various isoforms of nitric oxide synthases (NOS), which are partly constitutive and partly inducible. They all catalyse the five electron oxidation of the guanidine nitrogen of L-arginine to yield L-citrulline. The reaction requires a number of cofactors including NADPH, flavin and tetrahydrobiopterin. Whether tetrahydrobiopterin is synthesised in plants is not clear.

How is NO produced in plant shoots?

Until recently, evidence that NOS-type enzymes exits in plants was mostly based on activity measurements using ^{3}H-arginine, or immunological tests with commercial anti-NOS-antibodies, or on inhibitor studies *in vivo* using NOS-inhibitors which are mainly structural analogues of L-arginine. On the other hand, no mammalian type NOS sequence has been so far identified in *Arabidopsis*, and anti-NOS-antibodies have been shown to unspecifically cross-react with many plant enzymes (Butt *et al.* 2003). Most recently, however, a variant of the P-protein of the glycine decarboxylase complex (GDC) has been shown to have NOS activity and to be induced by plant pathogens (Chandok *et al.* 2003), and this finding may be a breakthrough in plant NO research.

Although the existence of a plant NOS is now well confirmed, plants - quite in contrast to animals - contain alternative enzymes catalysing NO production (Neill *et al.* 2003). Nitrate reductase is certainly the source for NO best studied. First reports on NO production and emission from plants date back almost 30 years now. They were obtained with soybean plants treated with photosynthetic inhibitor herbicides or other chemicals (Klepper 1979, 1990). It was suggested that this NO production was due to chemical reactions of accumulated nitrite with plant metabolites such as salicylate derivatives or to the chemical decomposition of HNO_2. NO was also observed to be the predominant compound evolved from accumulated nitrite during a purged *in vivo* assay with purified soybean NR (Klepper 1979). Experiments with ^{15}N-labelled nitrate showed that NO_x was produced form nitrate (Dean and Harper 1986). More recently it was shown that purified NR from maize also produced NO from nitrite plus NADH (Yamasaki *et al.* 1999, Yamasaki and Sakihama 2000), and unicellular green algae are also able to release NO when supported with nitrite (Mallick *et al.* 2000). Another potential source for NO production from nitrite is xanthine oxidase/dehydrogenase (XDH), which is also a MoCo-enzyme (Zangh *et al.* 1998, Godber *et al.* 2000). However, since XDH also depends on nitrite as a substrate, and since its NO production is limited to anoxic conditions, at least in above ground plant organs it probably plays no role for NO production under natural conditions.

And in roots?

Roots may receive nitrite not only from their endogenous NR, but also from soil microorganisms. Also, roots may become easily hypoxic or anoxic under water logging conditions, which favours nitrite production (Stoimenova et al. 2003a,b) and NO formation from nitrite (see above). So far, however, NO production by plant root systems has hardly been investigated. In preliminary experiments with tobacco we found that NO emission from aerated roots was below the detection limit of chemo-luminescence. However, under anoxia, NO emission rates from nitrate-fed roots were at least as high as from leaves (on a fresh weight basis). Interestingly, when feeding the roots with nitrite, we observed NO emission even when roots contained no detectable NR activity (not shown), confirming that in roots other reactions may participate in NO production. At least under anoxia, XDH may be one possible candidate for producing NO. However, in addition to XDH, roots appear to have yet another system able to produce NO. Stöhr and collaborators (summarised by Meyer and Stöhr 2003) showed that this NO production was catalysed by a newly described enzyme bound to the root PM, which was termed NI-NOR (nitrite:NO oxidoreductase) and which also depends on nitrite as a substrate. Whether an NOS-type enzyme exists also in roots is completely unknown.

NO emission from nitrate reductase

In leaves, NO emission from NR closely followed the NR activation state (Figure 1). NO emission was low in the dark (aerobic) and much higher in the light. Upon sudden light off, leaves often (but not always) produced a further transient increase in NO emission ("light-off peak", Figure 4). NO emission from leaves was at maximum under anoxia in the dark.
The factors that activated NR, like the 5'-AMP analogue AICAR, increased NO emission, whereas PP2A-inhibitors, which lead to a largely inactive NR prevented NO emission (Rockel et al. 2002). Thus, NO emission appeared to closely follow the NR activation state and the nitrite concentration in the tissue. As nitrite reduction was always blocked under anoxia, and NR was highly activated under these conditions, NO emission was always at maximum under anoxia because of the highly activated NR and the absence of nitrite reduction, which together led to massive nitrite accumulation in anoxic tissues (Rockel et al. 2002). Other reasons for the very high NO emission of plant tissues under anoxia might be the absence of reactive oxygen species (ROS), which can rapidly oxidise NO. Further evidence that NR (or nitrite) is a prerequisite for any NO formation in plants is based on the observation that plants grown on

ammonium (which do not express NR and NiR) and which, accordingly, do not accumulate nitrite, would not emit NO even under anoxia (Figure 4). The same holds for NR-deficient mutants (not shown). On the other side, a tobacco transformant having normal NR activities but expressing NiR in antisense orientation so that less than 5% of the normal nitrite reduction capacity remained, always accumulated nitrite and produced very high NO emission (Morot-Gaudry-Talarmain *et al.* 2002).

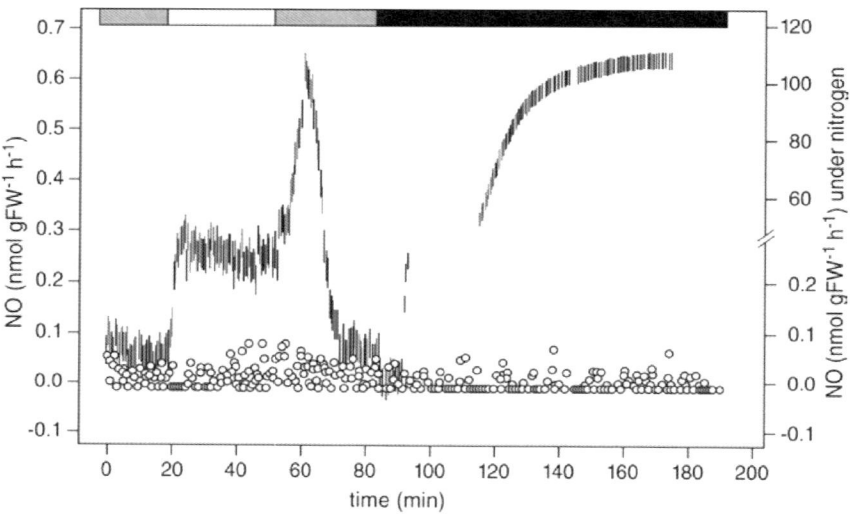

Figure 4. Typical pattern of NO-emission from detached tobacco leaves. The upper curve shows data from nitrate-fed plants. NO emission is low in the dark (grey or black bars on top of the figure), up to 10-fold higher in the light and increases transiently after light off. This is due to a transient overshoot of nitrate reductase leading to some nitrite accumulation, which stops after NR has been down regulated in the dark, requiring 5 -15 min. In contrast to NR, nitrite reduction stops immediately after light off. NO emission in the dark is drastically stimulated under anoxia, (black bar) because NR is activated, NiR does not work, leading to a strong accumulation of nitrite. Probably, the absence of reactive oxygen species under anoxia also contributes to the high NO-emission under dark-anoxic conditions (compare to Figure 5). Also shown (lower curve with open symbols) is the complete absence of NO-emission from leaves of ammonium-grown hydroponic tobacco plants, which contain no measurable NR-activity (W.M. Kaiser, unpublished data).

Reduction of nitrite to NO (Figure 1) is not the only side reaction catalysed by NR. Indeed, it has been shown that NR also catalyses the reduction of molecular oxygen to superoxide (Ruoff and Lillo 1990, Barber and Kay 1996, Yamasaki *et al.* 1999, Yamasaki and Sakihama 2000), but the capacity of that reaction appears as low as that for NO formation (Figure1). Still, NO emission from an aerated solution of NR is decreased upon addition of H_2O_2 (Figure 5) and is increased upon addition of catalase and SOD or under anoxia (not shown).

This indicates that upon simultaneous production of reactive oxygen species and of NO, both products react with each other, probably by forming the highly toxic peroxynitrite. This leads to a partial "quenching" of apparent NO emission from the solution. The observation is also important because in cells, many other reactions can produce reactive oxygen species and can thus interfere with the measurement of apparent NO production.

Although we are just beginning to explore the secrets of NO formation in plants, it seems obvious that they possess multiple NO-producing systems (Neill et al. 2003). With the exception of NOS, all of them depend finally on nitrite as a substrate, and therefore on NR or on exogenously produced nitrite from soil-borne microorganisms (see above).

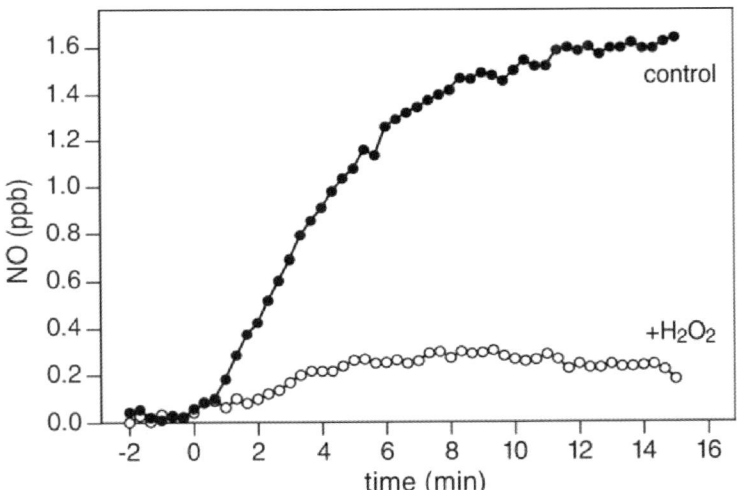

Figure 5. A stirred solution of purified NR (maize) emits NO into the gas phase after addition of NADH and nitrite at time zero. The apparent NO production is inhibited by addition of H_2O_2 (100 μM). It should be noted, that normal NR activity is not inhibited by H_2O_2 (not shown). (W.M. Kaiser, unpublished data).

Physiological roles of NO in plants

In spite of its relatively short half-life and high reactivity, NO appears to have multiple functions in plants, which are summarised in Table 2.

In animals, the physiological effects of NO result mainly from the activation of guanylate cyclase and the subsequent accumulation of cGMP. In plants, NO-treatment also lead to increased cGMP levels (Pfeiffer *et al.* 1994, Durner *et al.* 1998). NO appears involved in certain signalling cascades provoking the hypersensitive response (HR), which is a form of programmed cell death in plants serving to prevent growth and spread of the invading pathogen. Treatment of plants with NO-donors induced the expression of defence genes (Durner *et al.* 1998). In the HR, NO may have to be formed in parallel with reactive oxygen species (ROS) to fully express the symptoms (Delledonne *et al.* 1998). In all of the published literature on a role of NO as a second messenger in plant pathogen interactions, an NOS-like enzyme was considered as the source for NO, but clear evidence pro or contra a participation of NR or other NO-producing systems is still missing (see above).

In addition, NO appears as a signal affecting seed germination, deetiolation and hypocotyl elongation (Beligni and Lamattina 2001), cell growth and ethylene production (Leshem 2000). It binds to heme groups, *e.g.* of phytoglobines, and inhibits the iron sulphur centre of aconitase, preventing the conversion of citrate to isocitrate (Navarre *et al.* 2000). NO is also a potent inhibitor of the cytochrome oxidase pathway, but not of the alternative oxidase pathway in plant mitochondria (Millar and Day 1996, Millar *et al.* 2002), and may therefore affect energy metabolism.

NR-mediated NO formation appears also required for the abscisic acid-induced stomatal closure in *Arabidopsis, Pisum sativum* and *Vicia faba* (Desikan *et al.* 2002, Neill *et al.* 2002, Garcia-Mata and Lamattina 2003). Further suggestions concerning the role of NO in plants have been presented recently (Millar *et al.* 2002, Neill *et al.* 2003).

Table 2. Biological activities of nitric oxide in plants (adapted from Beligni and Lamattina 2001).

Seeds	Induction of germination
	Induction of respiration after imbibition
	Inhibition of aleuron cell death
Roots	Elongation
	Induction of lateral and adventitious roots
Tuber	Tuberisation
Hypocotyls	Inhibition of elongation under low light
Stems	Inhibition of internode elongation under low light
Leaves	Induction of deetiolation
	Delay of senescence
	Stomatal closure
	Leaf expansion
Plant-pathogen interactions	Induction of defence responses

OPEN QUESTIONS

Although NR is probably one of the best studied plant enzymes, details of its post-translational regulation are still unclear. We still do not know with certainty how the "auxiliary" enzymes which phosphorylate or dephosphorylate NR are regulated and whether this regulatory mechanism holds for all higher plants. So far, only one apparent exception has been described (Kandlbinder *et al.* 2000, Tsai *et al.* 2003). Details on the structure of the 14-3-3-P-NR complex and on the control of NR degradation are also unknown. Further, the mechanism of the one- electron transfer reactions from NADH to molecular oxygen or nitrite has not been studied in detail. With respect to the physiology and ecophysiology of nitrate reduction, it will be important to find out why and how nitrate reduction is beneficial for the survival of anoxia and how this is connected with the regulation of anaerobic metabolism. Last but not least, we still do not know how plants control NO production by several potentially interacting (at the product level) enzyme systems in order to channel NO into the right (signalling) pathways under various environmental conditions.

REFERENCES

Abd El-Baki, G.K., Siefritz F., Man H.M., Weiner H., Kaldenhoff R., Kaiser W.M. 2000. Nitrate reductase in *Zea mays* L. under salinity. - Plant Cell Environ. 23: 515-521.
Athwal G.S., Huber J.L., Huber S.C. 1998. Phosphorylated nitrate reductase and 14-3-3 proteins. Site of interaction, effects of ions, and evidence for an AMP-binding site on 14-3-3 proteins. - Plant Physiol. 118: 1041-1048.
Athwal G.S., Lombardo C.R., Huber J.L., Masters S.C., Fu H., Huber S.C. 2000. Modulation of 14-3-3-protein interaction with target peptides by physical and metabolic effectors. - Plant Cell Physiol. 41: 523-533.
Athwal G.S., Huber S.C. 2002. Divalent cations and polyamines bind to loop 8 of 14-3-3 proteins, modulating their interaction with phosphorylated nitrate reductase. - Plant J. 29: 119-129.
Bachmann M., Shiraishi N., Campbell W.H., Yoo B.C., Harmon A.C., Huber S.C. 1996a. Identification of Ser-543 as the major regulatory phosphorylation site in spinach leaf nitrate reductase. - Plant Cell 8: 505-517.
Bachmann M., Huber J.L., Liao P.C., Gage D.A., Huber S.C. 1996b. The inhibitor protein of phosphorylated nitrate reductase from spinach (*Spinacia oleracea*) leaves is a 14-3-3- protein. - FEBS Lett. 387: 127-131.
Bachmann M., Huber J.L., Athwal G.S., Wu K., Ferl R.J., Huber S.C. 1998. 14-3-3 proteins associate with the regulatory phosphorylation site of spinach leaf nitrate reductase in an isoform-specific manner and reduce dephosphorylation of Ser543 by endogenous protein phosphatases. - FEBS Lett. 398: 26-30.
Barber M.J., Kay C.J. 1996. Superoxide production during reduction of molecular oxygen by assimilatory nitrate reductase. - Arch. Biochem. Biophys. 326: 227-232.
Beligni M.V., Lamattina L. 1999. Nitric oxide protects against cellular damage produced by methylviologen herbicides in potato plants. - Nitric Oxide Biol. Chem. 3: 199-208.
Beligni M.V., Lamattina L. 2001. Nitric oxide: a non-traditional regulator of plant growth. - Trends Plant Sci. 11: 508-509.

Bloom A.J., Smart D.R., Nguyen D.T., Searles P.S. 2002. Nitrogen assimilation and growth of wheat under elevated carbon dioxide. - Proc. Natl. Acad. Sci. USA 99: 1730-1735.

Botrel A., Magne C., Kaiser W.M. 1996. Nitrate reduction, nitrite reduction and ammonium assimilation in barley roots in response to anoxia. - Plant Physiol. Biochem. 34: 645-652.

Botrel A., Kaiser W.M. 1997. Nitrate reductase activation state in roots in relation to the energy and carbohydrate status. - Planta 201: 496-501.

Britto D.T., Kronzucker H.J. 2001. Constancy of nitrogen turnover kinetics in the plant cell: insights into the integration of subcellular N-fluxes. - Planta 21: 175-181.

Britto D.T., Kronzucker H.J. 2003. The case for cytosolic NO_3^- heterostasis: a critique of a recently proposed model. - Plant Cell Environ. 26: 183-188.

Butt Y.K.C., Lum J.H.K., Lo S.C.L. 2003. Proteomic identification of plant proteins probed by mammalian nitric oxide synthase antibodies. - Planta 216: 762-771.

Campbell W.H. 2003. Molecular control of nitrate reductase and other enzymes involved in nitrate assimilation. - In: Foyer C.H., Noctor G. (Eds.) Photosynthetic nitrogen assimilation and associated carbon and respiratory metabolism. - Dordrecht, The Netherlands, Kluwer Academic Publishers, pp. 35-48.

Chandok M.R., Ytterberg A.J., van Wijk K.J., Klessig D.F. 2003. The pathogen-inducible nitric oxide synthase (iNOS) in plants is a variant of the P-protein of the glycine decarboxylase complex. - Cell 113: 469-482.

Dean J.V., Harper J.E. 1986. Nitric oxide and nitrous oxide production by soybean and winged bean during *in vivo* nitrate reductase assay. - Plant Physiol. 82: 718-723.

Delledonne M., Xia Y., Dixon R.A., Lamb C. 1998. Nitric oxide functions as a signal in plant disease resistance. - Nature 394: 585-588.

Deruere J., Jackson K., Garbers C., Söll D., DeLong A. 1999. The RCN1-encoded A subunit of protein phosphatase 2A increases phosphatase activity in vivo. - Plant J. 20: 389-399.

De Sousa C.A.F., Sodek L. 2002. Metabolic changes in soybean plants in response to waterlogging in the presence of nitrate. - Physiol. Mol. Biol. Plants 8: 97-104.

Desikan R., Griffiths R., Hancock J., Neill S. 2002. A new role for an old enzyme: Nitrate reductase mediated nitric oxide generation is required for abscisic acid-induced stomatal closure in *Arabidopsis thaliana*. - Proc. Natl. Acad. Sci. USA 99: 16314-16318.

Douglas P., Morris N., MacKintosh C. 1995. Identification of a regulatory phosphorylation site in the hinge 1 region of nitrate reductase from spinach (*Spinacia oleracea*) leaves. - FEBS Lett. 377: 113-117.

Douglas P., Moorhead G., Hong Y., Morrice N., Mackintosh C. 1998. Purification of a nitrate reductase kinase from *Spinacia oleracea* leaves, and its identification as a calmodulin-domain protein kinase. - Planta 206: 435-442.

Drew M.C. 1997. Oxygen deficiency and root metabolism: injury and acclimation under hypoxia and anoxia. - Annu. Rev. Plant Physiol. Plant Mol. Biol. 48: 223-250.

Durner J., Wendehenne D., Klessig D.F. 1998. Defense gene induction in tobacco by nitric oxide, cyclic GMP, and cyclic ADP-ribose. - Proc. Natl. Acad. Sci. USA 95: 10238-10333.

Durner J., Klessig D.F. 1999. Nitric oxide as a signal in plants. - Curr. Opin. Plant Biol. 2: 369-374.

Garcia-Mata C., Lamattina L. 2003. Abscisic acid, nitric oxide and stomatal closure - is nitrate reductase one of the missing links? - Trends Plant Sci. 8: 20-26.

Glaab J., Kaiser W.M. 1993. Rapid modulation of nitrate reductase in pea roots.- Planta 191: 173-179.

Glaab J., Kaiser W.M. 1995. Inactivation of nitrate reductase involves NR-protein phosphorylation and subsequent binding of an inhibitor protein. - Planta 195: 514-518.

Godber B., Doel J., Sapkota G., Blake D., Stevens C., Eisenthal R., Harrison R. 2000. Reduction of nitrite to nitric oxide catalyzed by xanthine oxidoreductase. - J. Biol. Chem. 275: 7757-7763.

Heuer B., Plaut Z., Federmann E. 1979. Nitrate and nitrite reduction in wheat leaves as affected by different types of water stress. - Physiol. Plant. 46: 318-323.

Huber J.L., Huber S.C., Campbell W.H., Redinbaugh M.G. 1992. Reversible light/dark modulation of spinach leaf nitrate reductase involves protein phosphorylation. - Arch. Biochem. Biophys. 296: 58-65.

Huber S.C., Kaiser W.M. 1996. 5-Aminoimidazol-4-carboxamide riboside activates nitrate reductase in darkened spinach and pea leaves. - Physiol. Plant. 98: 833-837.

Kaiser W.M., Förster J. 1989. Low CO_2 prevents nitrate reduction in leaves. - Plant Physiol. 91: 970-974.

Kaiser W.M., Brendle-Behnisch E. 1991. Rapid modulation of spinach leaf nitrate reductase activity by photosynthesis. I. Modulation *in vivo* by CO_2 availability. - Plant Physiol. 96: 363-367.

Kaiser W.M., Spill D. 1991. Rapid modulation of spinach leaf nitrate reductase by photosynthesis. II. In vitro modulation by ATP and AMP. - Plant Physiol. 96: 368-375.

Kaiser W.M., Huber S.C. 1994. Modulation of nitrate reductase in vivo and in vitro: Effects of phosphoprotein phosphatase inhibitors, free Mg^{2+} and 5'-AMP. - Planta 193: 358-364.

Kaiser W.M., Brendle-Behnisch E. 1995. Acid-base modulation of nitrate reductase in leaf tissues. - Planta 196: 1-6.

Kaiser W.M., Huber S.C. 1997. Correlation between apparent activation (phosphorylation) status of nitrate reductase (NR), NR hysteresis and degradation of NR protein.- J. Exp. Bot. 312: 1367-1374.

Kaiser W.M., Stoimenova M., Man H.M. 2002. What limits nitrate reduction in leaves? - In: Foyer C.H., Noctor G. (Eds.) Photosynthetic nitrogen assimilation and associated carbon and respiratory metabolism. - Dordrecht, The Netherlands, Kluwer Academic Publishers, pp. 63-70.

Kanamaru K., Wang R., Su W., Crawford N.M. 1999. Ser-543 in the hinge 1 region of Arabidopsis nitrate reductase is conditionally required for binding of 14-3-3 proteins and *in vitro* inhibition. - J. Biol. Chem. 274: 4160-4165.

Kandlbinder A., Weiner H., Kaiser W.M. 2000. Nitrate reductase from leaves of Ricinus (*Ricinus communis* L) and spinach (*Spinacia oleracea* L.) have different regulatory properties. - J. Exp. Bot. 51: 1099-1105.

Klepper L.A. 1979. Nitric oxide (NO) and nitrogen dioxide emissions from herbicide-treated soybean plants. - Atmos. Environ. 13: 537-542.

Klepper L.A. 1990. Comparison between NO_x evolution mechanisms of wild-type and nr1 mutant soybean leaves.- Plant Physiol. 93: 26-32.

Leclerc M.C. 1985. Variations saisonnières de la disponibilité du nitrate et de son utilisation par la végétation des dunes littorales de la Mediterranée. - Acta Oecol. 6: 87-106.

Lee R.B. 1979. The release of nitrite from barley roots in response to metabolic inhibitors, uncoupling agents and anoxia. - J. Exp. Bot. 30: 119-133.

Leshem Y.Y., Wills R.B.H., Ku V.V.V. 1998. Evidence for the function of the free radical gas - nitric oxide (NO) - as an endogenous maturation and senescence regulating factor in higher plants. - Plant Physiol. Biochem. 36: 825-833.

Leshem Y.Y. 2000. Nitric oxide in plants. Occurrence, function and use - Dordrecht, The Netherlands, Kluwer Academic Publishers.

Lillo C. 1994. Light regulation of nitrate reductase in higher plants. - Physiol. Plant. 62: 89-94.

Lillo C., Appenroth K.J. 2001. Light regulation of nitrate reductase in higher plants: Which photoreceptors are involved? - Plant Biol. 3: 455-465.

MacKintosh C. 1992. Regulation of spinach leaf nitrate reductase by reversible phosphorylation. - Biochim. Biophys. Acta 1137: 121-126.

MacKintosh C. 1998. Regulation of cytosolic enzymes in primary metabolism by reversible protein phosphorylation. - Curr. Opin. Plant Biol. 1: 224-229.

Mallick M., Mohn F.H., Soeder C.J. 2000. Evidence supporting nitrite-dependent NO release by the green microalga *Scenedesmus obliquus*. - J. Plant Physiol. 157: 40-46.

Man H.M., Abd El-Baki. G., Stegmann P., Weiner H., Kaiser W.M. 1999. The activation state of nitrate reductase is not always correlated with total nitrate reductase activity in leaves. - Planta 209: 462-468.

McMichael R.W., Bachmann M., Huber S.C. 1995. Spinach leaf sucrose phosphate synthase and nitrate reductase are phosphorylated/inactivated by multiple protein kinases in vitro. - Plant Physiol. 108: 1077-1082.

Meyer C., Stöhr C. 2003. Soluble and plasma membrane-bound enzymes involved in nitrate and nitrite metabolism. - In: Foyer C.H., Noctor G. (Eds.) Photosynthetic nitrogen assimilation and associated carbon and respiratory metabolism. - Dordrecht, The Netherlands, Kluwer Academic Publishers, pp. 49-62.

Millar A.H., Day D.A. 1996. Nitric oxide inhibits the cytochrome oxidase but not the alternative oxidase of plant mitochondria. - FEBS Lett. 398: 155-158.

Millar A.H., Day D.A., Mathieu C. 2002. Nitric oxide synthesis by plants and its potential impact on nitrogen and respiratory metabolism. - In: Foyer C.H., Noctor G. (Eds.) Photosynthetic nitrogen assimilation and associated carbon and respiratory metabolism. - Dordrecht, The Netherlands, Kluwer Academic Publishers, pp. 193-204.

Miller A.J., Smith S.J. 1996. Nitrate transport and compartmentation in cereal root cells. - J. Exp. Bot. 47: 843-854.

Moorhead G., Douglas P., Morrice N., Scarabel M., Aitken A., MacKintosh C. 1996. Phosphorylated nitrate reductase from spinach leaves is inhibited by 14-3-3 proteins and activated by fusiccocin. - Curr. Biol. 6: 1104-1113.

Morilla C.A., Boyer J.S., Hagemann R.H. 1973. Nitrate reductase activity and polyribosomal content of corn (*Zea mays* L.) having low leaf water potentials. - Plant Physiol. 51: 817-824.

Morot-Gaudry-Talarmain Y., Rockel P., Moureaux T., Quillere I., Leydecker M.T., Kaiser W.M., Morot-Gaudry J.F. 2002. Nitrite accumulation and nitric oxide emission in relation to cellular in nitrite reductase antisense tobacco. - Planta 215: 708-715.

Navarre D.A., Wendehenne D., Durner J., Noad R., Klessig D.F. 2000. Nitric oxide modulates the activity of tobacco aconitase. - Plant Physiol. 122: 573-582

Neill S.J., Desikan R., Clarke A., Hurst R.D., Hancock J.T. 2002. Hydrogen peroxide and nitric oxide as signalling molecules in plants. - J. Exp. Bot. 53: 1237-1247

Neill S.J., Desikan R., Hancock J.T. 2003. Nitric oxide signalling in plants. - New Phytol. 159: 11-35.

Petosa C., Masters S.C., Bankston L.A., Pohl J., Wang B., Fu H., Liddington R.C. 1998. 14-3-3-zeta binds to a phosphorylated raf peptide and an unphosphorylated peptide via its conserved amphipathic groove. - J. Biol. Chem. 273: 16035-16310.

Peuke A.D., Glaab J., Kaiser W.M., Jeschke W.D. 1996. The uptake and flow of C, N and ions between roots and shoots in *Ricinus communis* L. IV. Flow and metabolism of inorganic nitrogen and malate depending on nitrogen nutrition and salt stress. - J. Exp. Bot. 47: 377-385.

Pfeiffer S., Janistin B., Jessner G., Pichorner H., Ebermann R. 1994. Gaseous nitric oxide stimulates guanosine-3,5'-cyclic monophosphate (cGMP) formation in spruce needles. - Phytochemistry 36: 259-262.

Provan F., Askland L.M., Meyer C., Lillo C. 2000. Deletion of the nitrate reductase N-terminal domain still allows binding of 14-3-3 proteins but affects their inhibitory properties. - Plant Physiol. 123: 757-764.

Rockel P., Strube F., Rockel A., Wildt J., Kaiser W.M. 2002. Regulation of nitric oxide production by plant nitrate reductase in vivo and in vitro. - J. Exp. Bot. 53: 1-8.

Rockel P., Kaiser W.M. 2002. NO production in plants: Nitrate reductase versus nitric oxide synthase. - In: Esser K., Lüttge U., Beyschlag W., Hellwig F. (Eds.) Progress in Botany 63, - Berlin, Germany, Springer Verlag, pp. 246-257.

Ruoff P., Lillo C. (1990) Molecular oxygen as electron acceptor in the NADH-nitrate reductase system. - Biochem. Biophys. Res. Comm. 172: 1000-1005.

Scheible W.R., Gonzales-Fontes A., Morcuende R., Lauerer M, Geiger M., Glaab J., Gojon A., Schulze E.D., Caboche M., Stitt M. 1997.Tobacco mutants with a decreased number of functional nia-genes compensate by modifying the diurnal regulation of transcription, post-translational modification and turnover of nitrate reductase. - Planta 203: 304-319.

Shenolikar S. 1994. Protein serine/threonine phosphatases- new avenues for cell regulation. - Annu. Rev. Cell Biol. 10: 55-86.

Speer M., Brune A., Kaiser W.M. 1994. Replacement of nitrate by ammonium as N-source increases salt sensitivity of pea plants. I. Ion concentrations in roots and in leaves. - Plant Cell Environ. 17: 1215-1221.

Spill D., Kaiser W.M. 1994. Partial purification of two proteins (100 kDa and 67 kDa) participating in the ATP-dependent inactivation of spinach leaf nitrate reductase. - Planta 192: 183-188.

Stoimenova M., Hänsch R., Mendel R., Gimmler H., Kaiser W.M. 2003a. The role of nitrate reduction in the anoxic metabolism of roots I. Characterization of root morphology and normoxic metabolism of wild type tobacco and a transformant lacking root nitrate reductase. - Plant Soil 253: 145-153.

Stoimenova M., Libourel I.G.L., Ratcliffe R.G., Kaiser W.M. 2003b. The role of nitrate reduction for the anoxic metabolism of roots II. Anoxic metabolism of tobacco roots with or without nitrate reductase activity. - Plant Soil 253: 155-167.

Su W., Huber S.C., Crawford N.M. 1996. Identification *in vitro* of a post-translational regulatory site I the hinge-1 region of *Arabidopsis* nitrate reductase. - Plant Cell 8: 519-527.

Sudgen C., Donaghy P.G., Halford N.G., Hardie D.G. 1999. Two SNF1-related protein kinases from spinach leaf phosphorylate and inactivate 3-hydroxy-3-methylglutaryl-coenzyme A reductase, nitrate reductase, and sucrose phosphate synthase in vitro. - Plant Physiol. 120: 257-274.

Toroser D., Plaut Z., Huber S.C. 2000. Regulation of a plant SNF1-related protein kinase by glucose-6-phosphate. - Plant Physiol. 123: 403-412.

Tsai C.B., Kaiser W.M., Kaldenhoff, R. 2003. Molecular cloning and characterization of nitrate reductase from *Ricinus communis* L. heterologously expressed in *Pichia pastoris*. - Planta 217: 962-970.

Van der Leij M., Smith S.J., Miller A.J., 1998. Remobilisation of vacuolar stored nitrate in barley root cells. - Planta 205: 64-72.

Vincentz M., Moureaux T., Leydecker M.T., Vaucheret H., Caboche M. 1993. Regulation of nitrate and nitrite reductase expression in *Nicotiana plumbaginifolia* leaves by nitrogen and carbon metabolites. - Plant J. 3: 315-324.

Weiner H., Stitt M. 1993. Sucrose-phosphate synthase phosphatase, a type 2A protein phosphatase, changes its sensitivity toward inhibition by inorganic phosphate in spinach leaves. - FEBS Lett. 333: 159-164.

Weiner H., Kaiser W.M. 1999. 14-3-3 proteins control proteolysis of nitrate reductase in spinach leaves. - FEBS Lett. 455: 75-78.

Weiner H., Kaiser W.M. 2001. Antibodies to assess phosphorylation of spinach leaf nitrate reductase on serine 543 and its binding to 14-3-3- proteins. - J. Exp. Bot. 52: 1165-1172.

Yamasaki H., Sakihama Y., Takahashi S. 1999. An alternative pathway for nitric oxide production in plants: new features for an old enzyme. - Trends Plant Sci. 4: 128-129.

Yamasaki H., Sakihama Y. 2000. Simultaneous production of nitric oxide and peroxynitrite by plant nitrate reductase: in vitro evidence for the NR-dependent formation of active nitrogen species. - FEBS Lett. 486: 89-92.

Zangh Z., Naughton D., Winyard P.G., Benjamin N., Blake D.R., Symons M.C.R. 1998. Generation of nitric oxide by a nitrate reductase activity of xanthine oxidase: a potential pathway for nitric oxide synthase activity. - Biochem. Biophys. Res. Comm. 249: 767-772.

Chapter 8

INTEGRATED MOLECULAR ANALYSIS OF THE POLYAMINE METABOLIC PATHWAY IN ABIOTIC STRESS SIGNALLING

Alejandro Ferrando, Pedro Carrasco, Juan Cruz Cuevas, Teresa Altabella and Antonio F. Tiburcio

INTRODUCTION

Land plants experience constant fluctuations in the availability of water, thus they have evolved adaptive features to mine and absorb water through the root system, to prevent excessive transpiration water loss using cuticles and stomata on the shoot, and to adjust physiology and metabolism for continued growth and survival in the case of osmotic stress (Levitt 1972). Osmotic stress is broadly used to refer to situations where insufficient water availability limits plant growth and development; it can result from drought or from excessive salt in water. Chilling and freezing may also lead to osmotic stress due to reduced water absorption and cellular dehydration induced by ice formation (Zhu *et al.* 1997).

Global effects of desertification, soil salinisation, atmospheric CO_2 enrichment and effects of other pollutants are predicted to cause dramatic changes in the climatic conditions of arable lands in this century. These abiotic stresses together represent the primary cause of crop loss worldwide, reducing average yields for major crop plants by more than 50% (Bray *et al.* 2000). Breeding of crops and trees with enhanced tolerance to osmotic and other abiotic stresses is therefore one of the major goals of current developments in agronomy, forestry and environmental protection. Classical breeding approaches are time consuming and often yield unpredictable results. Therefore, it is a widely accepted consensus that a considerable improvement of stress tolerance traits requires the identification and modification of regulatory genes that play a key role in the control of plant stress responses (Zhu 2002).

Numerous signals and signal-like molecules have been identified and presumed, through some evidence, to function in plants as mediators of osmotic adaptation. These include several metabolites, proteins, or components of biochemical pathways that mediate ion homeostasis, osmolyte biosynthesis, toxic radical scavenging, water transport, and transducers of long-distance response coordination (Hasegawa *et al.* 2000).

With the complete *Arabidopsis* genome fully sequenced (The *Arabidopsis* Genome Initiative 2000), studies of individual genes and mechanisms are now rapidly being replaced by global analyses. Technological advances, summarised under the "genomics" label, make it possible to monitor the expression of many or even all genes in an organism simultaneously during the entire life cycle of the organism or in response to different stimuli. Recent transcriptome analysis in response to salt, drought and cold stress estimates that about 30% of the transcriptome (more than 7,000 genes) is potentially regulated by osmotic stress in *Arabidopsis* (Kreps *et al.* 2002). Because most aspects of plant physiology are impacted by stress, the large number of stress-regulated transcriptome changes underscores the difficulty of understanding the global context of a stress response (Kreps *et al.* 2002). Furthermore, although some of these individual stress-induced genes have been shown to affect stress tolerance in plants, their relationships to each other remains mostly unclear, and therefore we still know little of how these genes function in a stress signalling pathway (Hasegawa *et al.* 2000).

Although genomic technologies for determining global expression profile (*i.e.* gene chip microarrays) and proteomic developments such as the two-hybrid system enabling construction of protein interaction maps have proven invaluable, these large-scale data should be integrated into biological models in order to predict cellular behaviours that can be tested experimentally. This integrated genomic and proteomic approach has recently been implemented in yeast to explore the galactose metabolic pathway (Ideker *et al.* 2001).

The polyamines (PAs), spermidine (Spd) and spermine (Spm) and their diamine obligate precursor putrescine (Put), are small aliphatic amines that are ubiquitous in plant cells. The PA metabolic pathway is directly regulated by only half a dozen key enzymes and interacts with other metabolic pathways that are critical to plant development. Evidence gathered in recent years supports the involvement of PAs in many key physiological plant processes including cell division, morphogenic responses, senescence, secondary metabolism and stress responses. For reviews see Galston *et al.* (1997), Kumar *et al.* (1997), Malmberg *et al.* (1998) and Kaur-Sawhney *et al.* (2003). Thus the PA metabolic pathway has features that make it an excellent model for delineating interactions among networks of key pathways in plants. In this chapter we will 1) introduce the PA metabolic pathway and its relationship to N metabolism, 2) summarise our present knowledge regarding implications of the PA metabolic pathway in response to environmental stresses, 3) present an outlook of advanced genomic and proteomic approaches, and 4) discuss the necessity to undertake integrated molecular analyses in order to elucidate the function of the PA metabolic pathway in abiotic stress signalling.

THE POLYAMINE METABOLIC PATHWAY AND ITS RELATIONSHIP TO NITROGEN METABOLISM

The amount of free PAs present in plant tissue is the net result of its biosynthesis, conjugation with phenolic acids, binding to macromolecules, transport to other tissues, and its degradation during a given period of time. The diamine precursor Put is synthesised directly from Orn by Orn decarboxylase (ODC, EC 4.1.1.17) and indirectly from Arg by Arg decarboxylase (ADC, EC 4.1.1.9) and two other enzymes (Figure 1). The existence of two alternative pathways (ODC and ADC) for Put production in many plants complicates the situation regarding their metabolic regulation since the substrates (Orn and Arg) can also be inter-converted (Figure 1). The triamine Spd is formed from Put by the addition of an aminopropyl group derived from decarboxylated S-adenosyl-Met (SAM) in a reaction catalysed by Spd synthase (EC 2.5.1.16). The addition of another aminopropyl group to Spd gives rise to Spm by the action of Spm synthase (EC 2.5.1.16). Decarboxylated SAM is produced from SAM via SAM decarboxylase (SAMDC, EC 4.1.1.50). All three decarboxylases (ADC, ODC and SAMDC) have short half-lives (Cohen 1998), indicating that they are important metabolic control points in the cell. An overall up-regulation of PA biosynthesis often involves increases of both mRNA levels and respective enzyme activities (Tiburcio *et al.* 1997). In general, the levels of Spd are more tightly regulated than those of Put, with Spm being always a minor component of the PA pool (Bhatnagar *et al.* 2002).

A unique feature in plants is that SAM sits astride an important metabolic cross-road in the regulation of N metabolism (Figure 1). Thus SAM, in addition to participating in numerous transmethylation reactions, is also a precursor of ethylene via 1-aminopropyl-1-carboxylic acid (ACC) (Tiburcio *et al.* 1997). Furthermore, in some plants, the methyl moiety of SAM can be transferred to Put to form *N*-methyl-Put (Figure 1), which serves as a precursor of nicotine and other related alkaloids (Tiburcio *et al.* 1997).

In plants, PAs often occur as free molecular bases, but they can also be associated with small molecules like phenolic compounds (conjugated forms) and also to various macromolecules like proteins (bound forms) (Figure 1). The most common conjugated PAs are those which are covalently linked to cinnamic acids (reviewed by Martin-Tanguy 1987, 2001). Put hydroxycinnamoyl transferase involved in the conjugation of Put to phenolic compounds has been reported (Bagni and Tassoni 2001). Post-translation covalent linkage of PAs to proteins is catalysed by a class of enzymes known as transglutaminases (EC, 2.3.2.13), which have been localised both intra- and extra-cellularly (Serafinni-

Fracassini *et al.* (1995). Plant cells take up PAs by a rapid and active mechanism that shows biphasic kinetics and reaches saturation within a few minutes (Bagni and Torrigiani 1992). In whole plants, PAs can be translocated via the xylem (Bagni and Pistocchi 1991) and are also present in the phloem (Friedman *et al.* 1986). PAs are catabolised through the activity of one or more diamine oxidases (DAOs, E.C.1.4.3.6) and polyamine oxidases (PAOs, E.C. 1.5.3.3) (Smith 1985, Cohen 1998, Bagni and Tassoni 2001). DAO is believed to be loosely bound to the cell wall and can be released in the apoplast (Angelini *et al.* 1996, Møller and McPherson 1998). The best substrates for DAO are Put and cadaverine (Cad); however, it can also act on aromatic and aliphatic monoamines (Smith 1985). As a result of DAO action, Put is converted into Δ^1-pyrroline with the release of ammonia and hydrogen peroxide (Figure 1).

Pyrroline dehydrogenase then converts Δ^1-pyrroline to γ-amino-butyric acid (GABA), which enters the tricarboxylic acid (TCA) cycle via succinate after transamination and oxidation. This catabolic pathway recycles C and N from Put. An alternate route for GABA production in plants, especially under abiotic stress conditions, is directly from Glu through the action of cytosolic Glu decarboxylase (Shelp *et al.* 1999) (Figure 1).

Degradation of Spd by PAO yields Δ^1-pyrroline and 1,3 diaminopropane (Dap), whereas breakdown of Spm yields 1,3-aminopropylpyrroline (which subsequently gets converted to 1,5-diazabicyclononane), along with Dap and hydrogen peroxide (Smith 1985, Bouchereau *et al.* 1999, Bagni and Tassoni 2001). Dap is eventually converted into β-Ala (Figure 1). Thus, PA catabolism is not simply a degradation process but is also an important link between amino acid and C metabolism in plants.

INVOLVEMENT OF THE POLYAMINE METABOLIC PATHWAY ON PLANT STRESS RESPONSES

In higher plants, PA metabolism is responsive to external conditions. PAs have been involved in the signalling pathway of plant pathogen interactions (Martin-Tanguy 1987) and in the response to microbial symbionts important for plant nutrition (El Ghachtouli *et al.* 1996). However, the major changes in N and PA metabolism in different plants occur in response to various abiotic stress conditions like nutrients deficiency, osmotic stress, atmospheric pollutants or contamination by heavy metals (Flores 1991, Boucherau *et al.* 1999, Marco and Carrasco 2002, Groppa *et al.* 2003).

In animal cells, the increase in PA levels as a consequence of stress has been related to the dual ability of these compounds to interact with anions or cations. Binding of PAs to anions such as membrane phospholipids or nucleic acids

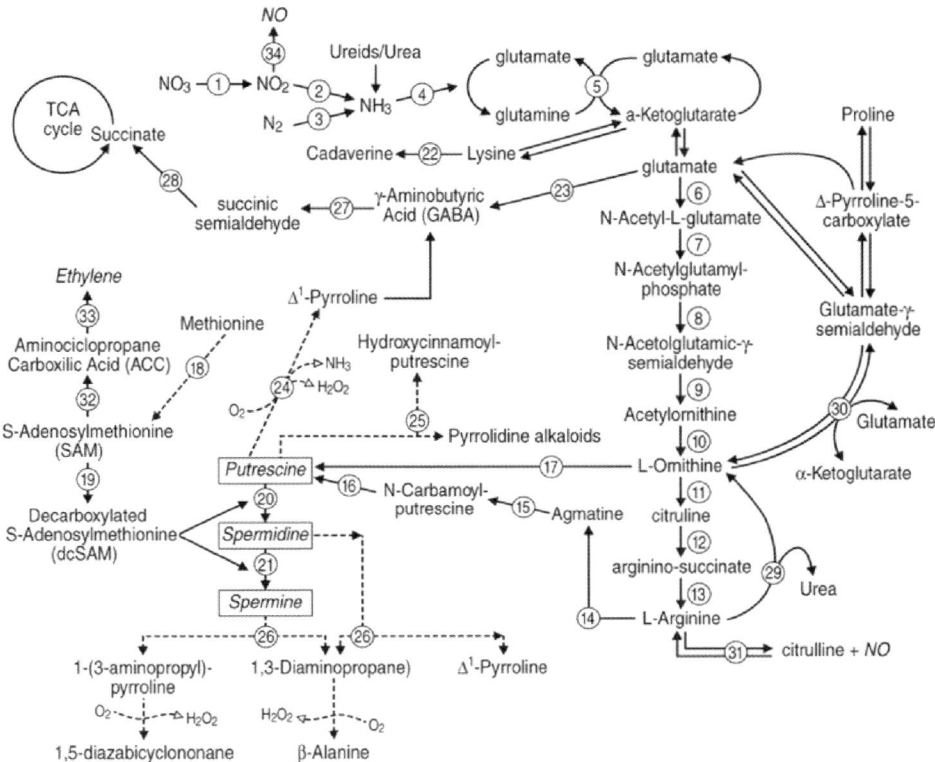

Figure 1. Polyamine metabolism and related pathways. The identified pathways and enzymes for biosynthesis, conjugation and degradation of the three main polyamines (Putrescine, Spermidine and Spermine) are shown. Catabolic and conjugation pathways are indicated by dashed lines while continuous lines show biosynthetic and related pathways. Numbers refer to the following enzymes: 1, nitrate reductase; 2, nitrite reductase; 3, nitrogenase; 4, glutamine synthetase; 5, glutamate synthase (GOGAT); 6, glutamate N-acetyltransferase; 7, acetylglutamate kinase; 8, N-acetyl-γ-glutamyl-phosphate reductase; 9, acetylornithine transaminase; 10, acetylornithine deacetylase; 11, ornithinecarbamoyl transferase; 12, argininosuccinate synthase; 13, argininosuccinate lyase; 14, arginine decarboxylase; 15, agmatine deiminase; 16, N-carbamoylputrescine amidohydrolase; 17, ornithine decarboxylase; 18, SAM synthetase; 19, SAM decarboxylase; 20, spermidine synthase; 21, spermine synthase; 22, lysine decarboxylase; 23, glutamate decarboxylase; 24, copper aminooxidase-diamine oxidase (DAO); 25, putrescine hydroxycinnamoyl transferase; 26, polyamine oxidase (PAO); 27, γ-aminobutyrate aminotransferase; 28, succinic semialdehyde dehydrogenase; 29, arginase; 30, ornithine aminotransferase; 31, nitric oxide synthase (NOS); 32, ACC synthase; 33, ACC oxidase; 34, nitrate reductase.

would protect those molecules from possible oxidative attacks. On the other hand, PA binding to cations such as Co^{2+}, Cu^{2+}, Fe^{2+} or Ni^{2+} would avoid formation of free radicals (Lovaas 1997). However, the knowledge about the reasons and physiological significance of the increase in PA levels by abiotic stresses in plants is scarce. Although differences in PA response under different abiotic stresses have been reported among and within species, in general, plants subjected to different types of stresses show a rapid and massive increase of Put levels. This response is mainly mediated by the ADC pathway, as shown by rapid labelling from Arg via agmatine and N-carbamoyl-Put, that correlates with increases of ADC activity but not of ODC (Figure 1), and inhibition of the response by application of the inhibitor DL-α-difluoromethylarginine (DFMA) (Flores and Galston 1984).

K^+ deficiency and osmotic stress

The accumulation of Put in leaves of K^+ deficient barley plants was first reported by Richards and Coleman (1952), and subsequent studies have established that this effect is common to different plant tissues (Flores 1991). In all cases, maximum accumulation of Put coincides with the appearance of several nutrient deficiency symptoms.

Osmotic treatments using sorbitol also induce high levels of Put and ADC activity in detached oat leaves (Flores and Galston 1984). However, Spd and Spm levels show a dramatic decrease. Similar results have been obtained in plants exposed to conventional water stress or to different osmotic agents (mannitol, proline, betaine, sucrose) (Flores and Galston 1984). These changes in PA levels are coincident with signs of stress, like wilting and protein loss, that can be retarded by treatment with Spm or DFMA-pre-treatment (Besford *et al.* 1993). Borrell *et al.* (1996) provided evidences for post-transductional regulation of ADC activity by Spm. Increases in Put levels and ADC activity have also been detected in *Arabidopsis* leaves exposed to osmotic stress, as a consequence of the transcriptional activation of the *ADC2* gene (Feirer *et al.* 1998, Soyka and Heyer 1999). On the other hand, when cell cultures of drought-tolerant alfalfa lines were exposed to water-deficit there was a trend towards increased accumulation of Spd and Spm and a loss of Put (Kuehn *et al.* 1990).

Salt stress

Salt stress differentially regulates ADC activity and its transcript accumulation in rice. The PA levels are higher in salt-resistant cultivars than in the salt-sensitive ones. Under stress conditions the PA pool in the salt-resistant cultivars is channelled to Spd and Spm. However, the sensitive cultivars under stress conditions only accumulate Put (Chattopadhyay *et al.* 1997). In tomato leaf explants subjected to salt stress a decrease in Put and Spd has been observed, while the levels of agmatine, Dap and Spm increase. The high levels of Dap indicate an activation of PA oxidation (Aziz *et al.* 1998). Moreover, there is an increase of proline accumulation which is correlated with the decrease in Put and Spd levels. Treatment with amino-guanidine, a DAO inhibitor, significantly inhibits proline accumulation, suggesting that in plant tissues under stress conditions, Put and proline are connected by a precursor-product relationship via the DAO and GABA metabolism (Bouchereau *et al.* 1999). These coordinated changes in PA and proline levels have also been described in *Brassica napus* under osmotic stress (Aziz and Larher 1995).

Heat and cold stress

PA metabolism is also altered by extreme temperatures. Plant cells have the ability to synthesise uncommon long-chained PAs (caldine, thermine) previously reported only in thermophilic bacteria, when challenged with heat stress. In bacteria, these uncommon long-chained PAs are apparently essential for continued protein synthesis at high temperatures. *Thermus thermophilus* produces a PA composed of five aminopropyl residues, which has been named caldopentamine (Oshima 1983, Hamana *et al.* 1985). This PA, as well as caldine and thermine, accumulate in pollen and cell cultures of heat-tolerant cotton genotypes (Kuehn *et al.* 1990, Phillips and Kuehn 1991). In rice callus heat stress has also been associated with the production of long-chained PAs (Roy and Ghosh 1996). It seems that, under heat stress, tolerant plants tend to maintain or to increase total PA (Spd and Spm) pools, at the expense of Put (Philipps and Kuehn 1991).

Exposure to low temperatures has been reported to induce Put accumulation in several species (reviewed by Martin-Tanguy 1987). In wheat an increase in Put has been described parallel to an increase in frost resistance (Racz *et al.* 1996). There are some data indicating that different plants respond to low temperature acclimation with a uniform and substantial increase of Spd (Martin-Tanguy 1987). Most of the published results support the view that Spd and Spm may inhibit chilling injury by retarding lipid peroxidation (Bouchereau *et al.* 1999).

Ozone stress

PAs have received considerable interest as possible protectants against ozone stress. Applying Put, Spd and Spm prevented leaf necrosis caused by ozone in tomato plants (Ormrod and Beckerson 1986). Studies with ozone sensitive and ozone tolerant tobacco cultivars have shown a rapid Put increase in the ozone tolerant cultivar Bel B (Langebartels *et al.* 1991). However, in the hypersensitive cultivar Bel W3, only a small increase was observed at a later stage when necrotic lesions had already been formed. Still, free Put accumulates mainly in undamaged tissues while conjugated Put is predominant in necrotic tissues of Bel W3 plants (van Buuren *et al.* 2002). Similarly, Put and Spm titers and ODC activity increased in potato leaves exposed to ozone (Reddy *et al.* 1993). Moreover, *SAMDC* mRNA levels are differentially affected by ozone in leaves and shoots of pea plants exposed to ozone (Marco and Carrasco 2002).

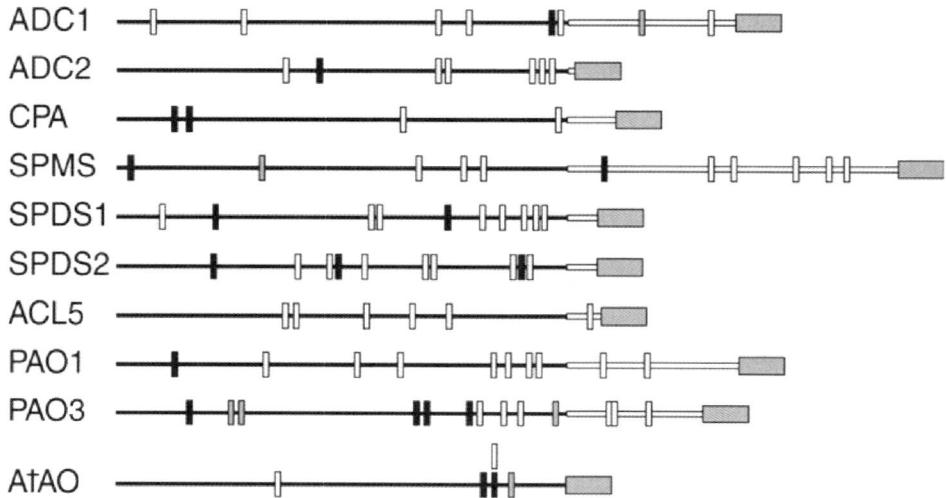

Figure 2. cis-regulatory elements in polyamine metabolism genes. Some regulatory *cis*-elements in promoters of polyamine metabolism genes were found using the PlantCare database of plant *cis*-acting regulatory elements. When searching for abiotic stress responsive elements, several heat-shock (HSE), drought responsive (DRE) and absicisic acid dependent elements (ABRE) were found as shown in the figure with white, grey or black vertical lines respectively. Horizontal black lines correspond to promoter sequences (1000 bp), white horizontal lines represent variable 5'UTR, and crossed boxes indicate the first 100 bp coding sequences. Diagrams are drawn to scale. The genes for which regulatory elements were found are *ADC1* (At2g16500), *ADC2* (At4g34710), *CPA* (At2g27450), *SPMS* (At5g53120), *SPDS1* (At1g23820), *SPDS2* (At1g70310), *ACL5* (At5g19530), *PAO1* (At1g65840), *PAO3* (At3g59050) and *AtAO* (At4g14940).

Polyamines as stress protectants

All these data indicate that PAs are somehow involved in the plant responses to stress challenges. Moreover, as shown in Figure 2, stress responsive elements are present in the promoters of several genes involved in PA metabolism. In most of the cases the existing data suggest that high levels of Spd and Spm, but not of Put, increase stress tolerance. In some plants feeding an excess of Put leads to loss of turgor and formation of necrotic areas, which mimic the symptoms of extreme salinisation (Flores 1991). Similar results have been obtained in transgenic tobacco plants overexpressing oat ADC, under the control of a tetracycline-inducible promoter (Masgrau *et al.* 1997). Diverse mechanisms have been suggested to explain the protective effect of Spm and Spd. Plants sensitive to drought (Wang *et al.* 1990) or salinity (Morgan and Drew 1997) increase ethylene biosynthesis when exposed to stress. The fact that Spd and Spm reduce ethylene biosynthesis by inhibiting ACC synthase (Davis *et al.* 1991) could explain their protective effect. On the other hand, both Spd and Spm can interact with membranes inhibiting the movement of phospholipids through the lipidic bilayer (Bratton 1994), stabilising molecular complexes at the thylakoid membranes (Besford *et al.* 1993) and/or inhibiting lipid peroxidation in osmotically-stressed oat leaves (Borrell *et al.* 1997).

THE ROLE OF THE POLYAMINE METABOLIC PATHWAY IN STRESS SIGNALLING

Although the main metabolic traits of the PA metabolic pathway are well defined (Figure 1), one of the major challenges consists in elucidating the regulatory processes involved in the modulation of this pathway under environmental stress conditions. Facing the paucity and the divergence of experimental data concerning the regulation of PA metabolism in response to abiotic stress and its involvement in cellular adaptive processes in plants, it seems necessary to undertake an integrated molecular analysis to elucidate the role of the PA metabolic pathway in abiotic stress signalling. Since genomic and proteomic tools are now becoming available in *Arabidopsis*, it is reasonable to use this model plant to study the PA response under abiotic stress. In the following sections an outlook of different genomic and proteomic technologies currently applicable to *Arabidopsis* will be discussed.

Genomic approaches

The complete sequence of *Arabidopsis* has already made an impact on plant research and much more remains to be learned from this model plant through the functional characterisation of mutations. Studies of other genomes, such as rice, are required to answer major questions about genome function and genome evolution, and to apply genomic information to practical problems such as increasing the yield and quality of food crops and other plant products (Delseny 2003). The most fundamental issue of comparative genomics is the number and variable number of functional plant genes. How many functional genes are there in any plant? To what extent do all plants contain the same genes? The number of genes inferred from the *Arabidopsis* genome sequence is about 26,000, whereas the number estimated for rice ranges from about 32,000 to 62,000. A large number of the genes inferred from the sequence have not been identified as transcripts, and it is not yet clear how many of the proposed genes are functional. Many genes may be transcribed rarely or under unusual circumstances (Schoof and Karlowski 2003).

Intelligent engineering of regulatory circuits will require detailed knowledge of signalling hierarchies and the impact of metabolic changes involved in stress responses. To this end, functional analysis is under way to determine the relationship between PA metabolism and other genes that participate in drought, salinity, ozone exposure, or other environmental stress responses. Analysis of plants carrying mutations in genes involved in PA metabolism is required to complement information obtained by gene discovery and expression profiling.

Global analysis of gene expression

An understanding of gene function begins with knowledge of when and where each gene is expressed during the process of plant development. Taken together, this information will become a platform from which the concerted gene action in the formation of tissues and organs can be elucidated. Further, examination of changes in gene expression that occur as consequence of environmental changes will illustrate the dynamic nature of gene regulation in plants and will facilitate the development of gene arrays useful for molecular diagnosis of plant physiological status in the field.

Microarrays were initially developed by Schena *et al.* (1995), and have now been widely applied in the field of Plant Biology. DNA microarrays are chips containing hundreds or thousands of gene snippets laid down in precise arrays that provide rapid snapshots of the expression of whole suites of genes. Two types of microarrays are commonly used, two-colour microarrays and oligonucleotide microarrays. In a two-colour microarray, collections of DNA samples (*i.e.* expressed sequence tag [ESTs] or other clones) are deposited onto a glass slide using robotics (Duggan *et al.* 1999, Wu *et al.* 2001). Oligonucleotide arrays are constructed from roughly 25-mer oligonucleotides synthesised *in situ* on a solid substrate (Lipschultz *et al.* 1999). This type of microarray requires exact sequence information and intricate bioinformatic design prior to the construction of the microarray. Oligonucleotide microarrays are highly consistent and offer sequence-specific detection of gene expression, which is especially important in the study of gene families. With both types of microarray analysis, data aggregation from multiple experiments is possible, allowing higher order analyses of transcript profiles. The general method in microarray analysis is 1) to isolate messenger RNAs (mRNAs) produced by a genome, 2) to convert mRNA into complementary DNA (cDNA), 3) to add a fluorescent tag to the cDNA for tracking purposes, and 4) to wash a solution of tagged cDNAs over a DNA microarray chip. Each DNA snippet on the chip will bind the cDNA from the corresponding gene, and by measuring the fluorescences arrayed on the chip, the profile of gene expression is revealed.

Recently, Matsuyama *et al.* (2002) have constructed a cDNA microarray to compare the response to different abiotic stresses for a set of *Arabidopsis* genes. Gene-specific cDNA probes for different PA biosynthetic genes were included in the array, expression of ADC2 being markedly induced by drought and wounding but not by ozone.

Knockout mutants

Since forward genetics relies on a phenotype arising from a single gene mutation, it is likely that a large number of genes may not be easily characterised using this approach. In order to identify functions for these genes, we need to develop a more sophisticated genetic toolkit for both forward and reverse genetic screens. Overexpression of natural or altered proteins can provide insights into families of genes that are collectively essential. A straightforward approach, albeit laborious, resembles strategic breeding and weds reverse and forward genetics. In this approach, the genome sequence is used to locate protein family members. Knockout mutations are found for all the genes in a family and the lines are crossed so that one plant contains knockout alleles of all the members of the gene family, creating a more robust phenotype. This strategic approach can lead to increased understanding of the functions of genes that exist in gene families, and is not possible without the entire sequence in hand.

The analysis of large T-DNA or insertional transposon mutant collections of *Arabidopsis* and rice provide essential resources for finding tagged mutations that can lead to defective stress tolerance responses (Winkler *et al.* 1998, Maes *et al.* 1999, Tissier *et al.* 1999). These populations can be surveyed using both forward and reverse genetic screens to isolate "knockout" mutants that are either tolerant or hypersensitive to stress. T-DNA activation-tagging collections are being generated in *Arabidopsis* backgrounds in order to isolate gain-of-function mutations that affect stress signalling (Weigel 2000).

Little is known about mutants affecting PA metabolism in plants. Mutants with high level of Put and high levels of ADC activity have been identified because of their abnormal floral morphology (Gerats *et al.* 1988) but the basis of the mutation remains uncharacterised. Screening for resistance to the SAMDC inhibitor methylglyoxal-bis (guanylhydrazone) (Malmberg and Rose 1987, Trull *et al.* 1992) or inhibitory concentrations of Spm (Mirza and Iqbal 1997), yielded mutants that showed reduced sensitivity to the respective agent, but these mutants have not been exploited for the analysis of PA function.

Watson *et al.* (1998) isolated EMS mutants of *Arabidopsis* with reduced ADC activity. The mutants fall into two complementation groups, *spe1* and *spe2*, and the strongest alleles within each group showed a reduction of ADC activity down to 23 and 36%, respectively. The double mutant *spe1-1 spe2-1* had lower ADC activity than each single mutant but still about 20% of the wild-type activity remained. Because two genes encoding ADC, *ADC1* and *ADC2*, are found in all members of the *Brassicaceae* studied to date except the basal genus

Aethionema (Galloway *et al.* 1998), the authors suggested that *spe1* and *spe2* might correspond to both ADC genes. The mutations have not been mapped and therefore it cannot be excluded that other functions, *e.g.* regulatory elements, are affected.

The mutant EN9 has been obtained by PCR-screening of an En-1-mutagenised *Arabidopsis* population (Baumann *et al.* 1998) for insertions at the *ADC2* locus. En-1 is an autonomous transposable element of maize and is able to transpose in the heterologous host *A thaliana* (ecotype Columbia) (Cardon *et al.* 1993). The 8.2 kb large transposon causes gene disruption or a frame shift when leaving a footprint during excision. In contrast to the mutants described by Watson *et al.* (1998), EN9 should therefore be regarded as a complete loss-of-function or knockout mutation of *ADC2*. The EN9 mutant line shows no obvious phenotype under normal growth conditions but is completely devoid of *ADC2* induction by osmotic stress as determined by incubating leaf discs (Soyka and Heyer 1999).

Overexpressions

Overexpression of natural or altered proteins can provide insights into families of genes that are collectively essential. Overexpression as well as antisense inhibition of biosynthetic enzymes has been employed to study PA function. This made use of either the constitutive 35S-promoter (Hamill *et al.* 1990, Descenzo and Minocha 1993, Bastola and Minocha 1995) or inducible promoters (Kumar *et al.* 1996, Masgrau *et al.* 1997). Overexpression of genes coding for specific enzymes of the PA biosynthetic and degradation pathways in transgenic plants is one of the approaches used to elucidate the functional role of PAs under stress (Kumar *et al.* 1997, Capell *et al.* 1998, Roy and Wu 2001). Several reports of transgenic plants showed that constitutive overexpression of PA biosynthetic genes was detrimental to the normal growth of transgenic plants (Kumar *et al.* 1997, Capell *et al.* 1998). The ABA-inducible promoter characterised by Su *et al.* (1998) was used to drive the expression of an oat *ADC* cDNA in rice (Roy and Wu 2001) under stress conditions. Those transgenic rice plants showed increased biomass production and a threefold higher level of Put accumulation than non-transformed plants under drought conditions. Similar results were obtained by transforming rice with *Tritordeum SAMDC* gene (Roy and Wu 2002), where exogenous ABA application and sodium chloride stress treatments resulted in different levels of SAMDC expression in transgenic rice lines. Accumulation of Spd and Spm under NaCl-stress conditions may contribute to increased stress tolerance due to enhanced biomass production as well as increased shoot length of transgenic rice plants as compared to non-transgenic plants. These results suggest that the regulation of the *ADC* and *SAMDC* genes, especially with respect to the response to salt or drought stress, may be coordinated in plants.

Proteomic approaches

A reciprocal benefit between genomics and proteomics can be expected for both fast-evolving areas. The "Proteome" was recently described as "the protein complement of the genome" (Wassinger *et al.* 1995), giving rise to the novel discipline of proteomics whose aim is the systematic characterisation of the proteome in a given organism, tissue, cell or subcellular compartment. The initial concept of proteomics as the identification of gene products in a given proteome, often associated with two-dimensional electrophoresis, is experiencing a rapid evolution. Not only large improvements in separation and identification methods are being reported, but also new perspectives in quantification and characterisation of protein-protein interactions are providing a multidimensional perception of how proteins execute their functions within the cellular context.

Yeast two-hybrid system

One approach to illustrate putative functions for unknown gene products is to identify their protein interacting partners or "interologs" (Walhout *et al.* 2000a). The establishment of a map of protein interactions ("interactome") may help establish clusters of interacting gene products. Based on this kind of protein network organisation, the identification of the protein partners for an unknown protein may assign a putative functional category to the uncharacterised protein (Tucker *et al.* 2001). The original development of the yeast two-hybrid system (Fields and Song 1989) has allowed the validation and standardisation of an amenable technique to characterise protein-protein interactions in living yeast cells. Two separated modules of a transcription factor, the DNA binding domain (DBD) and the transcriptional activator domain (AD) are translationally fused to interacting proteins X and Y generating the "bait" and "prey" constructs respectively. Upon X-Y interaction within the nucleus of the yeast cell, the hybrid transcription factor will be functionally reconstituted with the subsequent biological readout as the expression of reporter genes containing promoter-binding sites for the transcription factor. The strength of the idea relies on the powerful yeast molecular genetics, which allows a rapid and easy identification of coding sequences for interacting proteins when using cDNA libraries as a prey (Walhout *et al.* 2000b). This powerful technology also contains some drawbacks and limitations. Among the most disturbing consequences, it is the appearance of false positives due to spurious interactions the one centring most attention. Therefore any protein-protein interaction found by the yeast two-hybrid technology should be treated with caution until proven by alternative *in vivo* demonstrations. Also false negatives can rise due to intrinsic features of the system: improper folding, incorrect subcellular localisation, instability of hybrid proteins or absence of required post-translational modifications in yeast may lead to lack of interaction.

An immediate extension from the original application to study interactions between two known proteins was the identification of novel potential interactions at a large-scale, using either cDNA libraries fused to AD (library approach) or full length cDNAs as both "bait" and "prey" (matrix approach). These two-hybrid variants have been applied to diverse biological systems. Examples of the matrix approach extend from limited protein number in *Drosophila* (Finley and Brent 1994) to covering the whole proteome for the vaccinia virus (McCraith *et al.* 2000) and the baker's yeast (Uetz *et al.* 2000, Ito *et al.* 2001). The use of library screenings in a wide-proteome approach was initiated with the T7 phage (Bartel *et al.* 1996) and continued with the hepatitis C virus (Flajolet *et al.* 2000) and the bacteria *Helicobacter pylori* (Rain *et al.* 2001). Although the use of two-hybrid system is already a routine for *Arabidopsis* researchers, no global "interactome" analysis has been reported to date for plant systems. Parallel uses of both matrix and library approaches (Flajolet *et al.* 2000, Walhout *et al.* 2000a) indicate that the library strategy yields higher number of protein interactions. Another conclusion of the high-throughput assays is that large numbers of false positives preclude identification of as much as 90% of previously reported interactions. Therefore it seems evident the unavoidable request of independent validation for any reported protein-protein interaction based on the two-hybrid system.

Epitope-labelling technique

Among the different strategies to characterise protein interactions, the use of immunological methods is widely accepted. The coimmunopurification of proteins from total protein extracts represents unequivocal demonstration of *in vivo* protein associations. The same antibodies can provide information regarding tissue or subcellular localisation, data that may serve as an indirect evidence for protein association if overlapping signals are immunodetected.

The availability of appropriate antibodies is undoubtedly a crucial requirement for characterisation of protein associations by immunological procedures. One alternative to the laborious and uncertain practice of raising antibodies against purified proteins is the use of epitope-tagging methodology. This approach is particularly useful for available genomic coding sequences. In the highly favourable case of efficient yeast gene replacement, the complete genome sequence encouraged the development of successful chromosomal epitope tagging (Knop *et al.* 1999). The concept is based on the attachment of short genomic sequences coding for antigenic peptides (named "epitopes" of 6 to 20 amino acids length), to any cDNA of interest. The addition of a peptide sequence recognisable by a pre-existing antibody to a protein under study permits its surveillance by immunological methods readily after expressing the cloned coding sequence. In addition to the speed of the method, a series of additional advantages of epitope tagging to trace any protein of interest illustrate the power of this technique. For instance, the artificially attached epitope can be considered an outsider antigen within the total protein extract of cells that do not express the tagged protein. This represents an unbeatable negative control in terms of specificity when compared to the use of antibodies against peptides from the native protein. Moreover, the use of epitope tagging permits immunological discrimination between closely related proteins without risk of cross-reaction. Since the experimenter decides the location of the epitope within the investigated protein, suitable sites to avoid potential interferences in terms of localisation or function can be envisaged. Finally, the use of epitopes facilitates mild elution condition in protein purification methods by competition with the corresponding purified peptide. A large battery of peptide-antibody combinations is commercially available. For a recent review see Fritze and Anderson (2000). This panoply of suitable epitopes has been thoroughly exploited in yeast and animal cells (Jarvik and Telmer 1998).

In plant systems, the use of epitope-tagging technique has been modest (DeWitt and Sussman 1995, Boyes *et al.* 1998) mainly due to restrictions for working with transgenic plants. The recent development of an *Agrobacterium*-mediated technology to introduce intron-tagged epitope coding sequences in cultured *Arabidopsis* plant cells brings hope for a wider use of epitope-tagging technology in plant systems (Ferrando *et al.* 2000). This technology allows the detection, purification, and subcellular localisation of epitope-tagged proteins as early as five days after transformation. In addition, the high transformation rate achieved by means of *Agrobacterium* transformation, licenses cotransformation experiments whereby combinations of two differentially epitope-tagged proteins can be used for protein-protein interaction studies. This technique represents a fast and reliable means for the *in vivo* verification of protein interactions in plant cells (Ferrando *et al.* 2001). Production of transgenic cell lines expressing epitope-tagged proteins is also feasible by the use of intron-tagged epitope technology, thus providing unlimited source of transformed plant material for large-scale protein purification. In combination with the generation of large amount of tagged proteins, the purification of protein complexes by affinity chromatographic methods is a powerful technique to define multiproteic complex associations (Farràs *et al.* 2001).

In the laboratory, we are using some of these proteomic approaches to elucidate the formation of protein complexes between enzymes involved in PA biosynthesis in *Arabidopsis* (Panicot *et al.* 2002). Spd synthase SPDS2 was used as a "bait" using a cDNA library from *Arabidopsis* cell suspensions. SPDS2 was shown to interact in yeast with the functional homolog Spd synthase SPDS1, in addition to a novel Spm synthase named SPMS. Only heterodimers between these enzymes were found in the yeast two-hybrid system. In plant cells, *in vivo* evidence was demonstrated in *Arabidopsis* with the use of the intron-tagged epitope technique. Coexpression of hemaglutinin and c-Myc epitope-labelled proteins confirmed the presence of coimmunoprecipitating SPDS1-SPDS2 and SPDS2-SPMS heterodimers. In addition, the epitope-labeled proteins copurified associated to protein complexes in the range of 650-700kD. We have therefore suggested the formation of a metabolon involving at least the last two-steps of PA biosynthesis in *Arabidopsis* (Panicot *et al.* 2002). Further analysis of the identified protein complexes by mass spectrometry is expected to provide information about yet unknown regulatory subunits of SPDS-SPMS metabolon in the PA biosynthesis pathway in *Arabidopsis*. The use of proteomic techniques to unravel the formation of protein complexes in plant secondary metabolism is in progress. As an example, similar proteomic approaches have been employed to unravel the multi-enzymatic association in the flavonoid biosynthetic pathway (Burbulis and Winkel-Shirley 1999). By means of two-hybrid complemented with coimmunoprecipitation and protein retention studies, interactions among chalcone synthase (CHS), chalcone isomerase (CHI) and dihydroflavonol reductase (DFR) have been described.

Other advanced techniques

The full potential of this kind of approaches is achieved with the use of physico-chemical methods for peptide and protein identification. These combined approaches of affinity purification and protein identification by means of mass-spectrometry have been implemented for the characterisation of large multiproteic complex associations, as the spliceosome and the proteasome complexes (Neubauer *et al.* 1997, Verma *et al.* 2000). Compared to yeast or other eukaryotic systems, the proteomic analysis for plant systems is still in its infancy (van Wijk 2001). The most likely reason is the delayed completion of genome projects that are essential for protein identification. The fully sequenced *Arabidopsis* genome is giving boost to proteomic studies for higher plants, and the first reported proteomic analysis have appeared (Gallardo *et al.* 2001, Kruft *et al.* 2001, Peltier *et al.* 2001). Although limited, due to lack of complete genome sequences, studies in other plant systems are also feasible (Peltier *et al.* 2000, Huber *et al.* 2001).

In addition to the studies of physical protein interactions, the evidence of subcellular protein colocalisation can serve as a support to elucidate the formation of protein complexes. In this context, the epitope-tagging technology can also be exploited (Ferrando *et al.* 2000). The target proteins may also be fused to reporters such as fluorescent polypeptides to facilitate their cellular localisation (Quaedvlieg *et al.* 1998). A remarkable advance for protein interaction studies is the implementation of the fluorescence resonance energy transfer (FRET) technology as detailed by Kenworthy (2001) where the development of fluorescent proteins together with advances in confocal microscopy allow *in vivo* imaging of fusion-protein associations (Mas 2000, Kato *et al.* 2002).

INTEGRATED MOLECULAR ANALYSES - AN ORDER TO BE FILLED

In the second half of the twentieth century, advances in several scientific areas such as biochemistry, physiology, genetics, and cellular and molecular biology, have dramatically increased the knowledge of how the basic organisation of biological systems is accomplished. The remaining challenges for the twenty-first century await a deeper insight into the molecular mechanisms that govern those cellular pathways, to better understand the biological systems from a global perspective. This ambitious goal can be effectively achieved by taking advantage of novel and powerful technologies. The integration of advances in genetics, biophysics and biochemistry into novel scientific areas such as genomics and proteomics (*i.e.* two-hybrid, reverse genetics, imaging by fluorescence (or confocal) microscopy, DNA microarrays, real-time PCR, molecular mass fingerprints, *etc.*) allow, nowadays, a detailed and precise analysis of the molecules involved in basic cellular functions (DNA and proteins).

In the near future, the availability of full-genome microarrays will greatly expand our knowledge of the inter-connections and similarities among stress and defence response pathways, and will aid in the identification of genes previously unknown in stress responses. This progress will facilitate the fabrication of custom microarrays for more extensive studies of particular defence responses. An important facet of integrating the mass of new information from these experiments will be the development of databases that are specific to plant stress, which will be used to integrate the array information and provide it in a public repository. The availability of adequate data-mining, statistical and data presentation tools will also be integral to the understanding of functional genomic data. As the genomes of more crop species are fully sequenced and annotated, microarrays will be used to study stress responses in these plants. In the future, genes discovered in these experiments may provide new insights into stress resistance in multiple crop species.

Under a global perspective, we envisage that an integrated molecular analysis should be undertaken in order to draw any functional relationship among genes representing the PA metabolic network (Figure 3). By using the biochemically established metabolic pathway for PAs (Figure 1), further data should be added with regard to protein interactions, subcellular localisations, metabolite levels, enzymatic activities, protein expression levels, and coordinated gene expression (Figure 3). These global maps, gained by means of genomic, proteomic and metabolomic techniques, will then be used to design system perturbation strategies. These systematic perturbations will need to be performed at different levels: genetic (*e.g.* gene knockouts or overexpressions), biochemical (*e.g.* use of inhibitors) and environmental (*i.e.* abiotic stress), and the molecular consequences of these perturbations will be studied by genomics, proteomics and metabolomics (Figure 3). The acquired data after system perturbation will be integrated into the model and the system map will be refined. If necessary, further alterations of the systems will be performed and analysed again. As a result, further insights into the molecular mechanisms underlying the role of the PA metabolic pathway in abiotic stress signalling are envisaged.

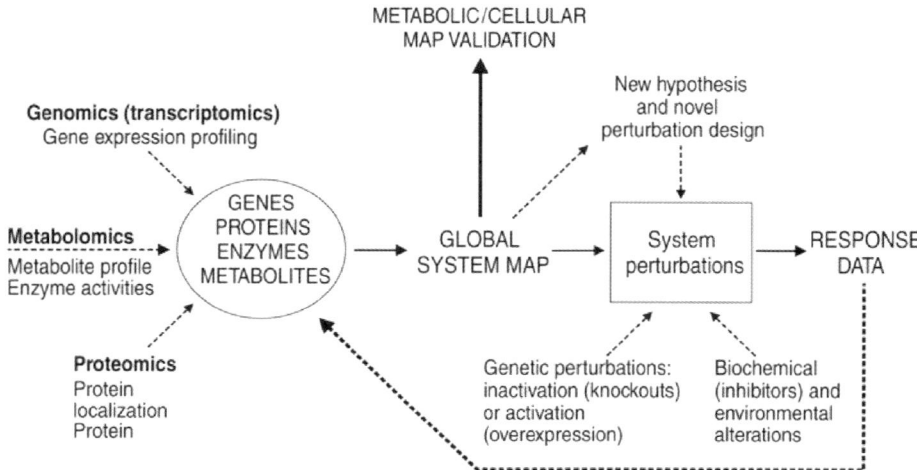

Figure 3. Integrated molecular analysis of metabolic pathways. Integrated molecular analysis based on genomics, proteomics and metabolomics techniques to validate cellular function maps related to metabolic pathways. After the model map has been established by the integrated analysis, the perturbations of the system allow further refinements of the model (see text for details).

ACKNOWLEDGEMENTS

The research has been supported by grants from Spanish CICYT-BIO2002-04459-C02-02 and from EC-QLK5-CT-2002-00841.

REFERENCES

Angelini R., Rea G., Federico R., D'Ovidio R. 1996. Spatial distribution and temporal accumulation of mRNA encoding diamine oxidase during lentil (*Lens culinaris* Medicus) seedling development. - Plant Sci. 119: 103-113.

Aziz A., Larher F. 1995. Changes in polyamine titters associated with the proline response and osmotic adjustment of rape leaf discs submitted to osmotic stress. - Plant Sci. 112: 175-186.

Aziz A., Martin-Tanguy J., Larher F. 1998. Stress-induced changes in polyamine and tyramine levels can regulate proline accumulation in tomato leaf discs treated with sodium chloride. - Physiol. Plant. 104: 195-202.

Bagni N., Pistocchi R. 1991. Uptake and transport of polyamine and inhibitors of polyamine metabolism in plants. - In: Slocum R.D., Flores H.E. (Ed.) Biochemistry and physiology of polyamines in plants. - Boca Raton, FL, CRC Press, pp. 105-120.

Bagni N., Torrigiani P. 1992. Polyamines: A new class of growth substances. - In: Karssen C.M., van Loon L.C., Vreugdenhil, D. (Eds.) Progress in plant growth regulation. - Dordrecht, The Netherlands, Kluwer Academic Publishers, pp. 264-275.

Bagni N., Tassoni A. 2001. Biosynthesis, oxidation and conjugation of aliphatic polyamines in plants. - Amino Acids 20:301-317.

Bartel P.L., Roecklein J.A., SenGupta D., Fields S. 1996. A protein linkage map of *Escherichia coli* bacteriophage T7. - Nature Genet. 12: 72-77.

Bastola D.R., Minocha S.C. 1995. Increased putrescine biosynthesis through transfer of mouse ornithine decarboxylase cDNA in carrot promotes somatic embryogenesis. - Plant Physiol. 109: 63-71.

Bhatnagar P., Minocha R., Minocha S. 2002. Genetic manipulation of the metabolism of polyamines in poplar cells. The regulation of putrescine catabolism. - Plant Physiol. 128: 1455-1469.

Baumann E., Lewald J., Saedler H., Schulz, B., Wisman E. 1998. Successful PCR-based reverse genetic screens using an En-1-mutagenised Arabidopsis thaliana population generated via single-seed descent. - Theor. Appl. Genet. 97: 729-734.

Besford R.T., Richardson C.M., Campos J.L., Tiburcio A.F. 1993. Effect of polyamines on stabilization of molecular complexes of thylakoid membranes of osmotically stressed oat leaves. - Planta 189: 201-206.

Borrell A., Besford R.T., Altabella T., Masgrau, C., Tiburcio A.F. 1996. Regulation of arginine decarboxylase by spermine in osmotically-stressed oat leaves. - Physiol. Plant. 98: 105-110.

Borrell A., Carbonell L., Farras R., Puig-Parellada P., Tiburcio A.F. 1997. Inhibition of lipid peroxidation by polyamines in senescent oat leaves in vivo. - Physiol. Plant. 99: 385-390.

Bouchereau A., Aziz A., Larher F., Martin-Tanguy J. 1999. Polyamines and environmental challenges: recent development. - Plant Sci. 140: 103-125.

Boyes D.C., Nam J., Dangl J.L. 1998. The *Arabidopsis thaliana* RPM1 disease resistance gene product is a peripheral plasma membrane protein that is degraded coincident with the hypersensitive response. - Proc. Natl. Acad. Sci. USA 95: 15849-15854.

Bratton D.L. 1994. Polyamine inhibition of transbilayer movement of plasma membrane phospholipids in the erythrocyte ghost. - J. Biol. Chem. 269: 22517-22523.

Bray E.A., Bailey-Serres J., Weretilnyk E. 2000. Responses to abiotic stresses. - In: Gruissen W. Buchannan B., Jones R. (Eds.) Responses to abiotic stresses - Rockville, Maryland American Society Plant Physiology, pp. 1158-1249.

Burbulis I.E., and Winkel-Shirley B. 1999. Interactions among enzymes of the Arabidopsis flavonoid biosynthetic pathway. - Proc. Natl. Acad. Sci. USA, 96: 12929-12934.

Capell T., Escobar C., Liu H., Burtin D., Lepri O., Christou P. 1998. Overexpression of the oat arginine decarboxylase cDNA in transgenic rice (*Oryza sativa* L.) affects normal development patterns in vitro and results in putrescine accumulation in transgenic plants. - Theor. Appl. Genet. 97: 246-254.

Cardon G.H., Frey M., Seadler H., Gierl A. 1993. Mobility of the maize transposable element En/Spm in *Arabidopsis thaliana*. - Plant J. 3: 773-784.

Chattopadhay M.K., Gupta S., Sengupta D.N., Ghosh B. 1997. Expression of arginine decarboxylase in seedlings of indica rice (*Oryza sativa* L.) cultivars as affected by salinity stress. - Plant Mol. Biol. 34: 477-483.

Cohen S.S. 1998. A guide to the polyamines. - New York, U.S.A., Oxford University Press.

Davis P.J., Rastogi R., Law, D.M. 1991. Polyamines and their metabolism in ripening tomato fruit. - In: Flores H.E., Arteca R.N. (Eds.) Polyamines and ethylene: Biosynthesis, physiology and interactions. - Rockville, Maryland, American Society Plant Physiology, pp. 112-125.

Delseny M. 2003. Towards an accurate sequence of the rice genome. - Curr. Op. Plant Biol. 6: 101-105.

Descenzo R.A., Minocha S.C. 1993. Modulation of cellular polyamines in tobacco by transfer and expression of mouse ornithine decarboxylase cDNA. - Plant Mol. Biol. 22: 113-127.

DeWitt N.D., Sussman M.R. 1995. Immunocytological localization of an epitope-tagged plasma membrane proton pump H^+-ATPase in phloem companion cells. - Plant Cell 7: 2053-2056.

Duggan D.J., Bittner M., Chen Y., Meltzer P., Trent J.M. 1999. Expression profiling using cDNA microarrays. - Nature Genet. 21:10-14.

El Gachtouli N., Martin-Tanguy J., Paynot M., Gianinazzi S. 1996. First report of the inhibition of arbuscular mycorrhizal infection of *Pisum sativum* by specific and irreversible inhibition of polyamine biosynthesis or by giberellic treatment. - FEBS Lett. 385: 189-192.

Farràs R., Ferrando A., Jasik J., Kleinow T., Okresz L., Tiburcio A. F., Salchert K., del Pozo C., Schell J., Koncz C. 2001. SKP1-SnRK protein kinase interactions mediate proteasomal binding of a plant SCF ubiquitin ligase. - EMBO J. 20: 2742-2756.

Feirer R.P., Hocking K.L., Woods P.J. 1998. Involvement of arginine decarboxylase in the response of *Arabidopsis thaliana* to osmotic stress. - J. Plant Physiol. 153: 733-738.

Ferrando A., Farràs R., Jasik J., Schell J., Koncz C. 2000. Intron-tagged epitope: a tool for facile detection and purification of proteins expressed in *Agrobacterium*-transformed plant cells. - Plant J. 22: 553-560.

Ferrando A., Koncz-Kalman Z., Farràs R., Tiburcio A.F., Schell J., Koncz C. 2001. Detection of in vivo protein interactions between Snf1-related kinase subunits with intron-tagged epitope-labelling in plants cells. - Nucleic Acids Res. 29: 3685-3693.

Fields S., Song O. 1989. A novel genetic system to detect protein-protein interactions. - Nature 340: 245-246.

Finley R.L.Jr. and Brent R. 1994. Interaction mating reveals binary and ternary connections between *Drosophila* cell cycle regulators. - Proc. Natl. Acad. Sci. USA 91: 12980-12984.

Flajolet M., Rotondo G., Daviet L., Bergametti F., Inchauspe G., Tiollais P., Transy C., Legrain P. 2000. A genomic approach of the hepatitis C virus generates a protein interaction map. - Gene 242: 369-379.

Flores H.E. 1991. Changes in polyamine metabolism in response to abiotic stress. - In: Slocum R., Flores H.E. (Eds.) The Biochemistry and physiology of polyamines in plants. - Boca Raton, FL, CRC Press, pp. 214-225.

Flores H.E., Galston A.W. 1984. Osmotic stress-induced polyamine accumulation in cereal leaves. I. Physiological parameters of the response. - Plant Physiol. 75: 102-109.

Friedman R., Levin N., Altman A. 1986. Presence and identification of polyamines in xylem and phloem exudates of plants. - Plant Physiol. 82: 1154-1157.

Fritze C.E. and Anderson T.R. 2000. Epitope tagging: general method for tracking recombinant proteins. - Meth. Enzymol. 327: 3-16.

Gallardo K., Job C., Groot S.P., Puype M., Demol H., Vandekerckhove J., Job D. 2001. Proteomic analysis of Arabidopsis seed germination and priming. - Plant Physiol. 126: 835-848.

Galloway G.L., Malmberg R.L., Price R.A. 1998. Phylogenetic utility of the nuclear gene arginine decarboxylase: An example from Brassicaceae. - Mol. Biol. Evol. 15: 1312-1320.

Galston A.W., Kaur-Sawhney R., Altabella T., Tiburcio A.F. 1997. Plant polyamines in reproductive activity and response to abiotic stress. - Bot. Acta 110: 197-207.

Gerats A.G.M., Kaye C., Collins C., Malmberg, R.L. 1988. Polyamine levels in petunia genotypes with normal and abnormal floral morphologies. - Plant Physiol. 86: 390-393.

Groppa M.D., Benavides M.P., Tomaro M.L. 2003. Polyamine metabolism in sunflower and wheat leaf discs under cadmium or copper stress. - Plant Sci. 164: 293–299.

Hamana K., Masahimo K., Onishi H., Akazawa T., Matsuzaki S. 1985. Polyamines in photosynthetic eubacteria and extreme halophyte archaebacteria. - J. Biochem. 97: 1653-1658.

Hamill J.D., Robins R.J., Parr A.J., Evans D.M., Furze J.M., Rohdes M.J.C. 1990. Overexpressing a yeast ornithine decarboxylase gene in transgenic roots of *Nicotiana rustica* can lead to enhanced nicotine accumulation. - Plant Mol. Biol. 15: 27-38.

Hasegawa P.M., Bressan R.A., Zhu J.K., Bohnert H.J. 2000. Plant cellular and molecular responses to high salinity. - Annu. Rev. Plant Physiol. Mol. Biol. 51: 463-499.

Huber C. Huber C.G, Timperio A.M., Zolla L. 2001. Isoforms of photosystem II antenna proteins in different plant species revealed by liquid chromatography-electrospray ionization mass spectrometry. - J. Biol. Chem. 276: 45755-45761.

Ideker T., Galitski T., Hood L. 2001. A new approach to decoding life: systems biology.- Annu. Rev. Genomics Hum. Genet. 2: 343-372.

Ito T., Tashiro K., Muta S., Ozawa R., Chiba T., Nishizawa M., Yamamoto K., Kuhara S., Sakaki Y. 2001. A comprehensive two-hybrid analysis to explore the yeast protein interactome. - Proc. Natl. Acad. Sci. USA 98: 4569-4574.

Jarvik J.W., Telmer C.A. 1998. Epitope tagging. - Annu. Rev. Genet. 32: 601-618.

Kato N., Pontier D., Lam E. 2002. Spectral profiling for the simultaneous observation of four distinct fluorescent proteins and detection of protein-protein interaction via Fluorescence Resonance Energy Transfer in tobacco leaf nuclei. - Plant Physiol. 129: 931-942.

Kaur-Sawhney R., Tiburcio A.F., Altabella T., Galston A.W. 2003. Polyamines in plants: an overview. - J. Cell Mol. Biol. 2: 1-12.

Kenworthy A.K. 2001. Imaging protein-protein interactions using fluorescence resonance energy transfer microscopy. - Methods 24: 289-296.

Knop M., Siegers K., Pereira G., Zachariae W., Winsor B., Nasmyth K., Schiebel E. 1999. Epitope tagging of yeast genes using a PCR-based strategy: more tags and improved practical routines. - Yeast 15: 963-972.

Kreps J.A., Wu Y., Chang H-S., Zhu T., Wang X., Harper J.F. 2002. Transcriptome changes for Arabidopsis in response to salt, osmotic and cold stress. - Plant Physiol. 130: 2129-2141.

Kruft V., Eubel H., Jänsch L., Werhahn W., Braun H.P. 2001. Proteomic approach to identify novel mitochondrial proteins in *Arabidopsis*. - Plant Physiol. 127: 1694-1710.

Kuehn G.D., Rodriguez-Garay B., Bagga S., Phillips G.C. 1990. Novel occurrence of uncommon polyamines in higher plants. - Plant Physiol. 94: 855-857.

Kumar A., Altabella T., Taylor M.A., Tiburcio A.F. 1997. Recent advances in polyamine research. - Trends Plant Sci. 2: 124-130.

Kumar A., Taylor M.A., Mad Arif S.A., Davies H.V. 1996. Potato plants expressing antisense and sense S-adenosylmethionine decarboxylase (SAMDC) transgenes show altered levels of polyamines and ethylene: Antisense plants display abnormal phenotypes. - Plant J. 9: 147-158.

Langebartels C., Kerner K.J., Leonardi S., Schraudner M., Trost M., Heller W., Sanderman, H. Jr. 1991. Biochemical plant responses to ozone I. Differential induction of polyamine and ethylene biosynthesis in tobacco. - Plant Physiol. 95: 882-889.

Levitt J. 1972. Responses of plants to environmental stresses. - New York, Academic Press.

Lipshultz R.J., Fodor S.P.A., Gingeras T.R., Lockhart D.J. 1999. High density synthetic oligonucleotide arrays. - Nature Genet. 21: 20-24.

Lovaas E. 1997. Antioxidative and metal-chelating effects of polyamines. - Adv. Pharmacol. 38: 119-149.

Maes T., De Keukeleire P., Gerats T. 1999. Plant tagnology. - Trends Plant Sci. 4: 90-96.

Malmberg R.L., Rose D.J. 1987. Biochemical genetics of resistance to MGBG in tobacco: mutants that alter SAM decarboxylase or polyamine ratios and floral morphology. - Mol. Gen. Genet. 207: 9-14.

Malmberg R.L., Watson M.B., Galloway G.L., Yu W. 1998. Molecular genetic analyses of plant polyamines. - CRC Crit. Rev. Plant Sci. 17: 199-224.

Mas P., Devlin P.F., Panda S., Kay S.A. 2000. Functional interaction of phytochrome B and cryptochrome 2. - Nature 408: 207-211.

Marco F., Carrasco P. 2002. Expression of the pea S-adenosylmethionine decarboxylase gene is involved in developmental and environmental responses. - Planta 214: 641–647.

Martin-Tanguy J. 1987. Hydroxycinnamic acids amides, hypersensitivity, flowering and sexual organogenesis in plants. - In: Von Wettstein D., Chua D. (Eds.) Plant molecular biology. - New York, Plenum Press, pp. 253-263.

Martin-Tanguy J. 2001. Metabolism and function of polyamines in plants: recent development (new approaches). - Plant Growth Regul. 34: 135-148.

Masgrau C., Altabella T., Farras R., Flores D., Thompson A.J., Besford R.T., Tiburcio A.F. 1997. Inducible overexpression of oat arginine decarboxylase in transgenic tobacco plants. - Plant J. 11: 465-473.

Matsuyama T., Tamaoki M., Nakajima N., Aono M., Kubo A., Moriya S., Ichihara T., Suzuki O., Saji H. 2002. cDNA microarray assessment for ozone-stressed *Arabidopsis thaliana*. - Environ. Pollut. 117: 191–194.

McCraith S., Holtzman T., Moss B., Fields S. 2000. Genome-wide analysis of vaccinia virus protein-protein interactions. - Proc. Natl. Acad. Sci. U.S.A. 97: 4879-4884.

Mirza J.I., Iqbal, M. 1997. Spermine-resistant mutants of Arabidopsis thaliana with developmental abnormalities. - Plant Growth Regul. 22: 151-156.

Moller S.G., McPherson M.J. 1998. Developmental expression and biochemical analysis of the Arabidopsis *ataoI* gene encoding a H_2O_2-generating diamine oxidase. - Plant J. 13: 781-791.

Morgan P.W., Drew M.C. 1997. Ethylene and plant response to stress. – Physiol. Plant. 100: 620-630.

Neubauer G., Gottschalk A., Fabrizio P., Seraphin B., Luhrmann R., Mann M. 1997. Identification of the proteins of the yeast U1 small nuclear ribonucleoprotein complex by mass spectrometry.- Proc. Natl. Acad. Sci. U.S.A. 94: 385-390.

Ormrod D.P., Beckerson, D.W. 1986. Polyamines as antiozonants for tomato. - HortScience 21: 1070-1071.

Oshima T.A. 1983. Novel polyamines in *Thermus thermophilus:* Isolation, identification and chemical synthesis. - Meth. Enzymol. 94: 401-411.

Panicot M., Minguet E., Ferrando A., Alcázar R., Blázquez M.A., Carbonell J., Altabella T., Koncz C., Tiburcio A.F. 2002. A polyamine metabolon involving aminopropyl transferases complexes in *Arabidopsis.* - Plant Cell 14: 2539-2551.

Peltier J.B., Friso G., Kalume D.E., Roepstorff P., Nilsson F., Adamska I., van Wijk K.J. 2000. Proteomics of the chloroplast: systematic identification and targeting analysis of lumenal and peripheral thylakoid proteins. - Plant Cell 12: 319-341.

Peltier J.B., Ytterberg J., Liberles D.A., Roepstorff P., van Wijk K.J. 2001 Identification of a 350-kDa ClpP protease complex with 10 different Clp isoforms in chloroplasts of *Arabidopsis thaliana.* - J. Biol. Chem. 276: 16318-16327.

Phillips G.C., Kuehn G.D. 1991. Uncommon polyamines in plants and other mechanisms. - In: Slocum R.D., Flores H.E. (Eds.) Biochemistry and physiology of polyamines in plants. - Boca Raton, FL, CRC Press, pp. 121-133.

Quaedvlieg N.E., Schlaman H.R., Admiraal P.C., Wijting S.E., Stougaard J., and Spaink H.P. 1998. Fusions between green fluorescent protein and beta-glucuronidase as sensitive and vital bifunctional reporters in plants. - Plant Mol. Biol. 37: 715-727.

Racz I., Kovacs M., Lasztity D., Veisz O., Szalai G., Paldi E. 1996. Effects of short-term and long-term low temperatures stress on polyamine biosynthesis in wheat genotypes with varying degrees of frost tolerance. - J. Plant Physiol. 148: 368-373.

Rain J.C. *et al.* 2001. The protein-protein interaction map of *Helicobacter pylori*. - Nature 409: 211-216.

Reddy G.N., Arteca R.N., Dai Y.R., Flores H.E., Negm F.B., Pell E.J. 1993. Changes in ethylene and polyamines in relation to mRNA levels of the large and small subunits of ribulose bisphosphate carboxylase/oxygenase in ozone-stressed potato foliage. - Plant Cell Environ. 16: 819-826.

Richards F.J., Coleman E.G. 1952. Occurrence of putrescine in potassium deficient barley. - Nature 170: 160-161.

Roy M., Ghosh B. 1996. Polyamines, both common and uncommon, under stress in rice (*Oryza sativa*) callus. - Physiol. Plant. 98: 196-200.

Roy M., Wu R. 2001. Arginine decarboxylase transgene expression and analysis of environmental stress tolerance in transgenic rice. - Plant Sci. 160: 869-875.

Roy M., Wu R. 2002. Overexpression of S-adenosylmethionine decarboxylase gene in rice increases polyamine level and enhances sodium chloride-stress tolerance. - Plant Sci. 163: 987-992.

Schena M., Shalon D., Davis R.W., Brown P.O. 1995. Quantitative monitoring of gene expression patterns with a complementary DNA microarray. - Science 270: 467-470.

Schoof H., Karlowski W.M. 2003. Comparison of rice and *Arabidopsis* annotation. - Curr. Op. Plant Biol. 6: 106-112.

Serafini-Fracassini D., Del Duca S., Beninati S. 1995. Plant transgutaminases. - Phytochemistry 40: 355-365.

Shelp B.J., Bown A.W., McLean M.D. 1999. Metabolism and functions of gamma-aminobutyric acid. - Trends Plant Sci. 4: 446-452.

Smith T.A. 1985. Polyamines. - Annu. Rev. Plant Physiol. 36: 117-143.

Soyka S., Heyer A. 1999. *Arabidopsis* knockout mutation of ADC2 gene reveals inducibility by osmotic stress. - FEBS Lett. 458: 219-223.

Su J., Shen Q., Ho T.-H.D., Wu, R. 1998. Dehydration-stress-regulated transgene expression in stably transformed rice plants. - Plant Physiol. 117: 913-922.

The *Arabidopsis* Genome Initiative. 2000. Analysis of the genome sequence of the flowering plant *Arabidopsis thaliana*. - Nature 408: 796-815.

Tiburcio A.F., Altabella T., Borrell A., Masgrau C. 1997. Polyamine metabolism and its regulation. - Physiol. Plant. 100: 664-674.

Tissier A.F., Marillonnet S., Klimyuk V., Patel K., Torres M.A., Murphy G., Jones J.D.G. 1999. Multiple independent defective suppressor-mutator transposon insertions in *Arabidopsis*: a tool for functional genomics. - Plant Cell 11: 1841-1852.

Trull M.C., Holaway B.L., Malmberg R.L. 1992. Development of stigmatoid anthers in a tobacco mutant - implications for regulation of stigma differentiation. - Can. J. Bot. 70: 2339-2346.

Tucker C.L., Gera J.F., Uetz P. 2001. Towards an understanding of complex protein networks. - Trends Cell Biol. 11: 102-106.

Uetz P. *et al.* 2000. A comprehensive analysis of protein-protein interactions in *Saccharomyces cerevisiae*. - Nature 403: 623-627.

Van Buuren M.L., Guidi L., Fornale S., Ghetti F., Franceschetti M., Soldatini F., Bagni N. 2002. Ozone-response mechanisms in tobacco: implications of polyamine metabolism. - New Phytol. 156: 389-398.

Van Wijk K.J. 2001. Challenges and prospects of plant proteomics. - Plant Physiol. 126: 501-508.

Verma R., Chen S., Feldman R., Schieltz D., Yates J., Dohmen J., and Deshaies R.J. 2000. Proteasomal proteomics: identification of nucleotide-sensitive proteasome-interacting proteins by mass spectrometric analysis of affinity-purified proteasomes. - Mol. Biol. Cell 11: 3425-3439.

Walhout A.J.M., Sordella R., Lu X., Hartley J.L., Temple G.F., Brasch M.A., Thierry-Mieg N., Vidal M. 2000a. Protein interaction mapping in *C. elegans* using proteins involved in vulval development. - Science 287: 116-122.

Walhout A.J.M., Boulton S.J., Vidal M. 2000b. Yeast two-hybrid systems and protein interaction mapping projects for yeast and worm. - Yeast 17: 88-94.

Wang C.Y., Wang C.Y., Wellburn A.R. 1990. Role of ethylene under stress conditions. - In: Alscher R., Cumming J. (Eds.) Stress responses in plants: Adaptation and acclimatation mechanisms. - New York, Wiley-Liss, pp. 147-173.

Wasinger V.C., Cordwell S.J., Cerpa-Poljak A., Yan J.X., Gooley A.A., Wilkins M.R., Duncan M.W., Harris R., Williams K.L., Humphery-Smith I. 1995. Progress with gene-product mapping of the Mollicutes: *Mycoplasma genitalium*. - Electrophoresis 16: 1090-1094.

Watson M.B., Emory K.K., Piatak R.M., Malmberg R.L. 1998. Arginine decarboxylase (polyamine synthesis) mutants of *Arabidopsis thaliana* exhibit altered root growth. - Plant J. 13: 231-239.
Weigel D. *et al*. 2000. Activation tagging in Arabidopsis. - Plant Physiol. 122: 1003-1013.
Winkler R.G., Frank M.R., Galbraith D.W., Feyereisen R., Feldman, K.A. 1998. Systematic reverse genetics of transfer-DNA-tagged lines of *Arabidopsis*. - Plant Physiol. 118: 743-750.
Wu S.-H., Ramonell K.M., Gollub J., Somerville S. 2001. Plant gene expression profiling with DNA microarrays. - Plant Physiol. Biochem. 39: 917-926.
Zhu J.K., Hasegawa P.M., Bressan R.A. 1997. Molecular aspects of osmotic stress in plants. - CRC Crit. Rev. Plant Sci. 16: 253-277.
Zhu J.K. 2002. Salt and drought stress signal transduction in plants. - Annu. Rev. Plant Biol. 53: 247-273.

Chapter 9

SIGNIFICANCE OF SECONDARY NITROGEN METABOLITES FOR FOOD QUALITY

Silvia Haneklaus and Ewald Schnug

INTRODUCTION

Nitrogen is essential for all organisms and a major constituent of amino and nucleic acids, proteins, proteids, nucleotides, enzymes and secondary compounds. In its functions N cannot be replaced by any other mineral as for instance the complementary features known for potassium and sodium (Marschner 1986). N deficiency in plants results in reduced protein synthesis and a diminished biological value of the crop. Characteristic symptoms of severe N deficiency are a stiff appearance together with apical chlorosis of the older leaves, which may proceed towards necrosis of the plant tissue. Plants show a poor degree of tillering, reduced growth rate and finally diminished crop yield. The symptomatological threshold for macroscopic symptoms ranges from 1.1 to 4.8% N and critical N concentrations vary between 1.2 and 5.9% N depending on crop type, stage of development and crop yield (Reuter *et al.* 1997).

Secondary metabolites contribute significantly to food quality, either as nutritives, or antinutritives. From N-containing secondary metabolites the structure of about 12,000 are alkaloids, 600 non-protein amino acids, 100 amines, 100 cyanogenic glycosides and 100 glucosinolates is known (Wink 1999a). It is the objective of this section to provide basic information about different aspects of food quality with special view to secondary compounds.

Food production implies food security, food safety and food quality. The Rome declaration on world food security "reaffirms the right of everyone to have access to safe and nutritious food, consistent with the right to adequate food and the fundamental right of everyone to be free from hunger" (World Food Summit 1996). Food safety regulations aim to protect consumers against food related diseases such as contamination with food borne pathogens, for example eggs and vegetables with *Salmonella*, fruit with *Cyclospora*, and frozen strawberries with the virus for hepatitis A, and naturally occurring mycotoxins

(Venkitanarayanan and Doyle 2001). Food or product quality comprises nutritional physiological quality, sensory quality and technical-physical quality (Anon 2003). The nutritional physiological quality is related to the content of energy, nutrients, fibres and secondary components as well as undesired substances (Anon 2003). The N supply is one important factor affecting food quality, both positively in terms of protein content and concentration of secondary metabolites, and negatively in terms of enrichment with nitrate, nitrite and amides.

Food quality, however, comprises much more aspects, which need to be considered. Numerous case studies have shown a relationship between food quality and the farming system (Anon 2003). The farming system is perhaps the most demanding standard for food quality. On this level not only nutritional aspects, but also the impacts the agricultural production system has on organisms that are part of the production system (*e.g.* marine environmental pollution by nitrogen and phosphorus) and other ecosystems are crucial parameters. This means that food quality is no longer restricted to the simple measurement of physical parameters, but is extended to an holistic approach, which implies the evaluation of all factors involved in food production (Anon 2003, Schnug 2003).

Intensive agricultural production systems are usually characterised by an excess of N, particularly on farms with high livestock densities, which may adversely affect not only crop quality, but also the environment. In general an N surplus results in increased levels of nitrate and amides in plants, enhances the susceptibility to pathogens and encourages N losses through leaching and volatile emissions, for instance NH_3 and N_2O, the latter being one of the contributors to global warming (Rolston and Venterea 2001). Since the change of the millennium there is an increasing striving by producers and consumers towards organic farming and organic products. Organically produced foodstuff requires a certification based on the EU regulation 2092/91/EWG for organic farming systems. Organic farming is a challenge to overcome problems of conventional farming systems such as health problems caused by pesticides or antibiotics, hormones etc. used in animal husbandry, environmental problems caused by pesticides and agrochemicals, pollution caused by animal manure and organic waste products and diminished biodiversity in ecosystems, including agricultural production as the use of synthetic pesticides and mineral N fertilisers is banned, and stocking density and the use of antibiotics in animal husbandry is strictly limited (Schmidt and Haccius 1998). Organically produced foodstuff may have significantly higher contents of secondary metabolites (Schlee 1992, Ebata *et al.* 1993, Brandt and Mølgaard 2001, Ren *et al.* 2001) and there are indications that organic food is beneficial for human health (Marckmann 2000). A common argument against organic farming is that it is archaic and does not take advantage of modern agricultural research. But it is the improved understanding of biological and physiological processes required in organic farming systems, for instance in order to maintain plant health by nutritive measures, which is an example for progress through know-how (Haneklaus et al. 2002).

The introduction of fast, processed and frozen foods in the 1950's had a lasting impact on dietary habits and caused a significant increase of obesity, type II diabetes, high blood pressure and heart disease (Kimbrell 2002). In order to overcome possible shortages brought about by an unbalanced diet the food industry offers nutraceuticals. Nutraceuticals comprise isolated nutrients, dietary supplements, diets, genetically engineered designer foods, herbal products and processed products (Andlauer and Fuerst 2002). A nutraceutical can be any substance that may be considered a food, or part of a food, and provides medical or health benefits including the prevention and treatment of disease (DeFelice 1992). Dietary supplements are products in the form of pills, tablets, capsules, liquids or powders that contain one or more nutritional substances in concentrated form (mainly vitamins, minerals and trace elements, but also secondary metabolites such as alliin, glutathione and chlorophyll). Dietary supplements and nutraceuticals need to be critically evaluated because their regular intake may support, or even encourage, malnutrition with as yet unknown consequences for health. It would be much better to promote the interest and consumption of 'authentic' food as provided by organic farming (Schnug 2003), rather than designer food, in order to take advantage of the whole range of compounds in natural food and their synergetic effects.

Phytopharmaceuticals are old, but recently rediscovered remedies for many medical disorders resulting from the side effects of environmentally dispersed chemicals, increasing resistance to antibiotics, carryover of chemicals such as antibiotics into the food chain and increasing costs for health care (Gruenwald 1998, Diaz-Cruz et al. 2003). The bioactive components in medicinal plants comprise the whole range of secondary metabolites and crop specific cultivation strategies, which include fertilisation, harvesting and processing techniques that are required for producing a consistently high level of bioactive constituents. Warranting a consistently high quality of the raw materials can be a problem, particularly if the active agent is unstable and decomposes after harvesting of the plant material, as is true for many secondary metabolites such as alliins and glucosinolates (Bloem et al. 2001a).

There are more than 100,000 known secondary plant compounds and for only a limited number of them the biochemical pathways, functions, nutritional and medicinal significance is known (Wink 1999b). Bioactive secondary plant compounds comprise various substances such as carotenoids, phytosterols, glucosinolates, flavonoids, phenolic acids, protease inhibitors, monoterpenes, phyto-oestrogens, sulphides, chlorophylls and roughages (Watzl and Leitzmann 1999). Humans take up about 1.5 g secondary compounds per day with their diet (Ames et al. 1990) and a causal relationship between consumption of secondary compounds and risk of degenerative diseases has been assumed (Steinmetz and

Potter 1996). Thus secondary plant compounds are to be considered to be invaluable for food quality.

The aim of this chapter is to outline the significance of individual N containing compounds for food quality and to provide examples for secondary N containing compounds used as nutraceuticals and phytopharmaceuticals with a special view to the impact of N fertilisation on the concentration of the secondary metabolites in different plants.

NITROGEN CONTAINING SECONDARY COMPOUNDS

So far, more than 100,000 secondary plant metabolites have been identified (Wink 1999a). N-free metabolites are, for instance, terpenes, polyketides, saponins and polyacetylenes and N-containing compounds are alkaloids, amines, cyanogenic glycosides, non-protein amino acids and glucosinolates (Wink 1999a). Wink and Waterman (1999) provide a chemotaxonomic classification of plant species in relation to molecular phylogeny. About 20 years ago, secondary compounds were classified as antinutritive because they were thought to reduce the utilisation of primary compounds so breeding and technological efforts were aimed at reducing their content (Watzl and Leitzmann 1999).

Often secondary metabolites are accumulated in plant tissues and concentrations of 1-3% dry weight have been determined (Wink, 1999a). Secondary compounds in plants usually have a pharmacological effect on humans (Watzl and Leitzmann 1999). Sullivan and Hagen (2001) provide detailed information about the psychotropic effect of alkaloids. Secondary metabolites are used as dyes (*e.g.* indigo, shikonin), flavours (*e.g.* vanilline, capsaicine, mustard oils), fragrances (*e.g.* rose oil, lavender oil), stimulants (*e.g.* caffeine, nicotine, ephedrine), hallucinogens (*e.g.* morphine, cocaine), insecticides (*e.g.* nicotine, pyrethrine), vertebrate and human poisons (*e.g.* strychnine, coniine), but their possible functions in plants have, until now, often been obscure (Wink 1999a). Secondary metabolites have been shown to be crucial for plant defence mechanisms against viruses, bacteria, fungi, competing plants (allelopathies) and herbivores, and are triggers for pollination and seed dispersal (Wink 1999a). Flowers, fruits and seeds are usually rich in secondary metabolites (Wink 1999a). Several secondary metabolites show a regular turnover in plant metabolism, for instance on a daily base as in case of nicotine (Wink 1999a).

Most secondary compounds are not part of the basic molecular structure of the plant cell (primary metabolism), but are formed in special tissues or plant parts at defined stages of plant growth (Mohr and Schopfer 1995). The precursors for secondary metabolites from primary metabolism are shown in a simplified scheme (Figure 1). The sites of biosynthesis of secondary metabolites are compartmentalised in the plant cell (Wink 1999a). Most pathways are localised in the cytoplasm, either completely or partially, but there are indications that some alkaloids such as coniine and caffeine are synthesised in the chloroplast (Wink 1999a). Coniine and amine formation has been localised in mitochondria (Wink 1999a). Storage can also be tissue and cell specific (Guern et al. 1987). Some alkaloids and glucosinolates are, for instance, stored in specific idioblasts (Wink 1999a). Besides physical compartment, glucosinolates can be chemically compartmentalised in such way that both substrate (glucosinolates) and enzyme (myrosinase) are in the cytosol whereby the myrosinase activity is triggered by the ascorbic acid concentration, which is linked to core metabolic processes (Halliwell cycle) of the plant (Schnug and Haneklaus 1994). For metabolic pathways of secondary N containing metabolites see for instance Luckner (1990), Mohr and Schopfer (1995), Rosa et al. (1997), Roberts and Strack (1999), Selmar (1999), Wink (1999b) and De Luca and Laflamme (2001).

Nutritive and non-nutritive components in food have the potential to modulate target functions in the body, which are relevant to well being and health and/or reduction of disease risk (Roberfroid 1998). Some crops such as bamboo, lima beans and cassava accumulate toxic, cyanogenic glucosides in the range of 8-300 mmol kg^{-1}, and maceration and evaporation of the cyanide is required before they are edible for humans and animals (Selmar 1999). Antinutritives are, for instance, substances which interfere with the metabolic utilisation of minerals (Berdanier 2002). The degradation of glucosinolates yields for example thiocyanates, isothiocyanates, cyclic sulphur compounds and nitriles, which are goitrogenic. So-called cabbage goiter or struma inhibits the iodine uptake of the thyroid gland in humans (Berdanier 2002). One of the most potent glucosinolates is progoitrin, found in leaves, buds, seeds and roots of *Brassica* crops (Berdanier 2002). Though glucosinolates are antinutritive, their potential as anticarcinogenics has been recognised (Watzl and Leitzmann 1999).

In the following sections an overview of N-containing secondary metabolites with a view to the nutritional quality of plant products will be given and the influence of the N supply on their content will be addressed.

ALKALOIDS

Coniine ($C_5H_{10}NC_3H_7$), the Socratic poison, was the first alkaloid that was synthetically produced by Albert Ladenburg in 1886. About 30 different alkaloids are used in medicine, *e.g.* atropine, morphine, chinine and reserpine. Alkaloids are a group of N containing bases, which mostly have drug characteristics. The number and toxicity of alkaloid containing plants and the populations of herbivores and pathogens increases as latitude decreases (Salmore and Hunter 2001).

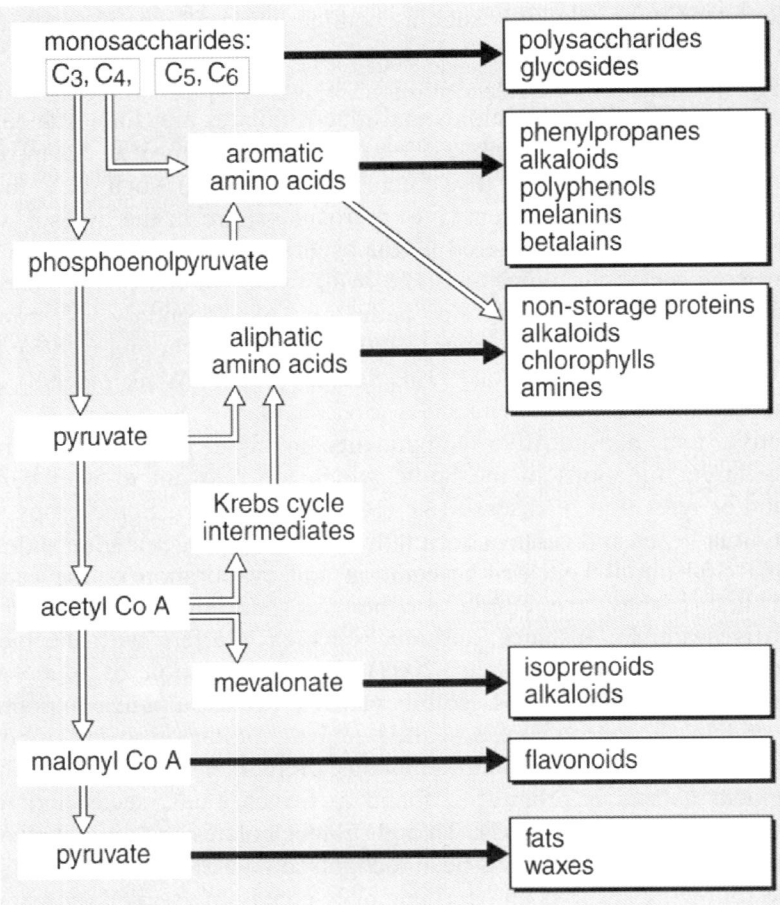

Figure 1. Groups of secondary compounds and their derivation from primary metabolism (Adapted from Mohr and Schopfer 1995).

N bound in alkaloids, non-protein amino acids and cyanogenic glycosides serve as an N reserve pool and N is reutilised during germination (Wink 1999a). Also, alkaloids and amines are effective defence mechanisms against herbivores and carnivores because they interfere with neuronal signal transduction (Wink and Schimmer 1999). The mode of actions of individual alkaloids is described in detail by Wink and Schimmer (1999). The antimicrobial activity of different alkaloids is summarised by Reichling (1999).

Phytoalexins are low molecular organic compounds rapidly synthesised by plants in response to infection or stress at the site of infection or stress. The phytoalexins of *Papaver somniferum* and *Thalictrum rugosum* are the alkaloids sanguinarine and berberine (Eilert *et al.* 1985, Funk *et al.* 1987). The phytoalexins of the *Brassica* family are synthesised as stress metabolites by cycling the thiocyanate component of indol glucosinolates after microbial infections or abiotic stress (Gross 1993). Their synthesis is triggered after infection by elicitors, either formed by the pathogen (*e.g.* substances of the cell wall of the fungus), or the host (*e.g.* breakdown products of the plant cell wall) (Mohr and Schopfer 1994). The immunity is generally of short duration and concentrated around the infected area. The involvement of phytoalexins in disease resistance depends upon the speed and magnitude with which they are produced and not on their selective toxicity or selectivity of elicitation (Kuć 1994).

Alkaloids are usually divided according to biogenetic principles, plant origin or basic heterocyclic ring systems (Figure 2, Franzke 1998).

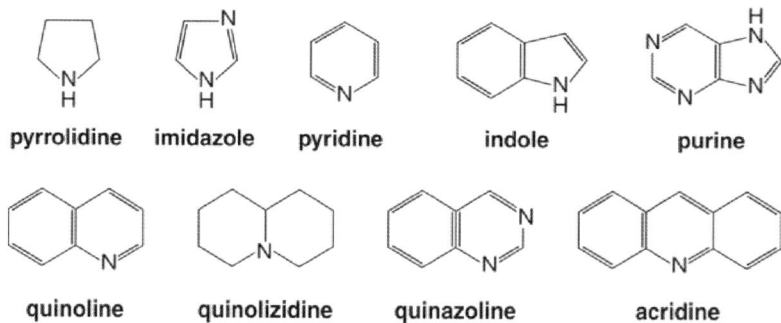

Figure 2. Heterocyclic ring systems of alkaloids (adapted from Franzke 1998).

Most alkaloids are derived from amino acids, *e.g.* lupinine and lupanine, which are bitter components of lupine, from lysine; ephedra, taxus and lunaria alkaloids from phenylalanine and d-tubicurarin from tryptophane. In the case of terpene and steroid alkaloids, N derives, for instance, from ammonia or ethanolamine (Franzke 1998). In Table 1 the basic components for alkaloid synthesis are listed.

In foodstuff the alkaloids caffeine, theobromine, theophylline, capsaicine, nicotine, piperine, chavicine and solanidine are of practical relevance (Franzke 1998). Piperine is the main alkaloid of pepper (Franzke 1998). Capsaicine is a phenylalkylamine, which occurs in paprika and pepper and causes the hot taste (Franzke 1998). For food safety solanum alkaloids in potatoes, aubergines and paprika are relevant (Franzke 1998). A concentration of more than 100 mg kg^{-1} fresh weight solanidin alkaloids in potatoes should not be exceeded due to the bitter taste and potential risk for human health; a dose of 25 mg is toxic and 400-500 mg are lethal for humans (Franzke 1998).

Table 1. Basic compounds for alkaloid synthesis, trivial names and plant origins of selected alkaloids (Oksmancaldentey *et al.* 1987, Luckner 1990, Franzke 1998, Roberts and Strack 1999, De Luca and Laflamme 2001).

Basic precursors	Alkaloid name/group based on origin	Family/genus/species
Histidine	pilocarpine	*Pilocarpus*
Lysine	lupinine, lupanine	*Lupinus*
Ornithine	hyoscamine, atropine	*Hyoscyamus, Atropa belladonna*
Phenylalanine	*Epehdra, Taxus* and *Lunaria* alkaloids	*Epehdra, Taxus, Lunaria*
Tyrosine	mecaline, colchicine	*Trichocereus pachanoi, Colchicum autumnale*
Tryptophane	*Vinca, Rauvolfia* and *Cataranthus* alkaloids	*Apocynaceae, Loganiaceae, Rubiaceae*
Anthranilic acid	quinoline, acridine and benzodiazepine alkaloids	*Rutaceae, Compositae*
Nicotinic acid	nicotine, ricinine, acalyphine	*Nicotiana, Ricinus communis, Acalypha indica*
Purine	caffeine, theobromine and theophylline	*Coffea arabica, Camellia sinensis, Ilex bonplandiana, Cola nitida, Theobroma cacao*
Polyketide	coniine	*Conium maculatum*
Phenylalkylamine	capsaicine	*Capsicum*
Terpenoid	aconitine	*Aconitum karacolicum*
Steroid	solanum alkaloids	*Solanum, Lycopersicon, Liliaceae, Asclepiadacea*

There are two basic concepts, which form the basis for the regulation of synthesis and allocation of secondary compounds. The carbon nutrient balance hypothesis formulated by Bryant *et al.* (1983) postulates that nutrient use and carbon allocation is balanced in such a way that an ample nutrient availability decreases the C/N ratio, favours rapid growth, more plasticity in allocation, and lower levels of N free secondary components (C, H, O). In comparison a low nutrient availability in the growth medium results in a higher C/N ratio, low growth rate, non plastic allocation patterns, and high levels of N free secondary components. Hoeft *et al.* (1996) conclude that their results fit to the carbon nutrient balance hypothesis because primary and secondary metabolites were not intercorrelated in their experiment. A high light intensity lowered the alkaloid content, and when the N supply was low, a higher proportion was bound in alkaloids; the total alkaloid content of *Tabernaemontana pachysiphon* increased, however, with fertilisation (Hoeft *et al*. 1996). In contrast, the concept of homeostasis assumes that complex physiological processes maintain a steady state in the plant organism. Lynds and Baldwin (1998) found a homeostasis in the allocation pattern of nicotine which was independent on N source and N fertiliser rate.

Gershenzon (1984) reviewed the literature on the influence of N supply and N fertilisation on the alkaloid content of different plant species and concluded that the effect of N fertilisation is related to the alkaloid type. Terpenoid derived alkaloids are not affected by N fertilisation, while those derived from amino acids show a clear increase. The N content of the alkaloid is another factor influencing the efficacy of N fertilisation in such a way that species producing N rich alkaloids show a consistent increase by N fertilisation (Gershenzon 1984).

The alkaloid content in lupines (*Lupinus* ssp.) may be as high as 4% in the dry matter (Aniszewski 1993). The quinolizidine alkaloid content depended on the variety and was about 40 times higher in bitter than in sweet lupines (Aniszewski *et al*. 2001). Also, the content of biogenic polyamines was significantly higher in bitter lupine varieties (Aniszewski *et al*. 2001). Aniszewski *et al*. (2001) suggest an internal balance of biogenic polyamines, quinolizidine alkaloids and basic amino acids in favour of the last in sweet lupines. The effect of N fertilisation on the alkaloid content in seeds of *Lupinus angustifolius* depended on climatic conditions and time of fertiliser application (Barlog 2002). On an average N fertilisation increased the alkaloid content by 9-17% (Barlog 2002).

Acock and Johnson (1998) found that an increasing N availability had no effect on the cocaine concentration in leaves of *Erythroxylum coca*, while dry weight of leaves increased. In comparison, N fertilisation increased the pinidine content in *Pinus ponderosa*, particularly under conditions of severe N deficiency, by 4.5 to 12 times and even 1.6 times when N supply was sufficient (Gerson and Kelsey 1999). Baricevic *et al*. (1999) determined the highest alkaloid content (hyosciamine and scopoloamine) of *Atropa belladonna* in a pot experiment when water supply was optimum and N supply was lowest; under water stress the highest N supply gave the highest alkaloid content. The isoquinoline content in rhizomes of *Sanguinaria canadensis* increased with its water content (Salmore and Hunter 2001). Gondola (2002) pointed out that N fertilisation of 200 kg N ha^{-1} increased the alkaloid content of tobacco from 1.46% to 2.61% compared to the control. A high N fertiliser rate compensated seasonal fluctuations in the alkaloid content (Gondola 2002). Infestation of *Lupinus albus* with *Cuscuta reflexa* caused a decline of the alkaloid quinolizidine content and synthesis by about 50% in the host (Baumel *et al*. 1995). During plant growth the N level decreased, but in infested plants the alkaloid content was 1.3-fold higher than in the noninfested plants, most of it found in *Cuscuta* (Baumel *et al.* 1995). Infestation apparently limited catabolic processes and rather promoted alkaloid synthesis in infested plants.

GLUCOSINOLATES

Glucosinolates are characteristic compounds of at least 15 dicotyledonous taxa. Out of these *Brassicaceae* are the most important agricultural crops. Glucosinolates act as attractants, repellents, insecticides, fungicides and antimicrobial protectors. The principle structure of a glucosinolate is given in Figure 3. About 80 different glucosinolates exist which consist of glucose, a sulphur containing group with an aglucon rest and a sulphate group (Watzl and Latzmann 1999). Alkenyl glucosinolates such as progoitrin and gluconapin have an aliphatic aglucon rest, while indole glucosinolates such as glucobrassicin and 4-hydroxyglucobrassicin in *Brassica napus* have an aromatic aglucon rest (Figure 3). Additional information on the characteristics of glucosinolate side chains is given by Underhill (1980), Larsen (1981), and Bjerg *et al.* (1987).

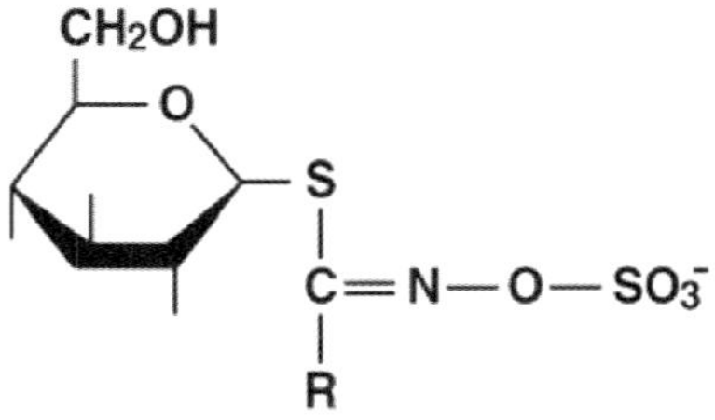

Figure 3. Basic structure of glucosinolates (Schnug 1990).

Glucosinolates are generally hydrolysed by the enzyme myrosinase, which is present in all glucosinolate containing plant parts. A proposed pathway for the recyclisation of sulphur (and nitrogen) under conditions of severe S deficiency is described by Schnug and Haneklaus (1994). Bones and Rossiter (1996) provide basic information about the biochemistry of the myrosinase/glucosinolate system. The degradation of glucosinolates results in the so called mustard oils, which are responsible for smell, taste and biological effect. Glucosinolates are vacuolar defence compounds (Wink and Schimmer 1999) of qualitative value (Rosenthal and Janzen 1979), which are effective against generalist insects at low tissue concentrations (Larsen *et al.* 1985). Isothiocyanates, the breakdown products after enzymatic cleavage of glucosinolates, may retard multiplication of spores, but do not hamper growth of fungal mycelium (Drobnica *et al.* 1967) and fungi may overcome the glucosinolate/ myrosinase system efficiently (Wu and Meijer 1999, Sexton and Howlett 2000).

Glucosinolates affect the flavour of spicy crops such as mustard and radish (Fenwick et al. 1983) and cause their pungency. Glucosinolates are antinutritive because of their goitrogenic effect (Berdanier 2002), which restricted for instance the use of extracted rape meal from single low oilseed rape cultivars in animal feeding. In human nutrition, glucosinolates are studied for their anticarcinogenic and antimicrobial potential (Watzl and Leitzmann 1999). In animal experimentation an anticarcinogenic effect could be attributed to isothiocyanates (phenethylisothiocyanate, bencylisothiocyanate, sulphoraphane), thiocyanates (bencylthiocyanate) and indoles (Watzl and Leitzmann 1999). Another beneficial effect of glucosinolates is their antimicrobial action, particularly that of bencylisothiocyanate (Watzl and Leitzmann 1999).

Secondary metabolites are often synthesised during maturation of crops. With a better degree of maturity, a higher concentration is expected in foods grown in organic farming systems (Mølgaard, 2000, Brandt et al. 2001). Broccoli is rich in sulphoraphane which, in animal feeding experimentation has shown a protective effect against chemically induced breast cancer (Zhang et al. 1994). Adam (2002) found sulphoraphane contents in broccoli that were two to six-fold higher in market products that were organically grown.

Generally, N fertilisation tended to result in lower glucosinolate contents (Rosa et al. 1997). Under field conditions the effect of N fertilisation on the glucosinolate content varied substantially between seasons (Asare and Scarisbrick 1995). Schnug (1989) found a distinct interaction between N and S fertilisation in the range of an insufficient N supply, whereby the alkenyl, but not the indole glucosinolate content in seeds of rapeseed increased with higher N and S rates. Kim et al. (2002) showed that N fertilisation increased particularly the alkenyl-glucosinolates, gluconapin and glucobrassicanapin, in *Brassica rapa*.

AMINES

Biogenic amines, except betains, result from the decarboxylation of amino acids (Franzke 1998). Amines may be divided into primary (R- NH_2), secondary (R_1, R_2-NH), tertiary (R_1, R_2, R_3N), and quarternary amines (R_1, R_2, R_3, R_4-N^+OH^-) (Luckner 1990). Polyamines such as spermidine and spermine comprise three or more amino groups (Luckner 1990). Biogenic amines are precursors of alkaloids and hormones, are neurotransmitters and components of phospholipids and vitamins (Altman and Levin 1993, Franzke 1998).

Polyamines can be found in all living organisms (Franzke 1998). Food products contain on average 20-40 mg kg^{-1} biogenic amines (Franzke 1998). Decarboxylation by bacteria can enhance the synthesis of amines so that fermented (sauerkraut) or contaminated products (rotten fish may contain up to 4,000 mg kg^{-1} histamine (Franzke 1998)) show distinctly higher amine concentrations (Shalaby 1996, Kalac et al. 2001).

Dietary polyamines have beneficial and detrimental effects on human health (Sørensen et al. 2001, Eliassen et al. 2002). Putrescine, spermine and spermidine are essential for cell proliferation, differentiation and renewal (Kalac et al. 2001, Eliassen et al. 2002, chapter 8). In comparison, histamine and tyramine may cause physiological disorders in humans, the first involved in food poisoning, the second possibly a trigger in the hypertensive crisis (Bauza et al. 1995, Shalaby 1996, Kalac et al. 2001). Polyamines stimulate growth and differentiation of tumours, and adversely effect the antitumour immune system (Eliassen et al. 2002). Secondary amines may react with nitrite, which yields carcinogenic nitrosamines (Silla Santos 1996).

Biogenic polyamines are positively charged cellular components, have a pH and cation effect (Friedman and Oshima 1989, Altman and Levin 1993), electrostatic binding to negatively charged components of nucleic acids and cell membrane phospholipids and have allelopathic functions in plants (Altman and Levin 1993). Biogenic polyamines function as N reserves in lupine (Aniszewski et al. 2001). The ratio of putrescine:spermidine:spermine was 100:10:1 in tobacco (Altman and Levin 1993). N fertilisation may increase the capsaicine content of pepper (Johnson and Decoteau 1996) and the histamine, putrescine and cadaverine content in red wines (Bauza et al. 1995). The form of N nutrition may also be of significance as NH_4^+ nutrition yielded a 100-fold increase in putrescine, and a moderate increase in other diamines and polyamines in soybean (Le Rudulier and Goas 1975). The intake of silage with elevated levels of amines by ruminants had a negative influence on health and physical condition due to a detrimental impact on ruminal and intestinal mucuous membranes (Krizek 1995). While the influence of N fertilisation was minor, suboptimal silage storage conditions increased the content of biogenic amines to values higher than 0.1% (Krizek 1995).

CYANOGENIC GLYCOSIDES

Cyanogenic glycosides are derived from the amino acids valine, isoleucine, leucine, phenylalanine and tyrosine, and the non-protein amino acid cyclopentenglycine (Selmar 1999). In Figure 4 the basic structures of cyanogenic glycosides and their precursors are given. So far more than 60 different structures of cyanogenic glycosides are known; basically they consist of α-hydroxynitriles, which are stabilised by glucose (Selmar 1999). Cyanogenic glycosides are stored in the vacuole and thus are separated from their hydrolytic enzymes (Wink and Schimmer 1999). After cell disrupture, β-glucosidase splits the glycoside into sugar and a nitrile moiety, which is further hydrolysed to hydrocyanic acid (HCN) and an aldehyde (Wink and Schimmer 1999). The high toxicity of cyanide makes these compounds an effective repellent against herbivores (Woodhead and Bernays 1977, Goodger et al. 2002). In comparison, most microorganisms are not sensitive against cyanide because of their ability to use the cyanide insensitive "alternative" respiratory pathway and their capability to detoxify cyanide, respectively (Selmar 1999). Further information about biochemical, physiological and nutritional aspects of cyanogenic glycosides is given by Selmar (1999).

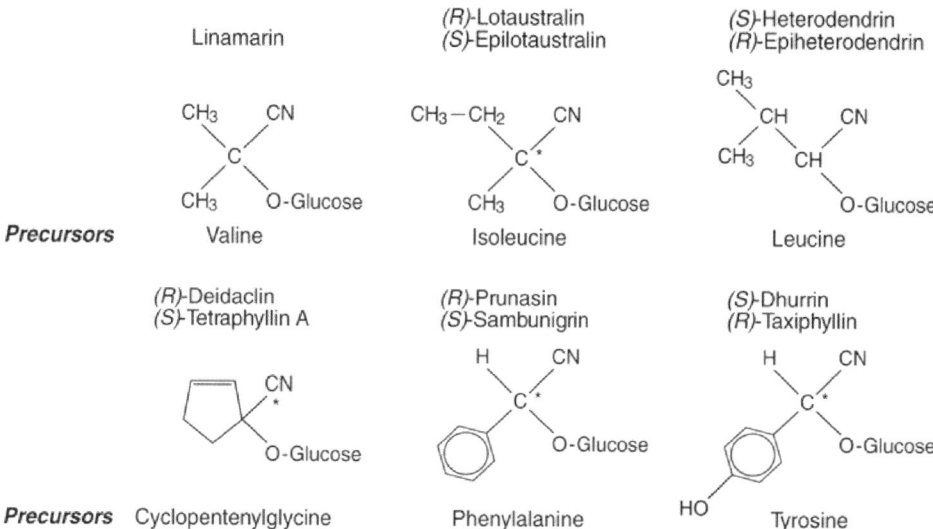

Figure 4. Basic structure of cyanogenic glycosides and their precursors (adapted from Selmar 1999)

More or less all plants contain cyanogenic glucosides and the cyanide content may be as low as 1 µg kg^{-1} in rice or as high as 2,100 µg kg^{-1} in plum juice (Lang 1990). Food processing of plants rich in cyanogenic glycosides, such as bamboo and cassava, by maceration for enzymatic cleavage and evaporation for final removal, is necessary in order to prevent toxic effects after consumption by humans or cattle (Selmar 1999).

Eucalyptus cladocalyx binds 15-20% of N in the cyanogenic glycoside prunasin (Gleadow and Woodrow 2000, Burns et al. 2002). The cyanogenic glycoside concentration was highest in young vegetative and generative plant tissue (Gleadow and Woodrow 2000). These results indicate that plant tissue, which is most susceptible to pathogens enriches cyanogenic glycosides. Plants grown in shade had lower prunasin concentrations relative to the N and chlorophyll content, also under conditions of a sufficient N supply, which shows that N was preferably used for the photosynthetic apparatus (Burns et al. 2002). Similar results were determined by Busk and Moller (2002) for the influence of N fertilisation on the dhurrin content of *Sorghum* leaves, which was more pronounced in full light than under shading conditions (Wheeler *et al.* 1990). By comparison, Goodger *et al.* (2002) found no relationship between foliar N content and prunasin concentration in two different *Eucalyptus polyanthernos* populations.

NON PROTEIN AMINO ACIDS

Non protein amino acids (NPAAs) are structural analogues to the 20 protein amino acids (Wink and Schimmer 1999). The structure of some simple NPAAs is given in Figure 5. In protein amino acids the amino group is linked to the α-carbon, while in NPAAs the amino group can be bound to the β-carbon as for instance in 3-alanine or to the γ-carbon like in 4-amino butyric acid (Figure 5). NPAAs are usually present in the free state (Luckner 1990). The formation of NPAAs is regulated either by specific pathways, the modification of primary amino acids or the modification of biosynthetic pathways leading to primary amino acids (Luckner 1990). About 900 amino acids, which are not linked to the primary metabolism, were identified and thus can be regarded as secondary metabolites (Lambein *et al.* 2001).

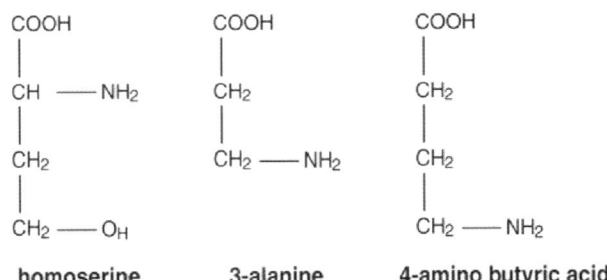

Figure 5. Structure of some simple non-protein amino acids (adapted from Selmar 1999).

Canavanine, a structural analogue of arginine is present in many legume crops and affects in particular the immune system, but is also discussed as a chemotherapeutic agent for human pancreas cancer (Lambein *et al.* 2001). Besides canavanine, mimosine and canaline are tested as anticancer metabolites (Lambein *et al.* 2001). Other NPAAs, for example glutamine derivates are tested in pharmacology (Brauner-Osborne *et al.* 1997).

Toxic NPAAs in plants may offer protection against herbivores (Rosenthal 2001). Possible effects of NPAAs on pests and diseases are, however, obscure at the moment. NPAAs function as storage pools for reduced N (Shelp *et al.* 1999).

Seeds of *Griffonia simplicifolia* and *Mucuna mutisiana* contained for example 14% 5-hydroxy-L-tryptophan and 8% L-DOPA (L-3,4-dihydroxyphenylalanine), respectively (Luckner 1990). Other functions of NPAAs are the provision of energy by, for example, creatine phosphate, and osmotic regulation, for example by betaines (Luckner 1990).

N fertilisation increased the content of δ-acetylornithine in rhizomes of *Bistorta bistortoides* (Lipson et al. 1996). *Leguminosae* are a major source for NPAAs such as albizine, canavanine and lathyrine (Wink and Waterman 1999), but the influence of the N nutritional status on their synthesis remains speculative.

NUTRACEUTICALS

The costs for nutrition related diseases are estimated to exceed currently 50 billion € in Germany (Anon 2003). The main reasons being malnutrition, overconsumption and physical inactivity (Anon 2003). The purchase of nutraceuticals suggests the compensation of these faults to consumers and the rapidly expanding market for nutraceuticals is at least economically highly beneficial for the food industry. A nutraceutical can be any substance that may be considered a food or part of a food and provides medical or health benefits including the prevention and treatment of disease (DeFelice 1992). Nutraceuticals are sold under "hard claims", which are claims related to activities against diseases and "soft claims" which are used to describe preventive health claims (Juretko 1999). Nutraceuticals comprise isolated nutrients, dietary supplements, diets, genetically engineered designer foods, herbal products and processed products (Andlauer and Fuerst 2002). The terms nutraceuticals or functional food are approved in Japan as FOSHU (Food for Specific Health Use) if the food or added ingredients have been acknowledged to provide potential health benefits and which comprises of carbohydrates, proteins, minerals and other components such as rice globulin and eucommia leaf glycoside (Goldberg 1994). The question, however, remains open whether nutraceuticals may not only promote malnutrition, but will also have adverse long-term effects, particularly through regular intake of high doses of bioactive compounds in the form of different products fortified with the same component.

There is substantial evidence from epidemiological studies that diets, which are rich in fruits and vegetables, may prevent cardiovascular diseases and cancer (Steinmetz and Potter 1996). Potentially anticarcinogenic compounds include *Allium* compounds, isothiocyanates and chlorophyll (Dashwood 1997, Watzl and Leitzmann 1999). The significance of secondary metabolites in the diet for cancer prevention is summarised by Weisburger and Butrum (2002). The same authors stress that dietary supplements are not as effective as the food, presumably because of synergistic effects among different components.

In this section, the significance of secondary metabolites as nutraceuticals will be discussed exemplary for alliin, glutathione and chlorophyll, all of them being part of a healthy diet and available as dietary supplements.

Alliins

Allium species contain four S-alk(en)yl-L-cysteine sulfoxides, namely S-1-propenyl-, S-2-propenyl, S-methyl- and S-propyl-L-cysteine sulphoxides (Block 1992). *Iso*-alliin is the main form in onions, while alliin is the predominant form in garlic (Kawakishi and Morimitsu 1994, Figure 6).

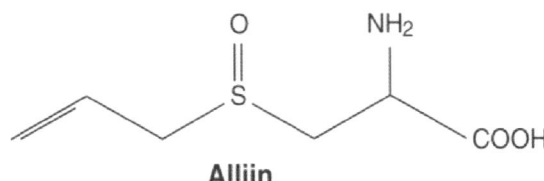

Alliin

Figure 6. Chemical structure of alliin (adapted from Watzl and Leitzmann 1999).

The characteristic flavour of *Allium* species is caused after the enzyme alliinase hydrolyses cysteine sulfoxides to form pyruvate, ammonia and S containing volatiles. In the intact cell alliin and related cysteine sulfoxides are located in the cytoplasm while the C-S lyase enzyme alliinase is localised in the vacuole (Lancaster and Collin 1981). Disruption of the cell releases the enzyme, which causes subsequent α,β-elimination of the sulfoxides, ultimately affording volatile and odorous low molecular weight organosulphur compounds (Block and Calvey 1994). The cysteine sulfoxide content of *Allium* species is an important quality parameter with regard to organoleptic features since it determines the taste and sharpness.

Alliin acts as an antioxidant by activating glutathione enzymes (see below) and has anticarcinogenic and antimicrobial effect (Watzl and Leitzmann 1999). *In vitro* the degradation of *iso*-alliin yields thiopropanal S-oxide that strongly inhibits the human platelet aggregation, which would explain the therapeutic effect of onions (*Allium cepa* L.) on vascular diseases such as thrombosis, arteriosclerosis, hyperlipidemia and rheumatic arthritis (Kawakishi and Morimitsu 1994). Garlic is used against arteriosclerosis, high blood pressure, and has been shown to have antibacterial, antifungal, antiviral and antiprotozoal activities. It also modulates the cardiovascular and immune system and has antioxidative and anticarcinogenic properties (Harris *et al.* 2001).

On average 21% of S, but only 0.9% of N were bound as *iso*-alliin in onion bulbs at the start of bulb growth (Bloem *et al.* 2003). Interactions between N and

S supply existed in such a way that an N and S fertilisation decreased total S and N contents in onion (Bloem *et al.* 2003). N rate and N form influenced the alliin content and flavour quality of onions (Gamiely *et al.* 1991, Coolong and Randle 2003). N fertilisation increased the synthesis of flavour precursors in onions (Randle 2000), but Bloem *et al.* (2003) found no significant influence of N fertilisation on the *iso*-alliin content in onion and the alliin content in garlic bulbs. There was, however, a tendency that higher N supply resulted in a decreased alliin content (Bloem *et al.* 2003).

Glutathione

The reduced glutathione tripeptide (L-γ-glutamyl-L-cysteinylglycine, Figure 7) is conventionally called glutathione. In plants glutathione is among others essential as an antioxidant, for the modulation of enzyme activity and the synthesis of phytochelatins (Foyer and Noctor 2001). Glutathione is part of the immune system in the human body (Galen 2000). Glutathione is an endogenous, water soluble, non-enzymatic antioxidant, which maintains exogenous antioxidants such as vitamins C and E in their reduced, active forms (Bray and Taylor 1994, Watzl and Leitzmann 1999). The enzymes glutathione peroxidases and glutathione reductase catalyse reactions by which glutathione eliminates free radicals and peroxides (Watzl and Leitzmann 1999). Consequently, glutathione is discussed as an anticarcinogenic (Bounous *et al.* 1991). A higher activity of glutathione-*S*-transferase induced for instance by sulphides, isothiocyanates and indole glucosinolates is one mechanism for inactivating carcinogenic substances (Watzl and Leitzmann 1999).

$$O=C(OH)-\overset{\alpha}{CH}(NH_2)-CH_2-CH_2-\overset{\gamma}{C}(=O)-NH-CH(CH_2SH)-C(=O)-NH-CH_2-C(=O)OH$$

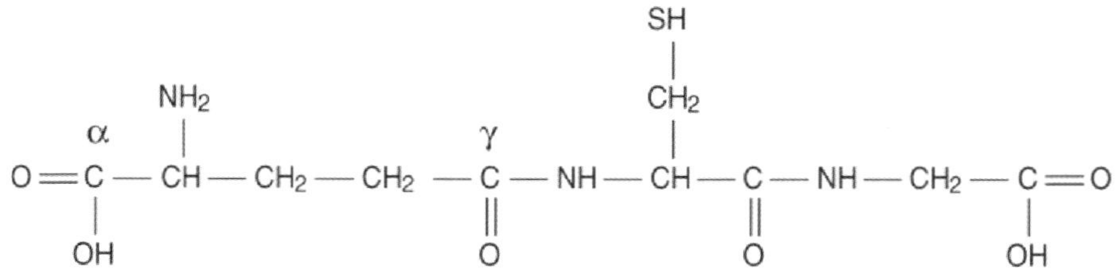

Figure 7. Principal structure of the reduced form of glutathione (adapted from Wonisch and Schaur 2001).

Asparagus (*Asparagus officinalis*), avocado (*Persea gratissima*) and watermelon (*Citrullus lanatus*) are particularly rich in glutathione with mean contents of 35, 31 and 280 mg glutathione kg^{-1} fresh weight, respectively (Saito *et al.* 2000). Young, developing leaves showed the highest glutathione content, which then decreased with plant age (Bielawski and Joy 1986). Shalaby (2003) found genotypical differences in the glutathione content of different white asparagus cultivars. N fertilisation had no significant influence on the glutathione content in asparagus spears, while S fertilisation significantly increased the glutathione content (Shalaby 2003).

Chlorophyll

Chlorophylls are health promoting phytochemicals (Ferruzzi *et al.* 2001). The chemical structure of chlorophyll a is given in Figure 8. After intake of chlorophyll, chlorophyllin is mediated in the gastrointestinal tract of humans whereby the magnesium atom is substituted by a metal, for example Cu, Co or Fe (Watzl and Leitzmann 1999); both components can impair the occurrence of cancer (Dashwood 1997). Egner *et al.* (2001, 2003) found that a mixture of semisynthetic derivates of chlorophyll, which is commercially available, reduced the urinary levels of an aflatoxin biomarker by 55% compared to the placebo group. The same authors conclude that a diet rich in chlorophyll might be efficient in cancer protection. Some mycotoxins cause cell membrane damage through increased lipid peroxidation. Chlorophyll and its derivates as well as other compounds (*e.g.* vitamins, carotenoids, phenolics), are discussed as being efficient against toxic effects of mycotoxins through acting as superoxide anion scavengers (Atroshi *et al.* 2002).

Photosynthetic N use efficiency determines plant and leaf N productivities (Garnier *et al.* 1995). Ercoli *et al.* (1993) found a close relationship between N and chlorophyll a and b content on a leaf area base in a N response experiment with maize (Figure 9). An increase of 1 mmol N m^{-2} was associated with an increase of 11 µmol chlorophyll a m^{-2} and 3.9 µmol chlorophyll b m^{-2} (Ercoli *et al.* 1993).

The relationship between N and chlorophyll a and b content, respectively was close with a coefficient of correlation of r = 0.91 and r = 0.87; the coefficient of correlation was r = 0.91 for the sum of chlorophyll a and b (Ercoli 1993). In comparison Yoder and Pettigrew-Crosby (1995) found only a coefficient of correlation of r = 0.66 for maple leaves under controlled N management. Ercoli *et al.* (1993) and Yoder and Pettigrew-Crosby (1995) worked with leaves of defined age and insertion, respectively. For undefined leaf material there is, however, no general relationship between chlorophyll content and specific leaf weight and N content (Baret and Fourty 1997), because there are many more environmental factors affecting the chlorophyll content of plant tissue than the N supply, such as drought, temperature, diseases and nutrient disorders (Haneklaus and Schnug 2002). Here, a fractionation of N bound in thylakoid membranes as pigment/protein complexes (chlorophyll a and b) and soluble proteins (photorespiratory enzymes) would be required to determine the relationship between N status and chlorophyll content as the proportion of total N in thylakoid membranes remains constant with increasing N per unit leaf area, while the proportion of soluble protein increases (Evans 1989).

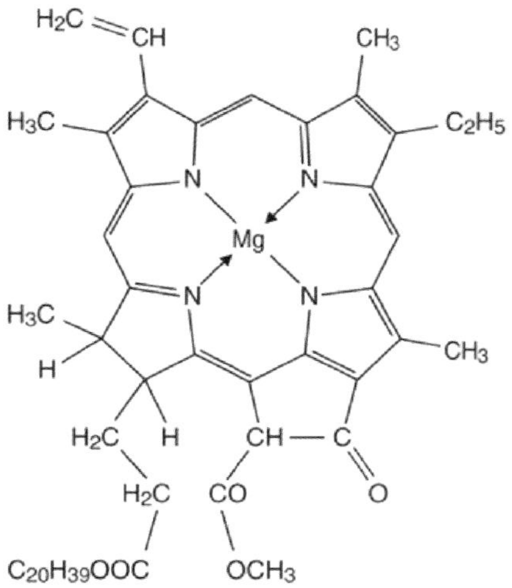

Figure 8. Chemical structure of chlorophyll a (adapted from Schroeter *et al.* 1995).

PHYTOPHARMACEUTICALS

Plants provide an invaluable source of secondary metabolites, which are used in their isolated form and which are indispensable in medical treatments, for example morphine, atropine and codeine. Heilmann and Bauer (1999) give an overview of new secondary compounds with pharmaceutical potential. The alkaloids ellipticine, 9-methoxyellipticine and olivacine were isolated from *Ochrosia* and *Aspidosperma* species and showed in vitro antitumour activity (Heilmann and Bauer 1999). Taxane derived compounds were already isolated in the 1960's. Paclitaxel is a diterpenoid with an alkaloidal side chain, which was first isolated from *Taxus brevifolia* (Heilmann and Bauer 1999). Paclitaxel was identified as being cyctotoxic to human epidermoid carcinoma cells and in leukaemia cell systems, but also other cancer types (Wani *et al.* 1971). The indolizidine alkaloid swainsonine is an important compound against viral infections such as HIV and is constituent of different locoweed species (Heilmann and Bauer 1999). Other antiviral N containing secondary metabolites are the α- and β-glucosidase I inhibitors, desoxynojirimycin and castanospermine from *Castanospermum australe* and the dimeric naphtylisoquinoline alkaloids, so-called michellamines from *Ancistrocladus korupensis* (Heilmann and Bauer 1999).

Besides the potential of secondary metabolites for pharmacology, several thousand plants are used in phytomedicine to treat human ailments and diseases (Wink 1999a) and their therapeutically active compounds claim among others to act as anticarcinogenic, antibiotic, antihypertensive and cholesterol reducing agents (Srivastava *et al.* 1995, Verhoeven *et al.* 1997). Phytopharmaceuticals are old remedies for many medical disorders. As long ago as the 12th century Hildegard von Bingen (1098 - 1183) described the medicinal action of 213 trees and plants and most of her knowledge is still valid today. The World Health Organisation (WHO) defines phytopharmaceuticals as medicine, which active compounds only consist of plant material for example plant powder, plant secretion, essential oil or plant extracts (Kern 1996). The registration of new products is a complicated process due to the lack of clinical investigations, which prove their medical activity. A herbal preparation is considered as one active compound of a drug, although it contains many different chemically defined constituents. Isolated constituents of herbal origin are not considered to be herbal remedies (Keller 1991). In Germany, herbal medicines have a special status set up by the Commission E, which investigated and affirmed the safety and effectiveness of phytomedicines, of in total, 363 herbal plants (Blumenthal *et al.* 1998). The Commission E handbook provides the decision basis for the registration of new and the re-registration of old products. Monodrugs receive preferential admission, while combination products, which contain parts of more than three different plants, are expected to receive no admission after 2004 (Karl 1999).

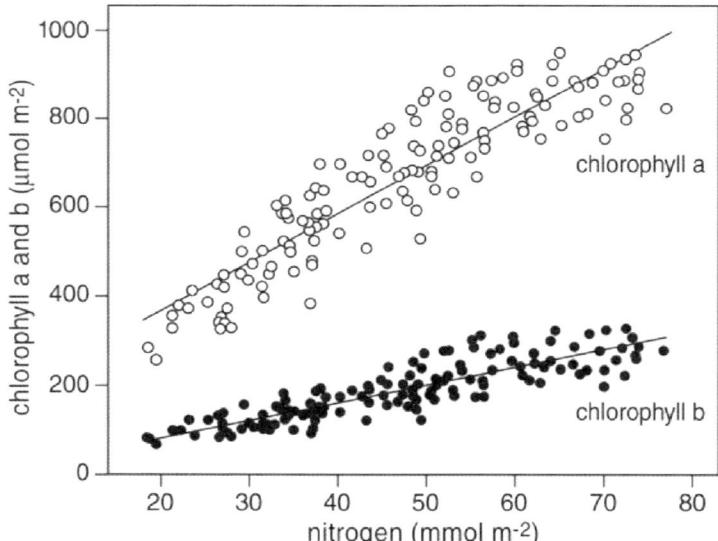

Figure 9. Relationship between N and chlorophyll a and b content in *Zea mays* in dependence on N fertiliser rates (adapted from Ercoli *et al.* 1993).

Today phytopharmaceuticals are recommended for primary healthcare (Benzi and Ceci 1997). In this section, the use of secondary N containing compounds such as phytopharmaceuticals will be discussed, exemplary for glucotropaeolin.

Glucotropaeolin

There is a great number of well known herbal plants such as horseradish, garlic and mustard with various bioactive compounds, such as glucosinolates, alliins, glutathione, but little is known about the specific effect of a compound and its interaction with other constituents. An oil extract from nasturtium (*Tropaeolum majus*) for example is 20 times more effective than the synthetically produced analogue (Winter 1954, Klesse and Lukoschek 1955).

Glucotropaeolin is the characteristic, aromatic glucosinolate found in *T. majus* which has been indicated for the treatment of scurvy, bronchitis, cystitis, pyelitis and as a general tonic and stimulant (Bardeau 1978). In the pharmaceutical industry it has long been recognised for its antimicrobial action against urethral infections (Dannenberg *et al*. 1956, Bergmann *et al*. 1966). The active component is benzylisothiocyanate, which is released after enzymatic cleavage of glucotropaeolin by myrosinase. Benzylisothiocyanate derived from *T. majus* may also have the ability to induce the synthesis of protective enzymes, which reduces the effects of chemical carcinogens (Fahey *et al*. 1997).

Variation of glucosinolate concentration within a plant is caused by genotypical differences and numerous environmental factors such as plant age, plant part, temperature, light intensity and plant pathogen interactions (Elliot and Stowe 1971, Rosenthal and Janzen 1979, Gershenzon 1984, Schnug 1987, Fieldsend and Milsford 1994, Rosa *et al*. 1997, Rosa and Rodrigues 1998). Besides these natural fluctuations, glucosinolates are prone to degradation after activation of the myrosinase enzyme following cellular disruption. This means that losses of the intact glucosinolate are inevitable during harvest and preservation of the crop (Bones and Rossiter 1996) and therefore a principal, major handicap for the pharmaceutical use of *T. majus*. The strongest exogenous factor influencing the glucosinolate content in both, vegetative and generative tissues of *Brassica* crops, proved to be, however, the sulphur supply (Schnug 1987, Walker and Booth 1994). N fertilisation did not significantly affect the glucotropaeolin content of *T. majus* (Figure 10, Bloem *et al*. 2001b).

S fertilisation, however, increased the glucotropaeolin content in leaves and stems by about 0.43 and 0.11 μmol kg^{-1} applied S, respectively. The glucotropaeolin content of seeds of *Tropaeolum majus* was almost doubled by S fertilisation from 33 to 56 μmol g^{-1} (Bloem *et al*. 2001b). The results reveal the significance of an optimum nutrient supply in order to achieve a consistently high crop quality, a criterion which is of particular relevance if the phytopharmaceuticals are directly prepared from the harvested plant material (Bloem *et al*. 2003).

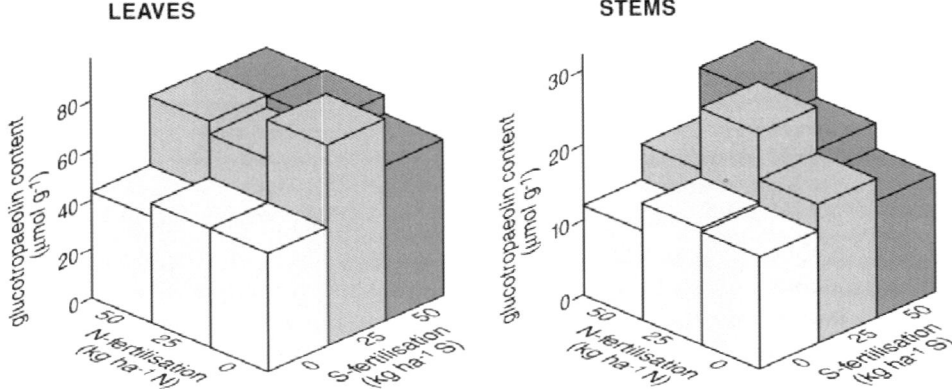

Figure 10. Influence of sulphur and nitrogen fertilisation on the glucotropaeolin content in leaves and stems of *T. majus* at the start of flowering (Bloem *et al.* 2001b).

CONCLUSIONS

The aspects of food safety and food quality are of prime interest in developed countries (Anon 2003). The significance of secondary metabolites in plants as protection against biotic and abiotic stress, as a criteria for food quality and as an invaluable potential for pharmacological and phytopharmaceutical use is meanwhile undisputed. For numerous secondary metabolites their role in plant and human metabolism is still unclear. The same applies to the influence of agronomic measures on the concentration of secondary metabolites in plants. A healthy diet is, in any case, not only based on high quality food, but is also closely linked to dietary habits (Read 2002). Nutraceuticals suggest a diet that balances possible deficiencies and provide consumers with a feeling of safety, but their consumption promotes the substitution of natural, manifold food.

ACKNOWLEDGEMENTS

The authors wish to express their most sincerest thanks to Herbert Daybell (Agrimedia, Bottesford, UK) for the linguistic revision of this contribution.

REFERENCES

Acock M.C., Johnson E.L. 1998. Modeling the effects of nitrogen on components of coca yield. - J. Plant Nutr. 21: 1501-1509.
Adam S. 2002. Vergleich des Gehaltes an Glucoraphanin in Broccoli aus konventionellem und aus ökologischem Anbau. - Bundesforschungsanstalt für Ernährung (Hrsg.): Jahresbericht 2001.

Altman A., Lewin N. 1993. Interactions of polyamines and nitrogen nutrition in plants. - Physiol. Plant. 89: 653-658.
Ames B.N., Profet M., Gold L.S. 1990. Dietary pesticides (99.99% all natural). - Proc. Natl. Acad. Sci. USA 7777-7781.
Andlauer W., Fuerst P. 2002. Nutraceuticals: a piece of history, present status and outlook. - Food Res. 35: 171-176.
Aniszewski T. 1993. Lupine: a potential crop in Finland. Studies on the ecology, productivity and quality of *Lupinus* ssp. - PhD thesis, University of Joensuu, Finland.
Aniszewski T., Ciesiolk D., Gulewicz K. 2001. Equilibrium between basic nitrogen compounds in lupin seeds with differentiated alkaloid content. - Phytochemistry 57: 43-50.
Anon 2003. Bewertung von Lebensmitteln verschiedener Produktionsverfahren. - Statusbericht des Senats der Bundesforschungsanstalten, 103 pp.
Asare E., Scarisbrick D.H. 1995. Rate of nitrogen and sulfur fertilizers on yield, yield components and seed quality of oilseed rape (*Brassica napus* L.). - Field Crops Res. 44: 41-46.
Atroshi F., Rizzo A., Westermarck T., Ali-Vehmas T. 2002. Antioxidant nutrients and mycotoxins. - Toxicology 180: 151-167.
Baret F., Fourty Th. 1997. Radiometric estimates of nitrogen status of leaves and canopies canopies. - In: Lemaire G. (Ed.) Diagnosis of the nitrogen status in crops. - Berlin, Germany, Springer Verlag, pp. 201-227.
Barlog P. 2002. Effect of magnesium and nitrogenous fertilisers on the growth and alkaloid content in *Lupinus angustifolius* L. - Aust. J. Agric. Res. 53: 671-676.
Baumel P., Jeschke W.D., Rath N., Czygan F.C., Proksch P. 1995. Modelling quinolozidine alkaloid net flows in *Lupinus albus* and between *Lupinus albus* and the parasite *Cuscuta reflexa* - new insights into the site of quinolizidone alkaloid synthesis. - J. Exp. Bot. 46: 1721-1730.
Bauza T., Blaise A., Mestres J.P., Teissedre P.L., Daumas F., Cabanis J.C. 1995. Biogenic amines contents and their variation parameters in Cotes Du Rhone, Vallee Du Rhone and Provence Wines. - Sciences des Aliments 15: 367-380.
Benzi G., Ceci A. 1997. Herbal medicines in European regulation. - Pharmacol. Res. Commun. 35: 355-362.
Berdanier, C.D. 2002. Food constituents. - In: Berdanier D., Feldman E.B., Flatt W.P., Jeor A.T.St. (Eds.) Handbook of nutrition and food. - Boca Raton, CRC Press, pp. 3-98.
Bergmann M., Lipsky H., Glawogger F. 1966. An antibiotic from *Tropaeolum majus* against urethral infections. - Med. Clin. 61: 1469-1472.
Bielawski W., Joy K.W. 1986. Reduced and oxidised glutathione and glutathione reductase activity in plant tissue of *Pisum sativum*. - Planta 169: 267-272.
Bjerg B., Kachlicki P.W., Larsen L.M., Sorensen H. 1987. Metabolism of glucosinolates. - Int. Rapeseed Congress, Poznan, Vol. 1/2: 496-506.
Block E. 1992. The organosulfur chemistry of the genus *Allium*. Implications for organic sulfur chemistry. - Angewandte Chemie 31: 1135-1178.
Block E., Calvey E.M. 1994. Facts and artifacts in *Allium* chemistry. - American Chemical Society Symposium Series ACS 564: 63-79.
Bloem E., Haneklaus S., Köhler S., Sator C., Schnug E. 2001a. Produktionstechnische Maßnahmen zur Erhöhung des Glucotropaeolingehaltes von Kapuzinerkresse (*Tropaeolum majus* L.). - In: NAROSSA: 7th Int. Conf. for Renewable Resources, Magdeburg, Germany, pp. 1-10.
Bloem E., Haneklaus S., Peplow E., Sator C., Köhler T., Schnug E. 2001b. The effect of sulphur and nitrogen fertilisation on the glucotropaeolin content in *Tropaeolum majus* (L.). - Proceedings der XXXVI. Vortragstagung Gewürz- und Heilpflanzen der DGQ, Jena, Germany, pp. 185-190.
Bloem E., Haneklaus S., Schnug E. 2003. Influence of nitrogen and sulphur fertilisation on the alliin content of onions (*Allium cepa* L.) and garlic (*Allium sativum* L.). - J. Plant Nutr. (*in press*).

Blumenthal M., Busse W.R., Goldberg A., Gruenwald J., Hall T., Riggins C.W., Rister R.S. 1998. The complete German commission E monographs: Therapeutic guide to herbal medicines. - Austin (TX), Boston (MA), American Botanical Council in cooperation with Integrative Medicine Communications.

Bones A.M., Rossiter J.T. 1996. The myrosinase-glucosinolate system, its organisation and biochemistry. - Physiol. Plant. 97: 194-208.

Bounos G., Batist G., Gold P. 1991. Whey proteins in cancer prevention. - Cancer Lett. 57: 91-94.

Brandt K, Mølgaard J.P. 2001. Organic agriculture: does it enhance or reduce the nutritional value of plant foods? - J. Sci. Food Agric. 81: 924-931.

Brauner-Osborne H., Nielson B., Stensbol T.B., Johanson T.N., Skjaerbaek, N. 1997. Molecular pharmacology of 4-substituted glutamic acid analogues at ionotropic and metatropic excitatory amino acid receptors. - Eur. J. Pharmacol. 335: R1-R3.

Bray T.M., Taylor C.G. 1994. Enhancement of tissue glutathione for antioxidant and immune functions in malnutrition. - Biochem. Pharmacol. 47: 2113-2123.

Bryant J.P., Chapin F.S., Klein D.R. 1983. Carbon nutrient balance of boreal plants in relation to vertebrate herbivory. - Oikos 40: 357-368.

Burns A.E., Gleadow R.M., Woodrow I.E. 2002. Light alters the allocation of nitrogen to cyanogenic glycosides in *Eucalyptus cladocalyx*. - Oecologia 133: 288-294.

Busk P.K., Moller B.L. 2002. Dhurrin synthesis in sorghum is regulated at the transcriptional level and induced by nitrogen fertilization in older plants. - Plant Physiol. 129: 1222-1231.

Coolong T.W., Randle W.M. 2003. Ammonium nitrate fertility levels influence flavour development in hydroponically grown 'Granex 33' onion. - J. Sci. Food Agric. 83: 477-482.

Dannenberg H., Stickl H., Wenzel F. 1956. About the antimicrobial component of *Tropaeolum majus*. - J. Physiol. Biochem. 303: 248-256.

Dashwood R.H. 1997. Chlorophylls as anticarcinogens (Review). - Int. J. Oncol. 10: 721-727.

DeFelice S.L. 1992. The nutraceutical initiative: a recommendation for U.S. economic and regulatory reforms. - Gen. Engin. News 12: 13-15.

De Luca V., Laflamme P. 2001. The expanding universe of alkaloid biosynthesis. - Curr. Opin. Plant Biol. 4: 225-233.

Diaz-Cruz M.S., de Alda M.J.L., Barcelo D. 2003. Environmental behavior and analysis of veterinary and human drugs in soils, sediments and sludge. - Trends Anal. Chem. 22: 340-351.

Drobnica L., Zemanová M., Nemec P., Antos K., Kristián P., Stullerová M., Knoppavá V., Nemec P. 1967. Antifungal activity of isothiocyanates and related compounds. I. Naturally occurring isothiocyanates and their analogues. - Appl. Microbiol. 15: 701-709.

Ebata J., Kawai K., Furukawa H. 1993. Inhibitory effects of dietary leafy vegetables on mutagens and on active oxygens. - In: Bronzetti G., Hayatsu H., deFlora S., Waters M.D., Shankel D.M. (Eds), Antimutagenesis and anticarcinogenesis mechanisms III. - New York, USA, Plenum Press, pp. 99-102.

Egner P.A. *et al.* 2001. Chlorophyllin intervention reduces aflatoxin-DNA adducts in individuals at high risk for liver cancer. - Proc. Natl. Acad. Sci. USA 98: 14601-14606.

Egner P.A., Munoz A., Kensler T.W. 2003. Chemoprevention with chlorophyllin in individuals exposed to dietary aflatoxin. - Mutation Res. 523: 209-21.

Eilert U., Kurz W.G., Constabel F. 1985. Stimulation of sanguinarine accumulation in *Papaver somniferum* cell cultures by fungal elicitors. - J. Plant Physiol. 119: 86-76.

Eliasen K.A., Reistad R., Risoen U., Rønning H.F. 2002. Dietary polyamines. - Food Chem. 78: 273-280.

Elliot M.C., Stowe B.B. 1971. Distribution and variation of indole glucosinolates in woad (*Isatis tinctoria* L.). - Plant Physiol. 48: 498-503.

Ercoli L., Mariotti M., Masoni A., Massantini F. 1993. Relationship between nitrogen and chlorophyll content and spectral properties in maize leaves. - Eur. J. Agron. 2: 113-117.

Fahey J.W., Zhang Y., Talalay P. 1997. Broccoli sprouts: An exceptionally rich source of inducers of enzymes that protect against chemical carcinogens. - Proc. Natl. Acad. Sci. USA 94: 10366-10372.

Fenwick G.R., Heaney R.K., Mullin W.J. 1983. Glucosinolates and their breakdown products in food and food plants. - CRC Crit. Rev. Food Sci. Nutr. 18: 123-201.

Ferruzzi M.G., Failla M.L., Schwartz S.J. 2001. Assessment of degradation and intestinal cell uptake of carotenoids and chlorophyll derivates from spinach puree using an in vitro digestion and caco-2 human cell model. - J. Agric. Food Chem. 49: 2082-2089.

Fieldsend J., Milford G.F.J. 1994. Changes in glucosinolates during crop development of single and double low genotypes of winter oilseed rape (*Brassica napus*): I. production and distribution in vegetative tissues and developing pods during development and potential role in the recycling of sulphur within the crop. - Ann. Appl. Biol. 124: 531-542.

Foyer C.H., Noctor G. 2001. The molecular biology and metabolism of glutathione. - In: Grill D., Tausz M, De Kok L.J. (Eds.) Significance of glutathione in plant adaptation to the environment. - Dordrecht, The Netherlands, Kluwer Acad. Publ., pp. 27-56.

Franzke C. 1998. Allgemeines Lehrbuch der Lebensmittelchemie. - Hamburg, Germany, Behr's Verlag.

Funk C., Gügler K., Brodelius P. 1987. Increased secondary product formation in plant cell suspension cultures after treatment with a yeast carbohydrate preparation (elicitor). - Phytochemistry 26: 401-405.

Galen S.M.D. 2000. What to do for your liver when you are on toxic overload. - J. Nat. Health 6: 7.

Gamiely S., Randle W.M., Mills H.A., Smittle D.A., Banna G.I. 1991. Onion plant growth, bulb quality, and water uptake following ammonium and nitrate nutrition. - Hort. Science 26: 1061-1063.

Garnier E., Gobin O., Poorter H. 1995. Nitrogen productivity depends on photosynthetic nitrogen use efficiency and on nitrogen allocation within the plant. - Ann. Bot. 76: 667-672.

Gershenzon J. 1984. Changes in the levels of plant secondary metabolites under water and nutrient stress. - In: Timmermann B.N., Steelink C., Loewus A. (Eds.) Plant adaptations to stress. - New York, Plenum Press, pp. 273-321.

Gerson E.A., Kelsey R.G. 1999. Piperidine alkaloids in nitrogen fertilized *Pinus ponderosa*. - J. Chem. Ecol. 25: 2027-2039.

Gleadow R.M., Woodrow I.E. 2000. Temporal and spatial variation in cyanogenic glycosides in *Eucalyptus cladocalyx*. - Tree Physiol. 20: 591-598.

Goldberg I. 1994. Functional foods: designer foods, pharmafoods, nutraceuticals. - Dordrecht, The Netherlands, Kluwer Academic Publishers.

Goodger J.Q.D., Capon R.J., Woodrow I.E. 2002. Cyanogenic polymorphism in *Eucalyptus polyanthemos* Schauer subsp vestita L. Johnson and K. Hill (*Myrtaceae*). - Biochem. Syst. Ecol. 30: 617-630.

Gondola I. 2002. Influence of crop year, N fertilization and genotype on the variability of some agronomic and chemical properties of Burley tobacco (*Nicotiana tabacum* L.). - Novenytermeles 51: 143-159.

Gross D. 1993. Phytoalexins of the *Brassicaceae*. - J. Plant Dis. Prot. 100: 433-442.

Gruenwald J. 1998. The emerging role of herbal medicine in health care in Europe. - Drug Inf. J. 32: 151-153.

Guern J., Renaudin J.P., Brown S.C. 1987. The compartimentation of secondary metabolites in plant cell cultures. - In: Constabel F., Vasil I. (Eds.) Cell culture and somatic cell genetics. - New York, Academic Press, pp. 43-46.

Haneklaus S., Schnug E. 2002. An agronomic, ecological and economic assessment of site-specific fertilisation. - FAL Agric. Res. 52: 123-133.

Haneklaus S., Bloem E., Schnug E. 2002. The Significance of Sulphur Induced Resistance (SIR) for Sustainable Agricultural Production Systems. - 13th International Reinhardsbrunn Symposium Modern Fungicides and Antifungal Compounds, Friedrichroda, Germany, pp. 365-372.

Harris J.C., Cottrell S.L., Plummer S., Lloyd, D. 2001. Antimicrobial properties of *Allium sativum* (garlic). - Appl. Microbiol. Biotech. 57: 282-286.

Heilmann J., Bauer R. 1999. New medical applications of plant secondary metabolites. - In: Wink M. (Ed.) Functions of plant secondary metabolites and their exploitation in biotechnology. - Ann. Plant Rev., Vol. 3, Boca Raton, CRC Press, pp. 274-310.

Hoeft M., Verpoorte R., Beck E. 1996. Growth and alkaloid contents in leaves of *Tabernaemontana pachysiphon* Stapf (Apocynaceae) as influenced by light intensity, water and nutrient supply. - Oecologia 107: 160-169.

Johnson C.D., Decoteau D.R. 1996. Nitrogen and potassium fertility affects Jalapeno pepper plant growth, pod yield, and pungency. - Hort. Science 31: 1119-1123.

Juretko A. 1999. Facts über Functional Food. - Lebensmitteltechnik 3: 29.

Kalac P., Krizek M., Spicka J. 2001. Biogenic amines in Sauerkraut. - In: Pfannhauser W., Fenwick G.R., Khokhar S. (Eds.) Biologically-active phytochemicals in food: analysis, metabolism, bioavailability and function. - Cambridge, United Kingdom, Royal Soc. Chem., pp. 217-220.

Karl J. 1999. Tendenzen der Arzneimittelzulassung. - Naturheilpraxis 1: 45-47.

Kawakishi S., Morimitsu Y. 1994. Sulfur chemistry of onions and inhibitory factors of the arachidonic- acid cascade. - American Chemical Society Symposium Series ACS 546: 120-127.

Keller K. 1991. Legal requirements for the use of phytopharmaceutical drugs in the Federal Republic of Germany. - J. Ethnopharmacol. 32: 225-229.

Kern A. 1996. Heilpflanzen im Garten und Brauchtum. - Examensarbeit am Institut für Botanik und Pharmazeutische Biologie der Universität Erlangen-Nürnberg.

Kimbrell A. 2002. Fatal harvest - the tragedy of industrial agriculture. - Washington, Island Press.

Klesse P., Lukoschek P. 1955. Untersuchungen über die akteriostatische Wirksamkeit einiger Senföle. - Arzneimittel Forsch. 5: 505-507.

Kim S.J., Matsuo T., Watanabe M., Watanabe Y. 2002. Effect of nitrogen and sulphur application on the glucosinolate content in vegetable turnip rape (*Brassica rapa* L.). - Soil Sci. Plant Nutr. 48: 43-49.

Krizek M. 1995. Possible ways of affecting biogenic-amine content in silages. - Vet. Med. (Prague) 40: 111-115.

Kuć J. 1994. Relevance of phytoalexins - a critical review. - Acta Hort. 381: 526-539.

Lambein F., Kuo Y.H., Ikegami F. 2001. Non-protein amino acids and food safety. - In: Pfannhauser W., Fenwick G.R., Khokhar S. (Eds.) Biologically-active phytochemicals in food: analysis, metabolism, bioavailability and function. - Cambridge, UK, Royal Soc. Chem., pp. 580-583.

Lancaster J.E., Collin H.A. 1981. Presence of alliinase in isolated vacuoles and alkyl cysteine sulphoxides in the cytoplasm of bulbs in onion (*Allium cepa*). - Plant Sci. Lett. 22: 169-176.

Lang I. 1990. Cyanogene Verbindungen in Nahrungs, Gewürz- und Genußmittelpflanzen sowie in Nahrungs- und Genussmitteln. - MSc, Faculty of Biology, University of Saarbrücken, Germany.

Larsen P.O. 1981. Glucosinolates. - In: E. Conn (Ed.), The biochemistry of plants, Toronto, Canada, Academic Press, pp. 501-525.

Larsen L.M., Nielsen J.K., Ploeger A., Sørensen H. 1985. Responses of some beetle species to varieties of oilseed rape and to pure glucosinolates. - In: Nijhoff M., Jungk W. (Eds.) Advances in the production and utilization of cruciferous crops. - Dordrecht, The Netherlands, Kluwer Acad. Publ., pp. 230-244.

Lipson D.A., Bowman W.D., Monson R.K. 1996. Luxury uptake and storage of nitrogen in the rhizomatous alpine herb, *Bistorta bistortoides*. - Ecology 77: 1277-1285.

Luckner M. 1990. Secondary metabolism in microorganisms, plants, and animals. - Jena, Germany, VEB Gustav Fischer Verlag.

Lynds G.Y., Baldwin I.T. 1998. Fire, nitrogen, and defense plasticity in *Nicotiana attentuata*. - Oecologia 115: 531-540.

Marckmann P. 2000. Organic foods and allergies, cancers and other common diseases - present knowledge and future research. - Proc. 13th Int. Federation of Organic Agriculture Movements (IFOAM) Sci. Conf., Basel, Switzerland, p. 312.

Marschner H. 1986. Mineral nutrition of higher plants. - London, UK, Academic Press.

Mohr H., Schopfer P. 1995. Plant Physiology. - Berlin, Germany, Springer Verlag.

Mølgaard J.P. 2000. Nutrients, secondary metabolites and foreign compounds in organic foods. - Proc. 13th Int. Federation of Organic Agriculture Movements (IFOAM) Sci. Conf., Basel, Switzerland, p. 313.

Oksmancaldentey K.M., Vuorela H., Isenegger M., Strauss A., Hiltunen R. 1987. Selection for high tropane alkaloid content in *Hyoscyamus muticus* plants. - Plant Breeding 99: 318-326.

Randle W.M. 2000. Increasing nitrogen concentration in hydroponic solutions affects onion flavor and bulb quality. - J. Am. Soc. Hortic. Sci. 125: 254-259.

Read M. 2002. The health-promoting diet throughout life: adults. - In: Berdanier D., Feldman E.B., Flatt W.P., Jeor A.T.St. (Eds.) Handbook of nutrition and food. - Boca Raton, CRC Press, pp. 299-317.

Reichling J. 1999. Plant-microbe interactions and secondary metabolites with antiviral, antibacterial and antifungal properties. - In: Wink M. (Ed.) Functions of plant secondary metabolites and their exploitation in biotechnology. - Ann. Plant Rev., Vol. 3, Boca Raton, CRC Press, pp. 187-273.

Ren H., Endo H., Hayashi T. 2001. Antioxidative and antimutagenic activities and polyphenol content of pesticide-free and organically cultivated green vegetables using water-soluble chitosan as a soil modifier and leaf surface spray. - J. Sci. Food Agric. 81: 1426-1432.

Reuter D.J., Robinson J.B., Peveril K.J., Price G.H., Lambert M.J. 1997. Guidelines for collecting, handling and analysing plant material. - In: Reuter D.J., Robinson J.B. (Eds.) Plant analysis, an interpretation manual. - Collingwood, Australia, Commonwealth Scientific and Industrial Research CSIRO Publishing.

Roberfroid M.B. 1998. Prebiotics and synbiotics: concepts and nutritional properties. - British J. Nutr. 80: 197-202.

Roberts M.F., Strack D. 1999. Biochemistry and physiology of alkaloids and betalains. - In: Wink M. (Ed.), Biochemistry of plant secondary metabolism, Ann. Plant Rev., Vol. 2, Boca Raton, CRC Press, pp. 17-78.

Rolsten D.E., Venterea R.T. 2001. Gaseous loss of oxides of nitrogen from the agricultural nitrogen cycle. - Proc. No 464, Int. Fert. Soc., York, UK.

Rosa E.A.S., Heaney R.K., Fenwick G.R., Portas C.A.M. 1997. Glucosinolates in crop plants. - Hortic. Rev. 19: 99-215.

Rosa E.A.S., Rodrigues P.M.F. 1998. The effect of light and temperature on glucosinolate concentration in the leaves and roots of cabbage seedlings. - J. Sci. Food Agric. 78: 208-212.

Rosenthal G.A., Janzen D.H. 1979. Herbivores: their interaction with secondary plant metabolites. - New York, Academic Press.

Rosenthal G.A. 2001. L-Canavanine: a higher plant insecticidal allelochemical. - Amino Acids 21: 319-330.

Saito M., Rai D.R., Masuda R. 2000. Effect of modified atmosphere packaging on glutathione and ascorbic acid content of asparagus spears. - J. Food Process. Preserv. 24: 243-251.

Salmore A.K., Hunter M.D. 2001. Elevational trends in defense chemistry, vegetation, and reproduction in *Sanguinaria canadensis*. - J. Chem. Ecol. 27: 1713-1727.

Schlee D. 1992. Ökologische Biochemie. - Jena, Germany, Gustav Fischer Verlag.

Schmidt, H., Haccius M. 1998. EU Regulation "Organic Farming". A legal and agro-ecological commentary on the EU's Council Regulation (EEC) No. 2091/91. - Weikersheim, Germany, Margraf Verlag.

Schnug E. 1987. Relations between sulfur-supply and glucosinolate content of 0- and 00 oilseed rape. - Proc. 7th Int. Rapeseed Congr., Poznan, Poland, pp. 682-686.

Schnug E. 1989. Quantitative und qualitative Aspekte der Diagnose und Therapie der Schwefelversorgung von Raps (*Brassica napus* L.) unter besonderer Berücksichtigung glucosinolatarmer Sorten. - Habilitationsschrift (Dsc thesis) Agrarwiss. Fakultät der Christian-Albrechts-Universität zu Kiel.

Schnug E. 1990. Glucosinolates - fundamental, environmental and agricultural aspects. - In: H. Rennenberg, C. Brunold, L. J. De Kok and I. Stulen (Eds) Sulfur nutrition and sulfur assimilation in higher plants. - The Hague, The Netherlands, SPB Academic Publishing, pp. 97-106.

Schnug E., Haneklaus S. 1994. Sulphur deficiency in *Brassica napus*: biochemistry, symptomatology morphogenesis. - Landbauforsch. Voelk., Special issue 144.

Schnug E. 2003. Organic grown crops in the South - challenges and implications. - Proc. Int. Assoc. of Agric. Students, World Congress 'Food Quality - a challenge for North and South', Leuven, Belgium, pp. 81-95.

Schroeter W., Lautenschlaeger K.H., Bibrack H. 1995. Taschenlexikon der Chemie. - Thun u. Frankfurt, Verlag Harri Deutsch.

Selmar D. 1999. Biosynthesis of cyanogenic glycosides, glucosinolates and nonprotein amino acids. - In: Wink M. (Ed.), Biochemistry of plant secondary metabolism, Ann. Plant Rev., Vol. 2, - Boca Raton, CRC Press, pp. 79-150.

Sexton A.C., Howlett B.J. 2000. Characterization of a cyanide hydratase gene in the phytopathogenic fungus *Leptosphaeria maculans*. - Mol. Gen. Genet. 263: 463-470.

Shalaby A.R. 1996. Significance of biogenic amines to food safety and human health. - Food Res. Intern. 29: 675-690.

Shalaby T. 2003. Genetical and nutritional influences on the spear quality of white asparagus (*Asparagus officinalis* L). - FAL Agric. Res. (*in press*).

Shelp B.J., Bown A.W., McLean M.D. 1999. Metabolism and functions of gamma-aminobutyric acid. - Trends Plant Sci. 4: 446-452.

Sørensen H., Sørensen J.C., Sørensen S. 2001. Phytochemicals in food: the plants as chemical factories. - In: Pfannhauser W., Fenwick G.R., Khokhar S. (Eds.) Biologically-active phytochemicals in food: analysis, metabolism, bioavailability and function. - Cambridge, UK, Royal Soc. Chem., pp. 1-12.

Srivastava K.C., Bordia A., Verma S.K. 1995. Garlic (*Allium sativum*) for disease prevention. - S. Afr. J. Sci. 91: 68-77.

Steinmetz K.A., Potter J.D. 1996. Vegetables, fruit, and cancer prevention: a review. - J. Am. Diet. Assoc. 96: 1027-1039.

Sullivan R.J., Hagen E.H. 2001. Psychotropic substance-seeking: evolutionary pathology or adaptation? - Addiction 97: 389-400.

Underhill E.W. 1980. Glucosinolates. - In: E. A. Bell and B. V. Charlwood (Eds) Encyclopedia of plant physiology, Vol. 8, Secondary plant products. - Berlin, Springer Verlag, pp. 493-511.

Venkitanarayanan K.S., Doyle M.P. 2001. Foodborne infections and infestations. - In:. Berdanier D., Feldman E.B., Flatt W.P., Jeor A.T.St. (Eds.) Handbook of nutrition and food. – Boca Raton, CRC Press, pp. 1135-1161.

Verhoeven D.T.H, Verhagen H., Goldbohm R.A., van der Brandt P.A., van Poppel G. 1997. A review of mechanisms underlying the anticarcinogenicity by *Brassica* vegetables. - Chemico-Biol. Interac. 103: 79-129.

Walker K.C., Booth E.J. 1994. Sulphur deficiency in Scotland and the effects of sulphur supplementation on yield and quality of oilseed rape. - Norw. J. Agric. Sci. 15: 97-104.

Wani M.C., Taylor H.L., Wall M.E., Coggon P., McPhail A.T. 1971. Plant antitumor agents. VI. The isolation and structure of taxol, a novel antileukemic and antitumor agent from *Taxus brevifolia*. - J. Am. Chem. Soc. 93: 2325-2327.

Watzl B., Leitzmann C. 1999. Bioaktive Substanzen in Lebensmitteln. - Stuttgart, Germany, Hippokrates Verlag, pp. 254.

Weisburger E.K., Butrum R. 2002. Chemoprevention of cancer in humans by dietary means. - In:. Berdanier D., Feldman E.B., Flatt W.P., Jeor A.T.St. (Eds.) Handbook of nutrition and food. - Boca Raton, CRC Press, pp. 1011-1028.

Wheeler J.L., Mulcahy C., Walcott J.J., Rapp G.G. 1990. Factors affecting the hydrogen-cyanide potential of forage sorghum. - Aust. J. Agric. Res. 41: 1093-1100.

Wink M. 1999a. Introduction: biochemistry, role and biotechnology of secondary metabolites. - In: Wink M. (Ed.) Functions of plant secondary metabolites and their exploitation in biotechnology. - Ann. Plant Rev., Vol. 3, Boca Raton, CRC Press, pp. 1-16.

Wink M. (Ed.) 1999b. Biochemistry of plant secondary metabolism. - Ann. Plant Rev., Vol. 2, Boca Raton, CRC Press.

Wink, M., Schimmer O. 1999. Modes of action of defensive secondary metabolites. - In: Wink M. (Ed.) Functions of plant secondary metabolites and their exploitation in biotechnology. - Ann. Plant Rev., Vol. 3, Boca Raton, CRC Press, pp 17-133.

Wink M., Waterman P.G. 1999. Chemotaxonomy in relation to molecular phylogeny of plants. - In: Wink M. (Ed.) Biochemistry of plant secondary metabolism, Ann. Plant Rev., Vol. 2 - Boca Raton, CRC Press, pp. 300-341.

Winter A.G. 1954. Untersuchungen über die der Natur der flüchtigen antibiotischen Wirkstoffe aus *Tropaeolum majus*. (Untersuchungen über Antibiotika aus höheren Pflanzen. X. Mitteilung). - Naturwissenschaften 41: 337-338.

Wonisch W., Schaur R.J. 2001. Chemistry of glutathione. - In: Grill D., Tausz M., De Kok L.J. (Eds.) Significance of glutathione in plant adaptation to the environment. - Dordrecht, The Netherlands, Kluwer Acad. Publ., pp. 13-26.

Woodhead S., Bernays E. 1977. Changes in release rates of cyanide in relation to palatability of *Sorghum* to insects. - Nature 279: 235-236.

World Food Summit 1996. Rome declaration on world food security, Rome, Italy.

Wu X.-M., Meijer J. 1999. In vitro degradation of intact glucosinolates by phytopathogenic fungi of Brassica. - Proc. 10th Int. Rapeseed Congress, Canberra (CD-ROM).

Yoder B.J., Pettigrew-Crosby R.E. 1995. Predicting nitrogen and chlorophyll content and concentrations from reflectance spectra (400-2500 nm) at leaf and canopy scales. - Remote Sensing Environ. 53: 199-211.

Zhang Y., Kensler T.W., Cho C.G., Posner G.H. 1994. Anticarcinogenic activities of sulforaphane and structurally related synthetic norbornyl isothiocyanates. - Proc. Natl. Acad. Sci. USA 91: 3147-3150.

Chapter 10

BIOTECHNOLOGY OF NITROGEN ACQUISITION IN RICE - IMPLICATIONS FOR FOOD SECURITY

Dev T. Britto and Herbert J. Kronzucker

INTRODUCTION

The intensive irrigated rice farms in the lowlands of southern Asia are the world's most important agricultural systems, covering some 79 million hectares, and accounting for approximately 70% of global rice production. While rice yields are sufficient to meet current human needs, they must increase by 60% within the next 30 years, if present projections for human population growth are correct (Ladha and Reddy 2003). Among the factors compromising the realisation of rice yield potential in the field, the supply and uptake rates of nitrogen are recognised as being among the foremost (Ladha and Reddy 2003). For instance, Horie *et al.* (1997) showed that the number of spikelets produced by rice plants was closely related to the plants' N status, and Witt *et al.* (1999) showed a similarly strong relationship between grain yield and plant N status.

Indeed, nitrogen use by rice plants was a key factor in the "green revolution" of the 1960s, wherein new varieties of rice could absorb and assimilate more nitrogen, without lodging, than their predecessors. Lodging resistance was due to the dwarf stature of these plants, and, in combination with other important traits such as high tillering, high harvest index, and dark green and erect leaves, allowed them to become much more productive from a grain-setting perspective (Khush 2001). However, the yield potential of existing high-yielding strains of rice is not achieved by most irrigated-rice farmers, who typically apply 86 to 138 kg N ha^{-1} crop^{-1} (Dobermann and Cassman 2002), even though an estimated 250 kg ha^{-1} crop^{-1} is required to reach near-record yields of 10 t ha^{-1}, assuming (conservatively) that 50% of the applied N is lost from the cropping system via volatilisation and other factors (Sheehy 2000).

Nevertheless, for both economic and environmental reasons, the required increase in rice production needs to be accompanied by a general decrease in fertiliser use (Khush 2001), which makes for a significantly greater challenge,

S. Amâncio and I. Stulen (eds.),
Nitrogen Acquisition and Assimilation in Higher Plants, 261-281.
© 2004 *Kluwer Academic Publishers. Printed in The Netherlands.*

particularly given the rather low efficiency of N capture by the crop. This challenge can be partially addressed by improvements in field management practices, such as through the enhancement of the formation and stability of nitrate-N in paddy soils, by selecting for increased O_2 exudation from rice roots (Kirk and Kronzucker 2000), or by alternating periods of flooding with drying periods that facilitate nitrification (Arth *et al.* 1998, Kronzucker *et al.* 2000). Such strategies could potentially reduce losses due to N volatilisation (the main route of N loss from the system) and the resulting co-presence of nitrate and ammonium in rice paddies might have a positive, synergistic effect on yield (Kronzucker *et al.* 1999, 2000). More generally, techniques of precision agriculture, in which the timing, dose, and depth of N application are coordinated with the requirements of the developing plant, appear promising (Cassman *et al.* 1998, Cassman 1999). However, while the implementation of precision agriculture is an important long-term goal, with potentially high returns, it is at present an expensive technology relative to the incomes of most Asian rice farmers (Cassman 1999, Robert 2002).

A different approach to the issue of nitrogen and rice yield, and the one which is the main subject of this chapter, involves an examination of the physiology of rice-nitrogen relations, at whole-plant, cellular, biochemical, and genetic levels of organisation. This information is vital to the implementation of conventional and recombinant-DNA breeding methods to increase both the yield of the rice crop, and its agronomic N-use efficiency (*i.e.* the yield increment relative to N fertiliser input). This chapter will discuss several important aspects of rice-nitrogen relations, focusing on four main topics: 1) biological N fixation, 2) N uptake, 3) N assimilation and, 4) points of intersection between photosynthesis and plant N use.

BIOLOGICAL NITROGEN FIXATION IN RICE

In many areas under rice cultivation, nitrogen is fixed by naturally-occurring diazotrophic bacteria, and can be directly or indirectly available to the crop. Biological nitrogen fixation (BNF) contributes an average of 30 kg N per crop (Roger and Ladha 1992), and can be particularly important when N applied as fertiliser is in short supply, as can be the case during the critical phases of flowering and grain set. Under greenhouse conditions, the amount of biologically fixed N taken up by rice can be as much as 80%, and is strongly dependent on the cultivar used (Ladha *et al.* 1998). While much of the BNF in rice paddies is accomplished by free-living cyanobacterial diazotrophs, or by the cyanobacterium *Anabaena* in association with the aquatic water fern *Azolla* (chapter 3), it has been of much recent interest that there exist diazotrophic rhizobia that endophytically associate with rice plants under natural conditions (Figure 1, Yanni *et al.* 2001).

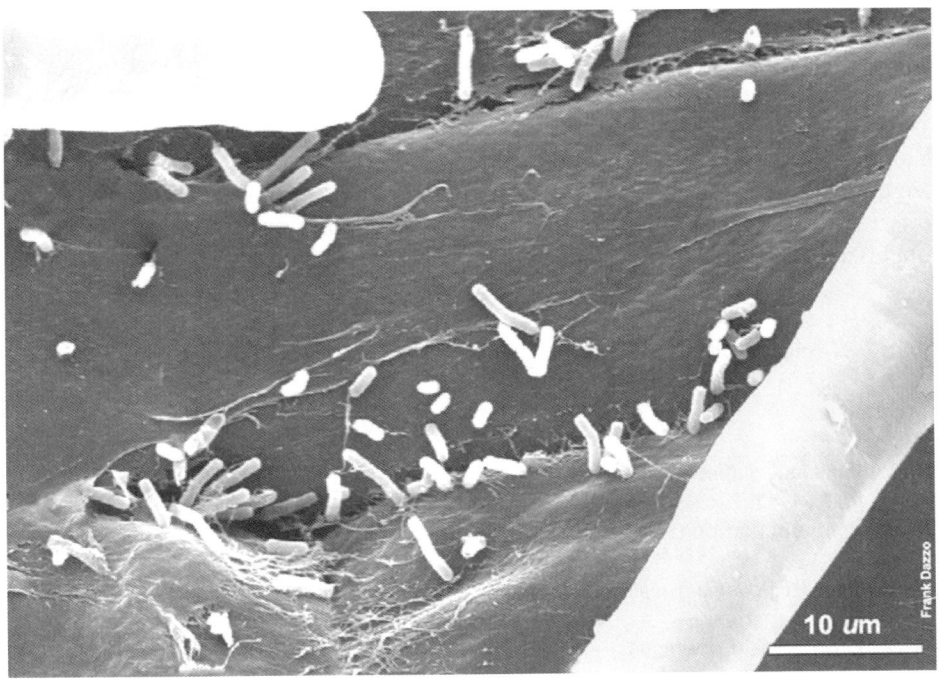

Figure 1. Scanning electron micrograph showing the "crack entry" mode of primary host infection into a rice root by cells of a rice-adapted endophytic strain of *Rhizobium leguminosarum* bv. trifolii in gnotobiotic culture. Bar scale equals 10 μm.

A cropping system consisting of a rotation of rice with berseem clover (*Trifolium alexandrinum*), which acts as a reservoir for *Rhizobium leguminosarum* bv. *trifolii*, has been in use in Egypt for centuries, and this system is only now being mechanistically unravelled. Recent field trials using this system, with specialised isolates of *Rhizobium leguminosarum*, have shown that substantial yield and agronomic NUE (N use efficiency) increases (of 30% or even higher) can result from bacterial inoculation, particularly when only 1/3 of the recommended dose of fertiliser N is used (Table 1, Yanni et al. 2001). However, in the case of this association, the yield increases could not be attributed to transfer of N from the bacterial partner to the rice plant, but resulted primarily from changes in root architecture that were stimulated by the presence of the colonising bacteria. Such changes can lead to a greater capacity of the plants to efficiently extract N and other nutrients from the soil. In addition, there is evidence that the colonising bacteria produce the hormones IAA and GA_7, and that this production is stimulated in the presence of root exudates (chapter 1, Yanni et al. 2001).

Although the rice/clover-*Rhizobium* rotation system entails a season in which rice is not grown, and hence cannot provide as much grain annually as can a double- or triple-cropping rice system, the fact that it occurs naturally is an indication that rice is genetically predisposed toward becoming the host partner in a tighter endophytic relationship with a diazotroph. This is especially true given that the associations appear to be very strain/variety specific, and heritable (Yanni *et al.* 2001). Therefore, while extremely challenging, the success of longer-range goals of engineering rice plants to behave like nodulating legumes, and requiring less fertiliser-N input while attaining maximal yields, is conceivable, and this area has attracted much research attention.

Table 1. N-related grain-yield enhancements in rice and wheat. Control value = 1.

Treatment	Yield enhancement	Reference
Rhizobium inoculation (rice, no applied N)	1.48	Yanni *et al.* 2001
Rhizobium inoculation (rice, 144 kg N ha^{-1})	1.41	Yanni *et al.* 2001
Transgenic PEPcase expression (rice)	1.22	Jiao *et al.* 2002
GS1 overexpression (wheat)	1.14	Habash *et al.* 2001

From legume to rice

In 1992, the International Rice Research Institute (IRRI) began spearheading a multi-center "Human Frontier" project with the principal long-term goal of bioengineering N fixation in rice. However, it should be noted that the sequence of steps involved in nodule formation, and subsequent symbiotic N fixation is lengthy, complex, and not completely understood. The current state of this knowledge is summarised in several recent reviews (Thies *et al.* 2001, Dey and Datta 2002, Trevaskis *et al.* 2002, Ladha and Reddy 2003,). In legumes (chapter 4), this sequence typically begins with the release of flavonoids and other signal compounds by the host plant which induce the expression of bacterial *nod* genes; *nod* factors produced by these bacterial genes include chito-lipo-oligosaccharides (CLOSes) which then, in turn, lead to the induction of "early nodulin" genes (*ENOD*s) in the root vascular tissue; this is followed by cell division and ultimately the formation of microfibril-entangled nodules, concomitant with increased auxin, gibberellin and cytokinin concentrations in these tissues (these hormones may be both of host plant and bacterial origin; Broughton *et al.* 2003). Once nodulation has advanced to a critical stage of

development, nitrogen fixation proceeds via an enzyme system encoded by *nif* genes (see below). In rice (Ladha and Reddy 2003), the scenario differs in several ways:
- Root exudates have only limited capacity, and only in some cultivars, to induce *nod* genes in diazotrophic bacteria (Reddy *et al.* 1998).
- No true nodule formation has been demonstrated in rice, either due to the presence of rhizobia or the addition of CLOS factors.
- Although rhizobia attach to root hairs, they do not infect them.

The primary mode of rhizobial entry into rice roots is via cracks at junctions between epidermal cells, or fissures formed transiently during lateral root emergence (Figure 1); however, unlike in legumes, this infection appears *nod*-independent, and classical infection threads are not formed (Ladha *et al.* 1998, Yanni *et al.* 2001).

Endophytic diazotrophs are seen in some cultivars (see above), including major varieties, such as IR-72 (Barraquio *et al.* 1997), but these are limited to intercellular (apoplastic) spaces including the xylem tissue (Cocking 2003) or cells undergoing lysis (Reddy *et al.* 1997), *i.e.* they are not endocytic; the thick cortical sclerenchyma in rice roots prevents deeper cross-sectional penetration of would-be endosymbionts.

Promising developments

Despite these major differences from legumes, and their associated impediments, however, bacterial attachment to rice roots (Terouchi and Syono 1990), and some morphological responses to rhizobial inoculation, have been documented in rice. Rhizobial plant growth substances have been shown to change lateral root architecture to favour short and thick laterals (Reddy *et al.* 1997), and root hair deformation, a process associated with early microsymbiont infection, has been demonstrated in at least one instance (Plazinski *et al.* 1985). It has also become clear that several salient aspects of the complex gene cascade required to achieve successful nodulation are in fact present in rice (Ladha and Reddy 2003). Particularly encouraging is the fact that homologues to legume *ENOD*s, such as *ENOD40* and *ENOD93*, have been discovered in rice, and rhizobial *nod* factors have been shown to induce at least one of these, *ENOD12* (Reddy *et al.* 1997); similarly, the expression of rice *ENOD40* protein in soybean revealed that its tissue localisation was identical to that of endogenous legume *ENOD*s, indicating that key regulatory features of these genes may be conserved in rice. A partial explanation for the presence of such genes in rice, and for their induction response to external stimuli, may lie in the fact that wild relatives of rice (although, typically, not in paddy-grown cultivated rices) possess the capacity to form mycorrhizal associations, and there appear to be substantial principal similarities at the genetic level between the formation of mycorrhizal associations and that of N-fixation symbioses (Hirsch and Kapulnik

1998). Further optimisation of a proposed rice-diazotroph symbiosis will involve increasing the export of fixed nitrogen to the plant host (Colnaghi *et al.* 1997), and decreasing the inhibitory effect that NH_4^+ exerts on the nitrogenase enzyme complex (Cheng *et al.* 1999).

Importantly, the above only pertains to successful "nodule" formation in rice. The possibly larger challenge in engineering N fixation in rice, however, lies in the successful incorporation, into the rice genome, of the 16 *nif* genes essential to nitrogenase activity. Plastids have been discussed as possibly the most suitable intracellular location for these genes (Merrick and Dixon 1984, Dixon *et al.* 2000), because the plastic genome and its transcription and translation signals most closely resemble those of N-fixing cyanobacterial prokaryotes (Whitfeld and Bottomley 1983), and because *nif* gene products appear to be more stable in the plastid environment (Merrick and Dixon 1984). In addition, the replacement of native *nif* gene promoters with plastid-specific promoters appears feasible. A necessity in any such gene transfer approach is that oxygen damage to the nitrogenase enzyme can be prevented (Dixon *et al.* 1997). In legumes, the leghemoglobin protein accomplishes the sequestration of oxygen, creating essentially an anaerobic environment for the proper operation of the enzyme (chapter 4). Possibly limiting *nif* expression to root plastids (*e.g.* amyloplasts), or diurnally regulating gene expression such that N fixation would occur only in the absence of photosynthetic oxygen evolution, *i.e.* at night, may be a partial solution to this problem. Another solution might be found through the expression in rice of the oxygen-tolerant nitrogenase from *Streptomyces thermoautotrophicus* (Ribbe *et al.* 1997).

While the above-stated problems are, in principle, surmountable, it will be likely to require many years of intensive research and development before a useful product makes it to the rice paddy. Enhancing rice N status by optimising associations between rice and naturally colonising endophytic bacteria in apoplastic spaces, may be more promising in the nearer future.

FEEDBACK LIMITATIONS IN PRIMARY N ACQUISITION

It appears plausible, and indeed has formed the rationale for many research initiatives, that plant N acquisition potential can be enhanced by the increased expression of primary nitrogen transport systems. This may be particularly attractive given that the initial cloning and characterisation of genes encoding multiple nitrate, ammonium, and amino acid transporters have been accomplished. Several reviews cataloguing these genes have been recently published (Forde 2000, Howitt and Udvardi 2000, von Wiren 2000, Glass *et al.* 2001, 2002). In chapters 1 and 2 of this book these genes are presented.

In the case of NH_4^+ transport, five homologues of the high-affinity AMT1 family have been identified in *Arabidopsis*. In rice, as many as ten representatives of this family may exist (Suenaga et al. 2003, chapter 2) and partial physiological characterisation of some of these members has begun recently (Kumar et al. 2003). As in the case of NO_3^- transport, the multiplicity of high-affinity NH_4^+-transporter genes appears to lend variability to the ion uptake characteristics of different cell types, at various stages of development. Low-affinity NH_4^+ transport is less well understood at the molecular level, although a gene named *amt2* may encode a transporter specific for this purpose. Alternatively, or additionally, NH_4^+ flux at high external concentrations might be mediated by non-selective cation channels, or by channels which are partially specific to other ions, such as some potassium channels (Kronzucker et al. 2001).

Despite these advances in the molecular genetics of nitrogen transport systems, the notion that plant N acquisition can be enhanced simply by overexpression of these primary N transporters is hampered by the fact that nitrogen influx across the plant cell plasma membrane is controlled by negative feedback processes (chapter 1). Typically, unidirectional influx (and net flux) of nitrogen into the plant are correlated negatively with plant N status (Figure 2), as has been seen in rice (Wang et al. 1993b, Kronzucker et al. 1998) and a variety of other plant species (Becking 1956, Morgan and Jackson 1988, Causin and Barneix 1993, Rawat et al. 1999). This indicates that plants already have excess flux capacity built into their transport systems, a condition that becomes particularly apparent when considering that nitrogen influx into the plant is accompanied by simultaneous efflux from the plant, even under conditions of N limitation, and that the ratio of efflux to influx tends to increase as influx increases (Lee and Clarkson 1986, Siddiqi et al. 1991, Wang et al. 1993a, Kronzucker et al. 1995a,b,c).

Because of this inherent excess flux capacity, it is likely that constitutive overexpression of transporters will need to be coordinated with the simultaneous overexpression of sink fluxes internal to the plant, to achieve the goal of enhanced plant growth and yield. These sink fluxes include metabolic fluxes into amino acid pools, subcellular sequestration fluxes, (*e.g.* into the vacuole), and fluxes from the root to the shoot. Another problem that may face those attempting to increase N acquisition by overexpression of N transporters is the poor correlation that is sometimes observed between mRNA levels for a transporter, and the level of transport activity measured, which implies the existence of regulatory mechanisms at the post-transcriptional or post-translational level (Rawat et al. 1999). As well, the multiplicity of transporters for a given ion appears to give the plant considerable ability to compensate for imposed changes in the expression patterns of individual transporters (Kaiser et al. 2002). This plasticity, and the set-points at which the concerted activities of the transport systems appear to be co-regulated, implies that there are high-level integrative controls in the plant that a molecular breeding program will have to consider (see below).

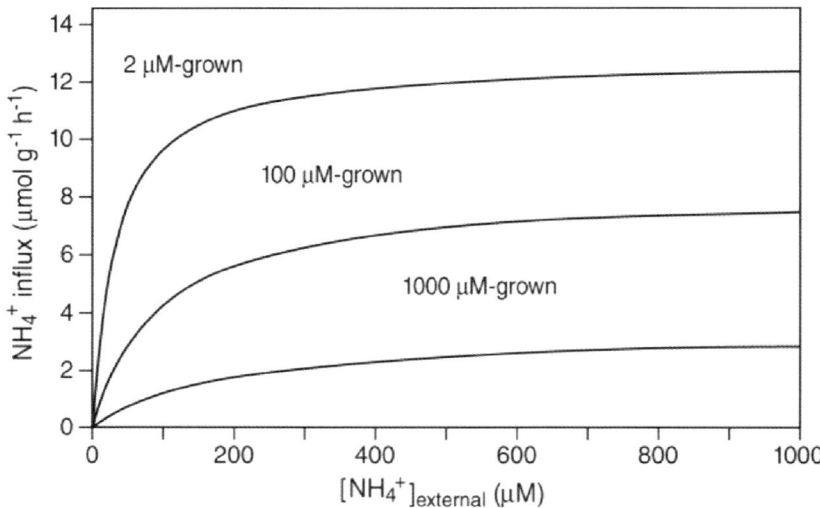

Figure 2. Isothermic patterns of NH_4^+ influx in rice, as influenced by nitrogen source and supply ($^{13}NH_4^+$ tracer data from Wang *et al.* 1993b).

TWO CLASSES OF FEEDBACK AGENTS

The chemical nature of the agents responsible for the negative feedback of plant N fluxes is not completely resolved, although it is essential for these agents, as well as the mechanisms by which they act, to be identified, if the prospect of altering N influx (or net N acquisition) is to be achieved. Therefore, this subject deserves further attention here. Feedback agents are likely to fall into two categories, the transported substrates themselves, or the N metabolites produced from these substrates (chapters 1 and 2). Correlative evidence that NH_4^+ accumulation in the cytosol of root cells feeds back negatively on NH_4^+ influx into these cells was found by Wang *et al.* (1993a), who grew rice plants hydroponically, in complete nutrient solutions with three NH_4^+ regimes (2, 100, and 1000 μM). As shown in Figure 2, the influx of NH_4^+ at a given external concentration declined with increasing steady-state concentration of NH_4^+ applied during growth; the cytosolic pool size of NH_4^+ was positively correlated with the external NH_4^+ concentration. This pattern has been seen in other plants, including white spruce (Kronzucker *et al.* 1995c) and *A. thaliana* (Rawat *et al.* 1999). However, other evidence shows that high tissue levels, or exogenous application, of certain amino acids, in particular glutamine and asparagine, are also correlated with decreased NH_4^+ influx across the plasma membrane (Lee *et al.* 1992, Causin and Barneix 1993, Wieneke and Roeb 1998, Rawat *et al.* 1999, chapter 2). In *Arabidopsis*, this down-regulation of influx was attributed mainly to a decrease in transcript abundance of *AMT1* effected by glutamine. An increase in cytosolic NH_4^+ concentration, achieved by blocking the assimilation of NH_4^+ by use of the glutamine synthetase inhibitor methionine sulfoximine (MSX), did have a small inhibitory effect on NH_4^+ influx while not down-regulating *AMT1* transcription (Rawat *et al.* 1999). However, this effect is by no means universal; indeed, in most cases MSX treatment results in both increased cytosolic NH_4^+ concentration and increased NH_4^+ influx (Wieneke and Roeb 1998). The difficulties in interpreting results obtained by use of MSX were discussed by Lee and Ayling (1993), Kronzucker *et al.* (1995d) and Glass *et al.* (1997).

In the case of NO_3^-, a situation similar to that shown in Figure 2 has been observed (Kronzucker *et al.* 1995b); again, there appears to be an inverse correlation between cytosolic NO_3^- pool size and NO_3^- influx. The possibility that nitrate down-regulates its own influx is also supported by experiments in which such down-regulation was observed in plants made deficient in nitrate reduction and assimilation (and hence deficient in downstream N metabolites), either through mutation or via the application of chemical inhibitors (Ingemarsson *et al.* 1987, Warner and Huffaker 1989, Mattsson *et al.* 1991, King *et al.* 1993). However, in some instances, such as over the course of induction of NO_3^- transport in white spruce, the highest NO_3^- influx was associated with the largest cytosolic NO_3^- pools (Kronzucker *et al.* 1995a). Alternative possibilities for down-regulatory agents include the nitrate pool in the vacuole, and the ammonium or amino acid pools in various compartments. The down-regulatory role of N metabolites on NO_3^- fluxes is supported by numerous studies (Cooper and Clarkson 1989, Lee *et al.* 1992, Imsande and Touraine 1994, Muller *et al.* 1995, Fraisier *et al.* 2000, Vidmar *et al.* 2000, Forde 2002, chapter 1). It is important to note, however, that in the cases of both NO_3^- and NH_4^+ fluxes, studies using exogenously applied amino acids as putative down-regulators are generally conducted against a backdrop of poorly understood uptake processes for the amino acids themselves. This is a subject that has only recently been making substantial progress (Persson and Nasholm 2003), but has potentially great importance given the high soil concentrations of amino acids in certain environments (Henry and Jefferies 2002).

REENGINEERING THE CARBON-NITROGEN INTERFACE

In the attempt to improve plant growth and yield via biotechnological means, it is crucial to consider what the factors are that limit these processes. Nitrate reductase (NR) was long considered to be the enzyme that limited nitrate assimilation and hence was thought to be pivotal to the growth response of plants to NO_3^- fertilisation (Srivastava and Shankar 1996). More recently, however, numerous studies involving increased or decreased amounts of expression of NR-encoding genes have called this idea into question, showing that such modifications result in little or no change in plant growth (Crawford 1995). It is possible that the transport and reduction of inorganic N is controlled at multiple points, in a regulatory matrix that maintains N turnover at high stringency, and resists attempts at artificial manipulation (Britto and Kronzucker 2001). This stringency may also apply in some cases where the enzyme glutamine synthetase (GS), which catalyses the critical step of inorganic N incorporation into amino acids (chapter 2), has been overexpressed, to result in no yield increases, or even a yield decrease (Vincent *et al.* 1997, Limami *et al.* 1999, Harrison *et al.* 2000, Fei *et al.* 2003).

Nevertheless, in some contrasting cases, alterations in the expression of GS and of glutamate dehydrogenase (GDH) (chapter 2) have resulted in substantial growth changes in some plant species (Table 1). In a particularly dramatic instance, a 42% increase in the cytosolic isoform of GS (GS1) in poplar trees resulted in an increase in height of 76% over controls after two months of growth, although this increase declined to 21% after six months (Gallardo *et al.* 1999). The potential for increasing GS activity to significantly increase yield in rice is suggested by the observation that GS1 overexpression experiments with another cereal grain, wheat, resulted in about 20% higher grain yield on a mass basis (Habash *et al.* 2001). GS1 overexpression also resulted in biomass increases in the legume *Lotus japonicus* (Hirel *et al.* 1997) and in tobacco (Fuentes *et al.* 2001) and, similarly, overexpression of GS2, the plastidic isoform, was shown to have positive effects on seedling biomass in tobacco (Migge *et al.* 2000). Although rice plants containing constructs for the overexpression of both GS1 and GS2 have been produced (Hanzawa *et al.* 2002, and Hoshida *et al.* 2000, respectively), the quantification of their effects on rice growth and yield are lacking. These measurements need to be made because of the success of this approach in other plants. Like GDH, the transgenic overexpression of which has conferred increased growth and yield on tobacco plants (Ameziane *et al.* 2000), GS is an enzyme that lies at the intersection of carbon and nitrogen metabolic pathways; this special property may be why modification of GS and GDH activity can overcome the restrictions on N metabolism that the regulatory matrix appears to impose on the flux through NR and other nitrogen fluxes (Britto and Kronzucker 2001).

GS and photorespiration

An intriguing possibility that emerges from the GS-overexpression work is that high activities of this enzyme may confer increased N-use efficiency by having a greater capacity to refix N released as NH_4^+ by the photorespiratory pathway in plants. Indeed, evidence from expression studies on GS1 in tobacco

(Oliveira et al. 2002), GS2 in tobacco (Kozaki and Takeba 1996), and GS2 in rice (Hoshida et al. 2000) has suggested that GS is a limiting factor in this pathway. In all of these studies, there was a positive correlation between GS activity and the amount of photorespiration (chapter 6). In conjunction with the increased growth rates reported in the tobacco plants, these findings are striking in that they appear to contradict the idea that photorespiration is a "wasteful process", in which both fixed C and N can be lost in substantial quantities from the plant (Ehleringer and Monson 1993, Mattsson et al. 1997). Such waste, however, might be minimised if GS overexpressors are able to refix the N released in the glycine decarboxylation step of photorespiration, and if C released in the same step can also be effectively refixed, possibly by phosphoenolpyruvate carboxylase (PEPcase, see below and chapter 6), which is up-regulated by glutamine levels; these in turn might be increased by higher GS activity (Migge et al. 1997, Vincent et al. 1997, Fuentes 2001, Suarez et al. 2003). Under such conditions, and under high light intensity, photorespiration may confer growth increases by relieving the photosystems of excitation pressure (Heber et al. 1996, Noctor et al. 2002) and of the associated production of reactive oxygen species. This might be particularly advantageous in rice, given the high irradiances in rice paddies, and the likelihood that stomata are closed during much of the day due to the low hydraulic conductivity of rice roots and the water deficit that results from this, even under flooded conditions (Miyamoto et al. 2001). The photosynthetic performance enhancement seen in GS2-overexpressing rice grown under saline conditions (Hoshida et al. 2000) may have a similar foundation. It is interesting to ask whether the advantages gleaned from optimisation of the GS systems may be further enhanced through alterations in primary carbon fixation. This possibility is discussed below.

NITROGEN AND PHOTOSYNTHESIS

Despite the claims that GS may catalyse a rate-limiting step in the photorespiratory pathway (Kozaki and Takeba 1996, Hoshida et al. 2000, Oliveira et al. 2002), however, this idea contradicts the theoretical basis of relative rates of Rubisco carboxylation and oxygenation activities, which are expected to be determined solely by the catalytic properties of the enzyme with respect to CO_2 and O_2, and by the partial pressures of these gases within the leaves (Brooks and Farquhar 1985). Therefore, the apparent causal relationship between GS and photorespiration, postulated by some authors (Kozaki and Takeba 1996, Olivera et al. 2002) requires reexamination. Because the substrate affinities of Rubisco for CO_2 and O_2 are essentially constant at a given temperature (Brooks and Farquhar 1985), it seems most likely that the parameter affected under GS overexpression is the CO_2/O_2 ratio in the leaf; a lower ratio would enhance the oxygenation of Rubisco, and hence increase the activity of the photorespiratory cycle. A likely explanation behind this change in CO_2/O_2 ratio is that GS overexpressors may display decrease in stomatal conductance, or increased control over stomatal function in general, as suggested by the observation that the most substantial gains in stress tolerance and/or biomass resulting from GS enhancement have been observed under salinity stress or drought stress conditions (Table 2, Hoshida *et al.* 2002, E.G. Kirby personal communication). Optimisation of stomatal control may be linked to GS for several reasons. First, high GS activities play a major role in the maintenance of low tissue NH_4^+ levels (Givan 1979, Schjoerring *et al.* 1993, Husted and Schjoerring 1995, Hoshida *et al.* 2000). Low tissue NH_4^+ is strongly associated with high plant potassium status (Britto and Kronzucker 2002), which in turn is critical for proper stomatal function (Smith and Stewart 1990, Hedrich *et al.* 2001). Also essential to stomatal movement are tissue malate pools (Marten 1993, Patonnier *et al.* 1999, Hedrich *et al.* 2001), which may be maintained at high levels in GS overexpressors due to the high glutamine/glutamate ratio in these plants, and the stimulated activity of PEPcase that is known to result from this elevated ratio (Golombek *et al.* 1999, Murchie *et al.* 2000, Pasqualini *et al.* 2001, see above).

Table 2. Effects of salt stress and GS2 expression on chlorophyll fluorescence in rice plants. The ratio F_v/F_m was used as an indicator of photosynthetic quantum yield, a declining ratio indicates increased stress on the photosynthetic apparatus. n.d. - not determined (data from Hoshida *et al.* 2000).

Plant genotype	F_v/F_m (days after transfer to 150 mM NaCl)				
	0	3	7	12	14
Control	0.72	0.73	0.63	0.28	0
GS2 overexpressors	0.72	0.72	0.70	0.75	0.65
GS2 underexpressors	0.72	0.54	0	n.d.	n.d

In rice, one of the most impressive gains in yield resulting from a single genetic alteration has been a 10-30% gain due to the overexpression of maize-specific PEPcase, with an additional 5-20% increase when the PEPcase-activating enzyme, pyruvate orthophosphate dikinase (PPDK), also from maize, was simultaneously overexpressed (Table 1, Ku *et al.* 2001, Jiao *et al.* 2002). A partial explanation of this yield enhancement is that transgenic plants display greater (up to 35%) photosynthetic capacity, possibly due to enhanced stomatal conductance which in turn leads to higher CO_2 concentrations in the leaf (Jiao *et* al. 2002, Ku *et al.* 2001). Also associated with this phenotype are increased carboxylation efficiency, a decreased CO_2 compensation point, and improved tolerance to photoinhibition as evidenced by improved maintenance of photochemical efficiency and photochemical quenching (Jiao *et al.* 2001, 2002, Huang *et al.* 2002). This body of work suggests that even a partial realisation of the C_4 photosynthetic pathway in rice (the decarboxylation aspect of the pathway not having been engineered in this case) can be of tremendous benefit to growth and yield, even in the absence of a true CO_2-concentrating mechanism. CO_2-concentrating mechanisms in C_4 plants are typically associated with Kranz anatomy, in which high concentrations of CO_2 are built up in the bundle sheath cells that are enriched in Rubisco. While the prospect of engineering such a dramatic anatomical change in rice appears distant (Evans and von Caemmerer 2000), the example of *Hydrilla verticillata*, a C_3-C_4 intermediate lacking Kranz anatomy (Voznesenskaya *et al.* 2001), demonstrates that it may not be necessary. Furthermore, the possibility that some varieties of rice may already possess partially favorable anatomy cannot be discounted, and a large-scale microscopic screening effort for such varieties, and wild species related to rice, which also show these anatomical features) is currently underway at IRRI (J. Sheehy, personal communication). In addition, it may be possible to produce an intracellular CO_2-concentrating mechanism by taking advantage of subcellular compartmentation and targeting a C_4-decarboxylating enzyme to the chloroplast. In fact, such a targeting has been accomplished in a study with rice, in which the phosphoenolpyruvate carboxykinase enzyme, which decarboxylates oxaloacetate to form phosphoenolpyruvate, releasing CO_2, was expressed at high activities in the chloroplast, by fusion of a Rubisco small subunit transit peptide with the structural protein (Suzuki *et al.* 2000). This study showed that decarboxylation of C_4 acids, and subsequent incorporation of released $^{14}CO_2$ into sucrose, occurred in the transformants at three times the rate of untransformed plants.

Given that yield increases have been observed through the overexpression of both GS and PEPcase/PPDK, the simultaneous overexpression of all three genes may lead to further yield gains. Moreover, these enzymes may work in a synergistic way, given that they all appear to be related to stomatal function. Optimisation of stomatal control in a triply transgenic plant might result from: 1) increased tissue K^+ due to lower tissue NH_4^+ resulting from increased GS activity, 2) increased tissue malate due to higher rates of C_4 acid production via PEPcase/PPDK, and 3) increased glutamine/glutamate ratios, due to high GS activity, which in turn reduce the malate inhibition of PEPcase/PPDK (Murchie *et al.* 2000). Such plants would also be expected to have particularly high agronomic (and photosynthetic) N use efficiency (NUE), due to more efficient recapture of NH_4^+ generated, in the leaves, by processes such as photorespiration and proteolysis, especially during senescence (Husted *et al.* 2002, Mattsson *et al.* 1997), and possibly due to a lower requirement for Rubisco, a feature typical of C_4 plants (Ehleringer and Monson 1993). A higher agronomic NUE may be consistent with the goal of achieving higher rice yields with less input of fertiliser N, and reduced loss associated with volatilisation and runoff.

However, maximisation of photosynthetic NUE in rice may not coincide with maximal grain yield. While C_4 plants have higher NUE under conditions of adequate soil nitrogen (Sage 2000), this does not necessarily translate into higher yields, because lower NUE plants may have a higher leaf N content under these conditions, a factor that strongly determines spikelet formation and grain filling (Peng *et al.* 1995, Horie 1997, Sheehy 2000). The importance to grain production of internal N reservoirs in the vegetative tissue of the plant is highlighted by the observation that the plant absorbs half of its requisite nitrogen by the time it has reached only one quarter of its maximum biomass (Sheehy *et al.* 1998). From a photosynthesis perspective, Rubisco is sometimes viewed as an enzyme present in excessive amounts in rice (Makino *et al.* 2000, Horton and Murchie 2000), but it is important to note that it acts as a nitrogen store which can be rapidly broken down and absorbed by the developing grain (Horton and Murchie 2000). It will be interesting to see, once a rice cultivar closer to C_4 plants is developed, whether Rubisco and N content in leaves will be lessened, and whether this will in fact result in a reduced yield (Sheehy 2000).

CONCLUDING REMARKS

At 450 megabase pairs, the rice genome is one of the most compact of the cereal genomes; its nuclear DNA content is one-fifth that of maize, and only 1/33 that of wheat (Arumuganathan and Earl 1991). The International Rice Genome Sequencing Project (IRGSP) was initiated in 1998 under the sponsorship of the Rice Genome Research Program in Tsukuba, Japan, and the preliminary sequences of both *indica* and *japonica* subspecies were reported for the first time in 2002 (Goff *et al.* 2002, Yu *et al.* 2002). Knowledge of this genome, and modern methods making possible its manipulation and its functional characterisation, will provide the basis for specifically enhancing or suppressing individual steps in the nitrogen acquisition pathway. However, it should be apparent that the components of this pathway interact in multiple and complex ways, and the mechanisms by which the pathway is regulated, both at the transcriptional level, and at the level of substrate-enzyme interactions, are still insufficiently resolved. Further complicating the situation is the often-significant multiplicity of closely related genes whose differential regulation as a function of developmental state, tissue type, and environmental conditions, is poorly characterised. Because of these considerations, single-trait manipulations can fail to produce the desired result of modulating the activity of the gene product, and can produce unpredictable results even in seemingly straightforward cases, such as that of NR.

Despite these complexities, much progress continues to be made in our understanding of both the genome and the nitrogen and carbon physiology of rice, and a conjunction of these fields will create bioengineering approaches that can powerfully complement traditional breeding efforts. Preliminary studies in non-rice as well as rice systems show that considerable growth and yield enhancements as well as improved tolerance to environmental stresses may be conferrable to new rice strains and may be feasible within the next several years. Examples of this include:
- the partial realisation of the C_4 syndrome in rice,
- the overexpression of glutamine synthetase,
- the optimisation of endophytic conditions for nitrogen-fixing bacteria.

Given the clear and urgent practical importance of this research endeavour, it is of paramount importance that all information in this area, as well as all plant lines produced, be shared among scientists engaged in this research, that collaborations be actively sought, and that information-sharing be overseen by the major non-profit international organisations, such as IRRI, that are already in place. Only through such means can the issues of hunger and poverty be maintained as the driving forces behind this research. Presently, there is a dangerous trend in which experimental results and materials are guarded by commercial or scientific oligarchies. This trend must be opposed, and the results of research efforts must be made publicly accessible (Khush 2001).

ACKNOWLEDGEMENTS

We thank Frank Dazzo and Youssef Yanni for their feedback and Frank Dazzo for micrograph provision, and thank the Natural Sciences and Engineering Council of Canada for financial support.

REFERENCES

Ameziane R., Bernhard K., Lightfoot D. 2000. Expression of the bacterial *gdhA* gene encoding a NADPH glutamate dehydrogenase in tobacco affects plant growth and development. - Plant Soil 221: 47-57.

Arth I., Frenzel P., Conrad R. 1998. Denitrification coupled to nitrification in the rhizosphere of rice. - Soil Biol. Biochem. 30: 509-515.

Arumuganathan K., Earle E.D. 1991. Nuclear DNA content of some important plant species. - Plant Mol. Biol. Rep. 9: 211-215.

Barraquio W.L., Revilla L., Ladha J.K. 1997. Isolation of endophytic diazotrophic bacteria from wetland rice. - Plant Soil 194: 15-24.

Becking J.H. 1956. On the mechanism of ammonium uptake by maize roots. - Acta Bot. Neerl. 5: 2-75.

Britto D.T., Kronzucker H.J. 2001. Constancy of nitrogen turnover kinetics in the plant cell: Insights into the integration of subcellular N fluxes. - Planta 213: 175-181.

Britto D.T., Kronzucker H.J. 2002. NH_4^+ toxicity in higher plants: a critical review. - J. Plant Physiol. 159: 567-584.

Brooks A., Farquhar G.D. 1985. Effect of temperature on the CO_2/O_2 specificity of ribulose-1,5-bisphosphate carboxylase oxygenase and the rate of respiration in the light - estimates from gas-exchange measurements on spinach. - Planta 165: 397-406.

Broughton W.J., Zhang F., Perret X., Staehelin C. 2003. Signals exchanged between legumes and *Rhizobium*: Agricultural uses and perspectives. - Plant Soil 252: 129-137.

Cassman K.G. 1999. Ecological intensification of cereal production systems: Yield potential, soil quality, and precision agriculture. - Proc. Natl. Acad. Sci. USA 96: 5952-5959.

Cassman K.G., Peng S., Olk D.C., Ladha J.K., Reichardt W., Dobermann A., Singh U. 1998. Opportunities for increased nitrogen-use efficiency from improved resource management in irrigated rice systems. - Field Crops Res. 56: 7-39.

Causin H.F., Barneix A.J. 1993. Regulation of NH_4^+ uptake in wheat plants - effect of root ammonium concentration and amino acids. - Plant Soil 151: 211-218.

Cheng J., Hipkin C.R., Gallon J.R. 1999. Effects of inorganic nitrogen compounds on the activity and synthesis of nitrogenase in *Gloeothece* (Nägeli) sp. ATCC 27152. - New Phytol. 141: 61-70.

Cocking E.C. 2003. Endophytic colonization of plant roots by nitrogen-fixing bacteria. - Plant Soil 252: 169-175.

Colnaghi R., Green A., He L.H., Rudnick P., Kennedy C. 1997. Strategies for increased ammonium production in free-living or plant associated nitrogen fixing bacteria. - Plant Soil 194: 145-154.

Cooper H.D., Clarkson D.T. 1989. Cycling of amino-nitrogen and other nutrients between shoots and roots in cereals - a possible mechanism integrating shoot and root in the regulation of nutrient uptake. - J. Exp. Bot. 40: 753-762.

Crawford N.M. 1995. Nitrate: nutrient and signal for plant growth. - Plant Cell 7: 859-868.

Dey M., Datta S.K. 2002. Promiscuity of hosting nitrogen fixation in rice: An overview from the legume perspective. - Crit Rev. Biotechnol. 22: 281-314.

Dixon R., Cheng Q., Shen G.F., Day A., Dowson-Day M. 1997. *Nif* gene transfer and expression in chloroplasts: Prospects and problems. - Plant Soil 194: 193-203.

Dixon R., Cheng Q., Day A. 2000. Prospects for constructing nitrogen-fixing cereals. - In: Ladha J.K., Reddy P.M. (Eds.) The quest for nitrogen fixation in rice. - Makati City, Philippines, International Rice Research Institute, pp. 327-336.

Dobermann A., Cassman K.G. 2002. Plant nutrient management for enhanced productivity in intensives grain production systems of the United States and Asia. - Plant Soil 247: 153-175.

Ehleringer J.R., Monson R.K. 1993. Evolutionary and ecological aspects of photosynthetic pathway variation. - Annu. Rev. Ecol. Syst. 24: 411-439.

Evans J.R., von Caemmerer S. 2000. Would C_4 rice produce more biomass than C_3 rice? - In: Sheehy J.E., Mitchell P.L., Hardy B. (Eds.) Redesigning rice photosynthesis to increase yield. - Amsterdam, The Netherlands, Elsevier, pp. 53-72.

Fei H.M., Chaillou S., Hirel B., Mahon J.D., Vessey J.K. 2003. Overexpression of a soybean cytosolic glutamine synthetase gene linked to organ-specific promoters in pea plants grown in different concentrations of nitrate. - Planta 216: 467-474.

Forde B.G. 2000. Nitrate transporters in plants: structure, function and regulation. Biochim. Biophys. Acta 1465: 219-235.

Fraisier V., Gojon A., Tillard P., Daniel-Vedele F. 2000. Constitutive expression of a putative high-affinity nitrate transporter in *Nicotiana plumbaginifolia*: evidence for post-transcriptional regulation by a reduced nitrogen source. - Plant J. 23: 489-496.

Fuentes S.I., Allen D.J., Ortiz-Lopez A., Hernandez G. 2001. Over-expression of cytosolic glutamine synthetase increases photosynthesis and growth at low nitrogen concentrations. - J. Exp. Bot. 52: 1071-1081.

Gallardo F., Fu J.M., Canton F.R., Garcia-Gutierrez A., Canovas F.M., Kirby E.G. 1999. Expression of a conifer glutamine synthetase gene in transgenic poplar. - Planta 210: 19-26.

Givan C.V. 1979. Metabolic detoxification of ammonia in tissues of higher plants. - Phytochemistry 18: 375-382.

Glass A.D.M., Erner Y., Kronzucker H.J., Schjoerring J.K., Siddiqi M.Y., Wang M.Y. Ammonium fluxes into plant roots: Energetics, kinetics and regulation. - Z. Pflanz. Bodenkunde 160: 261-268.

Glass A.D.M., Britto D.T., Kaiser B.N., Kronzucker H.J., Kumar A., Okamoto M., Rawat S.R., Siddiqi M.Y., Silim S.M., Vidmar J.J., Zhuo D. 2001. Nitrogen transport in plants, with an emphasis on the regulation of fluxes to match plant demand. - J. Plant. Nutr. Soil Sci. 164: 199-207.

Glass A.D.M., Britto D.T., Kaiser B.N., Kinghorn J.R., Kronzucker H.J., Kumar A., Okamoto M., Rawat S., Siddiqi M.Y., Unkles S.E., Vidmar J.J. 2002. The regulation of nitrate and ammonium transport systems in plants. - J. Exp. Bot. 53: 855-864.

Goff S.A. *et al.* 2002. A draft sequence of the rice genome (*Oryza sativa* L. ssp japonica). - Science 296: 92-100.

Golombek S., Heim U., Horstmann C., Wobus U., Weber H. 1999. Phosphoenolpyruvate carboxylase in developing seeds of *Vicia faba* L.: Gene expression and metabolic regulation. - Planta 208: 66-72.

Habash D.Z., Massiah A.J., Rong H.L., Wallsgrove R.M., Leigh R.A. 2001. The role of cytosolic glutamine synthetase in wheat. - Ann. Appl. Biol. 138: 83-89.

Hanzawa S., Matumura S., Hayakawa T., Yamaya T. 2002. Transgenic rice expressing either or CDNA for the cytosolic glutamine synthetase (GS1) under the control of GS1 promoter from rice. - Plant Cell Physiol. 43: S71-S71.

Harrison J., Brugiere N., Phillipson B., Ferrario-Mery S., Becker T., Limami A., Hirel B. 2000. Manipulating the pathway of ammonia assimilation through genetic engineering and breeding: consequences to plant physiology and plant development. - Plant Soil 221: 81-93.

Hashem M.A. 2001. Problems and prospects of cyanobacterial biofertilizer for rice cultivation. - Aust. J. Plant Physiol. 28: 881-888.

Heber U., Bligny R., Streb P., Douce R. 1996. Photorespiration is essential for the protection of the photosynthetic apparatus of C3 plants against photoinactivation under sunlight. - Bot. Acta 109: 307-315.

Hedrich R., Neimanis S., Savchenko G., Felle H.H., Kaiser W.M., Heber U. 2001. Changes in apoplastic pH and membrane potential in leaves in relation to stomatal responses to CO_2, malate, abscisic acid or interruption of water supply. - Planta 213: 594-601

Henry H.A.L., Jefferies R.L. 2002. Free amino acid, ammonium and nitrate concentrations in soil solutions of a grazed coastal marsh in relation to plant growth. - Plant Cell Environ. 25: 665-675.

Hirel B. *et al.* 1997. Manipulating the pathway of ammonia assimilation in transgenic non-legumes and legumes. - Z. Pflanz. Bodenkunde 160: 283-290.

Hirsch A.M., Kapulnik Y. 1998. Signal transduction pathways in mycorrhizal associations: Comparisons with the *Rhizobium*-legume symbiosis. - Fungal Gen. Biol. 23: 205-212.

Horie T., Ohnishi M., Angus J.F., Lewin L.G., Tsukaguchi T., Matano T. 1997. Physiological characteristics of high-yielding rice inferred from cross-location experiments. - Field Crops Res. 52: 55-67.

Horton P., Murchie E.H. 2000. C_4 photosynthesis in rice: some lessons from studies of C_3 photosynthesis in field-grown rice. - In: Sheehy J.E., Mitchell P.L., Hardy B. (Eds.) Redesigning rice photosynthesis to increase yield. - Amsterdam, The Netherlands, Elsevier, pp. 127-144.

Hoshida H., Tanaka Y., Hibino T., Hayashi Y., Tanaka A., Takabe T., Takabe T. 2000. Enhanced tolerance to salt stress in transgenic rice that overexpresses chloroplast glutamine synthetase. - Plant Mol. Biol. 43: 103-111.

Howitt S.M., Udvardi M.K. 2000. Structure, function and regulation of ammonium transporters in plants. - Biochim. Biophys. Acta 1465: 152-170.

Huang X.Q., Jiao D.M., Chi W., Ku M.S.B. 2002. Characteristics of CO_2 exchange and chlorophyll fluorescence of transgenic rice with C-4 genes. - Acta Bot. Sin. 44: 405-412.

Husted S., Schjoerring J.K. 1995. Apoplastic pH and ammonium concentration in leaves of *Brassic napus* L. - Plant Physiol. 109: 1453-1460.

Husted S., Mattsson M., Mollers C., Wallbraun M., Schjoerring J.K. 2002. Photorespiratory NH_4^+ production in leaves of wild-type and glutamine synthetase 2 antisense oilseed rape. - Plant Physiol. 130: 989-998.

Imsande J., Touraine B. 1994. N demand and the regulation of nitrate uptake. - Plant Physiol. 105: 3-7.

Ingemarsson B., Oscarson P., Ugglas M.A., Larrson C-M. 1987. Nitrogen utilization in *Lemna* II. Studies of nitrate uptake using ^{13}N-nitrate. - Plant Physiol. 85: 860-864.

Jiao D.M., Li X., Huang X.Q., Wei C., Kuang T.Y., Maurice K.S.B. 2001. The characteristics of CO_2 assimilation of photosynthesis and chlorophyll fluorescence in transgenic PEPC rice. - Chinese Sci. Bull. 46: 1080-1084.

Jiao D.M., Huang X.Q., Li X., Chi W., Kuang T.Y., Zhang Q.D., Ku M.S.B., Cho D.H. 2002. Photosynthetic characteristics and tolerance to photo-oxidation of transgenic rice expressing C-4 photosynthesis enzymes. - Photosynth. Res. 72: 85-93.

Kaiser B.N., Rawat S.R., Siddiqi M.Y., Masle J., Glass A.D.M. 2002. Functional analysis of an *Arabidopsis* T-DNA "Knockout" of the high-affinity NH_4^+ transporter *AtAMT1;1*. - Plant Physiol. 130: 1263-1275.

Khush G.S. 2001. Green revolution: The way forward. - Nat. Rev. Gen. 2: 815-822.

King B.J., Siddiqi M.Y., Ruth T.J., Warner R.L., Glass A.D.M. 1993. Feedback-regulation of nitrate influx in barley roots by nitrate, nitrite, and ammonium. - Plant Physiol. 102: 1279-1286.

Kirk G.J.D., Kronzucker H.J. 2000. Nitrogen uptake by rice roots. - In: Kirk G.J.D., Olk D.C. (Eds.) Carbon and nitrogen dynamics in flooded soils. - Los Banos, Philippines, International Rice Research Institute, pp. 147-162.

Kozaki A., Takeba G. 1996. Photorespiration protects C_3 plants from photooxidation. - Nature 384: 557-560.

Kronzucker H.J., Glass A.D.M., Siddiqi M.Y. 1995a. Nitrate induction in spruce - an approach using compartmental analysis. - Planta 196: 683-690.

Kronzucker H.J., Siddiqi M.Y., Glass A.D.M. 1995b. Compartmentation and flux characteristics of nitrate in spruce. - Planta 196: 674-682.

Kronzucker H.J., Siddiqi M.Y., Glass A.D.M. 1995c. Compartmentation and flux characteristics of ammonium in spruce. - Planta 196: 691-698.

Kronzucker H.J., Siddiqi M.Y., Glass A.D.M. 1995d. Analysis of $^{13}NH_4^+$ efflux in spruce roots - a test-case for phase identification in compartmental analysis. - Plant Physiol. 109: 481-490.

Kronzucker H.J., Siddiqi M.Y., Glass A.D.M., Kirk G.J.D. 1999. Nitrate-ammonium synergism in rice: A subcellular analysis. - Plant Physiol. 119: 1041-1046.

Kronzucker H.J., Glass A.D.M., Siddiqi M.Y., Kirk G.J.D. 2000. Comparative kinetic analysis of ammonium and nitrate acquisition by tropical lowland rice: implications for rice cultivation and yield potential. - New Phytol. 145: 471-476.

Kronzucker H.J., Britto D.T., Davenport R.J., Tester M. 2001. Ammonium toxicity and the real cost of transport. - Trends Plant Sci. 6: 335-337.

Kumar A., Silim S.N., Okamoto M., Siddiqi M.Y., Glass A.D.M. 2003. Differential expression of three members of the *AMT1* gene family encoding putative high-affinity NH_4^+ transporters in roots of *Oryza sativa* subspecies indica. - Plant Cell Environ. 26: 907-914.

Ku M.S.B., Cho D.H., Li X., Jiao D.M., Pinto M., Miyao M., Matsuoka M. 2001. Introduction of genes encoding C_4 photosynthesis enzymes into rice plants: physiological consequences. - In: Goode J.A., Chadwick D. (Eds.) Rice biotechnology: Improving yield, stress tolerance and grain quality, Novartis Foundation Symposium 236. - Chichester, U.K., Wiley, pp. 100-116.

Ladha J.K., Reddy P.M. 2003. Nitrogen fixation in rice systems: state of knowledge and future prospects. - Plant Soil 252: 151-167.

Ladha J.K., Kirk G.J.D., Bennett J., Peng S., Reddy C.K., Reddy, P.M., Singh U. 1998. Opportunities for increased nitrogen-use efficiency from improved lowland rice germplasm. - Field Crops Res. 56: 41-71.

Lee R.B., Clarkson D.T. 1986. ^{13}N studies of nitrate fluxes in barley roots. 1. Compartmental analysis from measurements of ^{13}N efflux. - J. Exp. Bot. 37: 1753-1767.

Lee R.B., Ayling S.M. 1993. The effect of methionine sulfoximine on the absorption of ammonium by maize and barley roots over short periods. - J. Exp. Bot. 44: 53-63.

Lee R.B., Purves J.V., Ratcliffe R.G., Saker L.R. 1992. Nitrogen assimilation and the control of ammonium and nitrate absorption by maize roots. - J. Exp. Bot. 43: 1385-1396.

Limami A., Phillipson B., Ameziane R., Pernollet N., Jiang Q.J., Poy R., Deleens E., Chaumont-Bonnet M., Gresshoff P.M., Hirel B. 1999. Does root glutamine synthetase control plant biomass production in *Lotus japonicus* L.? - Planta 209: 495-502.

Makino A., Nakano H., Mae T., Shimada T., Yamamoto N. 2000. Photosynthesis, plant growth and N allocation in transgenic rice plants with decreased Rubisco under CO_2 enrichment. - J. Exp. Bot. 51: 383-389.

Marten I., Busch H., Raschke K., Hedrich R. 1993. Modulation and block of the plasma-membrane anion channel of guard-cells by stilbene derivatives. - Eur. Biophys. J. Biophy. 21: 403-408.

Mattsson M., Johansson E., Lundborg T., Larsson M., Larsson C.-M. 1991. Nitrogen-utilization in N-limited barley during vegetative and generative growth. 1. Growth and nitrate uptake kinetics in vegetative cultures grown at different relative addition rates of nitrate-N. - J. Exp. Bot. 42: 197-205.

Mattsson M., Hausler R.E., Leegood R.C., Lea P.J., Schjoerring J.K.. 1997. Leaf-atmosphere NH_3 exchange in barley mutants with reduced activities of glutamine synthetase. - Plant Physiol. 114: 1307-1312.

Merrick M., Dickson R. 1984. Why dont plants fix nitrogen? - Trends Biotechnol 2: 162-166.

Migge A., Carrayol E., Kunz C., Hirel B., Fock H., Becker T. 1997. The expression of the tobacco genes encoding plastidic glutamine synthetase or ferredoxin-dependent glutamate synthase does not depend on the rate of nitrate reduction, and is unaffected by suppression of photorespiration. - J. Exp. Bot. 48: 1175-1184.

Migge A., Carrayol E., Hirel B., Becker T.W. 2000. Leaf-specific overexpression of plastidic glutamine synthetase stimulates the growth of transgenic tobacco seedlings. - Planta 210: 252-260.

Miyamoto N., Steudle E., Hirasawa T., Lafitte R. 2001. Hydraulic conductivity of rice roots. - J. Exp. Bot. 52: 1835-1846.

Morgan M.A., Jackson W.A. 1988. Inward and outward movement of ammonium in root systems - transient responses during recovery from nitrogen deprivation in presence of ammonium. - J. Exp. Bot. 39: 179-191.

Muller B., Tillard P., Touraine B. 1995. Nitrate fluxes in soybean seedling roots and their response to amino acids - an approach using ^{15}N. - Plant Cell Environ. 18: 1267-1279.

Murchie E.H., Ferrario-Mery S., Valadier M.H., Foyer C.H. 2000. Short-term nitrogen-induced modulation of phosphoenolpyruvate carboxylase in tobacco and maize leaves. - J. Exp. Bot. 51: 1349-1356.

Noctor G., Veljovic-Jovanovic S., Driscoll S., Novitskaya L., Foyer C.H. 2002. Drought and oxidative load in the leaves of C_3 plants: a predominant role for photorespiration? - Ann. Bot. 89: 841-850.

Oliveira I.C., Brears T., Knight T.J., Clark A., Coruzzi G.M. 2002. Overexpression of cytosolic glutamine synthetase. Relation to nitrogen, light, and photorespiration. - Plant Physiol. 129: 1170-1180.

Pasqualini S., Ederli L., Piccioni C., Batini P., Bellucci M., Arcioni S., Antonielli M. 2001. Metabolic regulation and gene expression of root phosphoenolpyruvate carboxylase by different nitrogen sources. - Plant Cell Environ. 24: 439-447.

Patonnier M.P., Peltier J.P., Marigo G. 1999. Drought-induced increase in xylem malate and mannitol concentrations and closure of *Fraxinus excelsior* L. stomata. - J. Exp. Bot. 50: 1223-1229.

Peng S.B., Cassman K.G., Kropff M.J. 1995. Relationship between leaf photosynthesis and nitrogen content of field-grown rice in tropics. - Crop Sci. 35: 1627-1630.

Persson J., Nasholm T. 2003. Regulation of amino acid uptake by carbon and nitrogen in *Pinus sylvestris*. - Planta 217: 309-315.

Plazinski J., Innes R.W., Rolfe B.G. 1985. Expression of *Rhizobium trifolii* early nodulation genes on maize and rice plants. - J. Bacteriol. 163: 812-815.

Rawat S.R., Silim S.N., Kronzucker H.J., Siddiqi M.Y., Glass A.D.M. 1999. *AtAMT1* gene expression and NH_4^+ uptake in roots of *Arabidopsis thaliana*: Evidence for regulation by root glutamine levels. - Plant J. 19: 143-152.

Reddy P.M., Ladha J.K., So R., Hernandez R., Dazzo F.B., Angeles O.R., Ramos M.C., de Bruijn F.J. 1997. Rhizobial communication with rice: induction of phenotypic changes, mode of invasion and extent of colonization in roots. - Plant Soil 194: 81-98.

Reddy P.M., Ladha J.K., Ramos M.C., Maillet F., Hernandez R.J., Torrizo L.B., Oliva N.P., Datta S.K., Datta K. 1998. *Rhizobial* lipochitooligosaccharide nodulation factors activate expression of the legume early nodulin gene *ENOD12* in rice. - Plant J. 14: 693-702.

Ribbe M., Gadkari D., Meyer O. 1997. N_2 fixation by *Streptomyces thermoautotrophicus* involves a molybdenum-dinitrogenase and a manganese-superoxide oxidoreductase that couple N_2 reduction to the oxidation of superoxide produced from O_2 by a molybdenum-CO dehydrogenase. - J. Biol. Chem. 272: 26627-26633.

Robert P.C. 2002. Precision agriculture: a challenge for crop nutrition management. - Plant Soil 247: 143-149.

Roger P., Ladha J.K. 1992. Biological nitrogen fixation in wetland rice fields: Estimation and contribution to nitrogen balance. - Plant Soil 141: 41–55.

Sage R. 2000. C_3 versus C_4 in rice: ecophysiological perspectives. - In: Sheehy J.E., Mitchell P.L., Hardy B. (Eds.) Redesigning rice photosynthesis to increase yield. - Amsterdam, The Netherlands, Elsevier, pp. 13-35.

Schjoerring J.K., Kyllingsbaek A., Mortensen J.V., Byskovnielsen S. 1993. Field investigations of ammonia exchange between barley plants and the atmosphere. 2. Nitrogen reallocation, free ammonium content and activities of ammonium-assimilating enzymes in different leaves. - Plant Cell Environ. 16: 169-178.

Sheehy J.E., Dionora M.J.A., Mitchell P.L., Peng S., Cassman K.G., Lemaire G., Williams R.L. 1998. Critical nitrogen concentrations: implications for high-yielding rice (*Oryza sativa* L.) cultivars in the tropics. - Field Crops Res. 59: 31-41.

Sheehy J.E. 2000. Limits to yield for C_3 and C_4 rice: An agronomist's view. - In: Sheehy J.E., Mitchell P.L., Hardy B. (Eds.) Redesigning rice photosynthesis to increase yield. - Amsterdam, The Netherlands, Elsevier, pp. 39-52.

Siddiqi M.Y., Glass A.D.M., Ruth T.J. 1991. Studies of the uptake of nitrate in barley. 3. Compartmentation of NO_3^-. - J. Exp. Bot. 42: 1455-1463.

Smith S., Stewart G.R. 1990. Effect of potassium levels on the stomatal behavior of the hemiparasite *Striga hermonthica*. - Plant Physiol. 94: 1472-1476.

Srivastava H.S., Shankar N. 1996. Molecular biology and biotechnology of higher plant nitrate reductases. - Curr. Sci. India 71: 702-709.

Suarez R., Marquez J., Shishkova S., Hernandez G. 2003. Overexpression of alfalfa cytosolic glutamine synthetase in nodules and flowers of transgenic *Lotus japonicus* plants. - Physiol. Plant. 117: 326-336.

Suenaga A., Moriya K., Sonoda Y., Ikeda A., von Wiren N., Hayakawa T., Yamaguchi J., Yamaya T. 2003. Constitutive expression of a novel-type ammonium transporter *OsAMT2* in rice plants. - Plant Cell Physiol. 44: 206-211.

Suzuki S., Murai N., Burnell J.N., Arai M. 2000. Changes in photosynthetic carbon flow in transgenic rice plants that express C4-type phosphoenolpyruvate carboxykinase from *Urochloa panicoides*. - Plant Physiol. 124: 163-172.

Terouchi N., Syono K. 1990. *Rhizobium* attachment and curling in asparagus rice and oat plants. - Plant Cell Physiol. 31: 119-128.

Thies J.E., Holmes E.M., Vachot A. 2001. Application of molecular techniques to studies in *Rhizobium* ecology: A review. - Aus. J. Exp. Agri. 41: 299-319.

Trevaskis B., Colebatch G., Desbrosses G., Wandrey M., Wienkoop S., Saalbach G., Udvardi M. 2002. Differentiation of plant cells during symbiotic nitrogen fixation. - Comp. Func. Gen. 3: 151-157.

Vidmar J.J., Zhuo D., Siddiqi M.Y., Schjoerring J.K., Touraine B., Glass A.D.M. 2000. Regulation of high-affinity nitrate transporter genes and high-affinity nitrate influx by nitrogen pools in roots of barley. - Plant Physiol. 123: 307-318.

Vincent R., Fraisier V., Chaillou S., Limami M.A., Deleens E., Phillipson B., Douat C., Boutin J.P., Hirel B. 1997. Overexpression of a soybean gene encoding cytosolic glutamine synthetase in shoots of transgenic *Lotus corniculatus* L plants triggers changes in ammonium assimilation and plant development. - Planta 201: 424-433.

von Wiren N., Gazzarrini S., Gojon A., Frommer W.B. 2000. The molecular physiology of ammonium uptake and retrieval. - Curr. Opin. Plant Biol. 3: 254-261.

Voznesenskaya E.V., Franceschi V.R., Kiirats O., Freitag H., Edwards G.E. 2001. Kranz anatomy is not essential for terrestrial C-4 plant photosynthesis. - Nature 414: 543-546.

Wang M.Y., Siddiqi M.Y., Ruth T.J., Glass A.D.M. 1993a. Ammonium uptake by rice roots. 1. Fluxes and subcellular distribution of $^{13}NH_4^+$. - Plant Physiol. 103: 1249-1258.

Wang M.Y., Siddiqi M.Y., Ruth T.J., Glass A.D.M. 1993b. Ammonium uptake by rice roots. 2. Kinetics of $^{13}NH_4^+$ influx across the plasmalemma. - Plant Physiol. 103: 1259-1267.

Warner R.L., Huffaker R.C. 1989. Nitrate transport is independent of NADH and NAD(P)H nitrate reductases in barley seedlings. - Plant Physiol. 91: 947-953.

Wieneke J., Roeb G.W. 1998. Effect of methionine sulphoximine on ^{13}N-ammonium fluxes in the roots of barley and squash seedlings. - Z. Pflanz. Bodenkunde 161: 1-7.

Whitfeld P.R., Bottomley W. 1983. Organization and structure of chloroplast genes. - Annu. Rev. Plant Physiol. 34: 297-310.

Witt C. *et al.* 1999. Internal nutrient efficiencies of irrigated lowland rice in tropical and subtropical Asia. - Field Crops Res. 63: 113-138.

Yanni Y.G. *et al.* 2001. The beneficial plant growth-promoting association of *Rhizobium leguminosarum* bv. trifolii with rice roots. - Aust. J. Plant Physiol. 28: 845-870.

Yu J. *et al.* 2002. A draft sequence of the rice genome (*Oryza sativa* L. ssp indica). - Science 296: 79-92.

INDEX

14-3-3 binding, 42, 170, 187-189
14-3-3 proteins, 42, 164, 170, 187

abiotic stress, 207, 209, 210, 212, 214, 215, 217, 223, 224, 237, 252
absicisic acid (ABA), 22, 200, 214, 218
Acacia, 104, 105
Acalypha indica, 238
ACC, oxidase / synthase, 24, 209, 211, 215
Acetobacter, 100
acetylglutamate kinase, 211
acetylornithine deacetylase, 211
acetylornithine transaminase, 211
Achromobacter, 25
acid rain, 65, 66
acid soils, 35, 73, 82
acidification, 69, 70, 72, 76, 84, 191, 194
Aconitum karacolicum, 238
actinomycetes, 101, 102
actinorhizal plants, 1, 100-103, 115, 117, 120
adventitious roots, 200
Aethionema, 218
age, 41, 247, 251
agmatine deiminase, 211
agriculture, 67, 68, 89, 99, 103, 121, 122, 232, 240, 261, 262
Agrobacterium, 221
agronomic traits, 52, 53
alanine (Ala), 17, 18, 39, 172, 210, 244
alfalfa, 42, 45, 51, 108, 111, 119, 172, 212
alga(e), 88, 154, 156-158, 196
aliphatic amines, 208, 210
alkaloid(s), 175, 209, 231, 234-241, 249
allantoic acid, 119
allelopathies, 234, 242
alliin, 233, 246, 247
Allium, 245, 246
Allorhizobium, 105

Alnus, 102
aluminum (Al) species, 72, 79, 88
amide(s), 35, 119, 139, 232
amine(s), 231, 234-236, 241, 242
amino acid(s), 15, 17, 18, 35, 36, 39, 42, 43, 45, 46, 49, 51, 66, 76, 78, 82, 83, 87, 88, 119, 139, 155, 157, 159, 168, 169, 172-176, 210, 220, 234, 236, 237, 239, 241, 242, 244, 266-270
amino acid transporters, 119, 266
amino acid uptake, 76, 82
aminocyclopropane carboxylate deaminase, 23
ammonia (NH_3), 17, 18, 35, 48, 51, 67, 68, 80, 81, 83, 85-89, 99, 103, 117, 119, 121, 137, 154, 177, 210, 232, 237, 246
ammonium (NH_4^+), 1, 2, 6, 10, 21, 35-41, 44-51, 53, 66-68, 72-74, 76, 78, 79-82, 84, 86, 88, 89, 99, 117-119, 136, 137, 139, 155-157, 165, 168, 172, 173, 176, 189, 242, 262, 266, 268-270, 272, 273
ammonium assimilation, 2, 35, 37, 41, 47, 49, 50, 53, 78
ammonium detoxification, 173
ammonium flux(es), 267-269
ammonium transporter genes, 36, 37, 266, 269
ammonium transporters, 36, 37, 266, 269
ammonium uptake, 1, 35, 36, 37, 74, 84, 137
Anabaena, 262
Ancistrocladus korupensis, 249
angiosperms, 100, 101
animal(s), 65, 67, 103, 136, 174, 175, 186, 189, 190, 196, 199, 210, 221, 232, 235, 241
anoxia, 8, 190, 191, 193-199, 201
anthocyanins, 176
anthropogenic N, 67, 89

anticarcinogenic compounds, 235, 241, 244-247, 249
antimicrobial action / activity, 236, 240, 241, 251
antioxidant(s), 121, 246, 247
apical meristem, 44, 116
Apocynaceae, 238
apoplasm / apoplast, 1, 3, 5, 36, 81, 138, 210
apoplastic NH_y, 81, 137
apoplastic space / transport, 41, 80, 266
aquatic ferns, 101, 262
Arabidopsis CLAVATA gene, 117
Arabidopsis genome, 158, 177, 208, 215, 222
Arabidopsis mutants, 10, 21, 217, 218
Arabidopsis thaliana, 4, 5, 9-11, 14-21, 23, 24, 36, 37, 40, 42-44, 46-49, 117, 151, 152, 154-156, 158-161, 163, 164, 166, 167, 169-173, 175, 176, 187, 196, 200, 208, 212, 215, 217, 218, 220-222, 266, 268, 269
arbuscular mycorrhizal (AM) fungi, 110, 111
arginase, 211
arginine (Arg), 84, 106, 196, 209, 211, 212, 244
arginine decarboxylase (ADC), 209, 211, 212, 215, 217, 218
argininosuccinate lyase, 211
aromatic amino acid(s), 153, 168, 172, 175, 176
Asclepiadacea, 238
ascorbic acid, 235
asparagine (Asn), 17, 18, 39, 119, 154, 169, 172, 173, 268
asparagine biosynthesis, 119, 173
asparagine synthetase (genes), 172-174
Asparagus officinalis, 247
aspartate (Asp), 17, 18, 84, 119, 169, 172-174, 187, 188
aspartate kinase (AK), 172, 174, 175
Aspergillus, 10, 155
Aspidosperma, 249
Astragalus, 105
atmosphere, 67, 72, 79, 99, 101, 121, 136, 165, 192
atmospheric CO_2, 16, 192, 207

atmospheric nitrogen, 67, 103, 121
atmospheric nitrogen dioxide (NO_2), 67, 68, 72, 82
atmospheric nitrogen oxide (NO), 67, 68, 72, 74
atmospheric nitrogen oxides (NO_x), 67, 69
atmospheric pollutants, 210
atmospheric reduced nitrogen (NH_y), 67-69, 81
ATP sulfurylase, 18
ATPase, 8, 26, 82
Atropa belladonna, 238, 239
autotrophic C fixation, 66
autotrophic nitrifying bacteria, 73
auxin(s), 19, 21, 23, 25, 159, 264
auxin signalling, 21, 25
auxin-resistant mutants, 21
Azolla, 262
Azorhizobium, 105
Azospirillum, 23, 24, 100
Azotobacter, 100

Bacillus radicicola, 104
bacteria / bacterium, 22-24, 35, 38, 48, 100, 101, 103-108, 113, 115, 116, 119, 153, 154, 175, 213, 220, 234, 241, 263, 266, 275
bacteroids, 22, 103, 114-120
barley, 4, 14, 17, 20, 40, 41, 43, 44, 51, 141, 156, 159, 160, 164, 167, 171, 175, 212
bean(s), 17, 40, 42, 104, 161, 163, 167, 235
betaine(s), 212, 245
biodiversity, 65, 232
biomass, 13, 25, 51, 52, 71, 218, 270, 272, 274
birch, 159, 163, 167, 168
Bistorta bistortoides, 245
blue light (receptors), 149, 151, 156, 157, 160, 166, 167, 171, 177
Bradyrhizobium, 105
Brassica, 24, 25, 170, 213, 235, 237, 240, 241, 251
Brassica napus, 24, 25, 170, 213, 240
Brassica rapa, 241
Brassicaceae, 217, 240

Brunchorstia pinea, 88
bryophytes, 101
bud burst, 76
bundle sheath cells / strands, 43, 48, 50, 51, 273

C/N ratio, 71, 73, 76, 84, 238
C_2 cycle, 176
C_3 (plants), 36, 50, 51, 273
C_3-C_4 intermediate, 273
C_4 (plants), 36, 50, 51, 273-275
cadaverine, 210, 242
cadmium (Cd), 48, 49
caffeine, 234, 235, 237, 238
calcium (Ca), 47, 72, 86, 108, 110-112, 164, 189, 191
calcium kinases, 189
calmodulin, 47
Calvin cycle (enzymes), 83, 139, 153, 176
Camellia sinensis, 238
Canavalia lineata, 170
Candida utilis, 38
canola, 18
canopy, 72, 74, 79, 80, 85, 88, 89, 143
Capsicum, 238
carbohydrate(s), 16, 48, 49, 51, 76, 77, 84, 86, 87, 139, 140, 158, 160, 169, 173, 245
carbon (C), 1, 7, 15, 16, 38, 39, 47, 49-51, 66, 71, 73, 75-78, 80, 82-84, 86, 88, 103, 106, 115, 116, 119, 135, 136, 138-142, 145, 149, 150, 154, 158, 159, 161, 165, 173, 175, 176, 188-192, 209, 210, 238, 246, 270, 271, 275
carbon assimilation / fixation, 83, 88, 136, 139, 176, 271
carbon metabolism, 7, 145, 149, 158, 189, 191, 210
carbon sensing, 16
Casparian strip, 3, 41, 46
Castanospermum australe, 249
Casuarina, 102, 103
catabolite repression, 155
catalase, 199
Cataranthus, 238

cation(s), 71-74, 80, 86, 89, 164, 186, 187, 210, 212
cation channels, 267
cation exchange, 72, 73
Ceasalpinioideae, 103
cell(s), 1-3, 5, 8, 36, 37, 40-42, 44, 46-48, 50, 54, 76, 80, 100-103, 108, 109, 113, 114, 116, 119-121, 135, 151, 152, 156, 157, 159, 164, 166, 175-177, 190, 194, 199-210, 212, 213, 219, 221, 234, 237, 242, 246, 248, 249, 263, 264, 267, 268, 275
cell membrane, 36, 242, 248
cell wall, 3, 80, 101-103, 113, 114, 210, 237
central cylinder, 3, 5, 37, 41, 44, 54
central tissue, 114-116, 119
central vascular bundle, 103
cereal(s), 39, 103, 270, 274
chalcone isomerase, 222
chalcone synthase, 222
chemotaxonomic classification, 234
chickpea, 105
Chlamydomonas reinhardtii, 10
chlorate resistant mutants, 10, 160, 166
Chlorella, 47, 51, 157
chlorophyll, 35, 83, 85, 87, 139, 143, 167, 233, 243, 245, 246, 248-250, 272
chlorophyllin, 85, 248
chloroplast(s), 36, 38-40, 42, 43, 47, 48, 51, 54, 100, 150, 151, 153, 154, 157, 159, 161, 165, 173-175, 177, 193, 235, 273
chloroplastic genome, 104
chromophore, 151
chromosome(s), 52, 53, 105
circadian rhythms / control, 161, 171
citrulline, 119, 196
Citrullus lanatus, 247
climate(s), 67, 70, 74, 103, 105, 136
clover, 104, 118, 119
CO_2, 50, 70, 73, 142, 144, 153, 165, 177, 189, 192, 193, 271, 272
Coffea arabica, 238
coimmunoprecipitation / purification, 188, 220-222
Cola nitida, 238

Colchicum autumnale, 238
cold stress, 86, 87, 208, 213
compartmental analysis, 192
Compositae, 238
conifer(s), 35, 50, 68, 69, 72, 82, 84, 88
coniferous forests, 67, 69
Conium maculatum, 238
Corallina elongata, 156
cortex, 3, 5, 10, 41, 44, 46, 54, 108, 113, 115
cortical cell(s), 3, 102, 103, 108, 109, 113
cortical parenchyma, 3, 41
cortical sclerenchyma, 265
Cortinarius, 76
cotyledons, 160, 167, 175
coumarines, 176
cowpea, 104, 119
critical load (CL), 69, 89
crop(s), 1, 52, 54, 65, 66, 82, 89, 103, 207, 215, 223, 231-233, 235, 241, 251, 261, 262, 263
cryptochrome (genes), 149, 151, 152, 163, 171
Cuscuta reflexa, 239
cuticle, 79-81, 87, 88, 207
cyanide, 118, 121, 235, 242, 243
cyanobacteria, 100, 101, 103, 153, 154
cyanogenic glucoside(s), 231, 234-236, 242, 243
cyclic sulphur compounds, 235
Cyclospora, 231
cystathionine γ-synthase, 175
cysteine sulfoxide(s), 246
cytochrome oxidase pathway, 200
cytokinin(s), 23, 141, 264
cytoplasm, 5, 8, 100, 115, 117, 152, 235, 246
cytosol, 7, 15, 36, 39, 120, 153, 157, 164, 165, 174, 192, 235, 268
cytosolic acidification, 191, 194
cytosolic ammonium, 269
cytosolic nitrate, 2, 4, 7, 192, 269
cytosolic sugar-P levels, 190

DAHPS synthase, 172, 175
decarboxylases, 209, 212, 214

defence, 116, 176, 200, 223, 234, 236, 240
denitrification, 72, 74
desertification, 207
development, 2, 3, 19, 22, 24, 26, 40, 45, 52, 53, 103, 115, 116, 121, 135, 149-151, 177, 207, 216, 219-223, 231, 265-267
diamine oxidases (DAO), 210, 211, 213
diazotroph(s), 262, 264, 266
diazotrophic prokaryotes, 100, 121
dicot(s), 2, 7, 40, 150, 160, 240
dihydroflavonol reductase, 222
diurnal regulation / variations, 13, 14, 16, 158, 167, 173
DMI (does not make infections) genes / mutants, 111, 112, 117
DNA, 52, 105, 112, 113, 152, 174, 216, 217, 219, 223, 274
DNA (cDNA), 37, 42, 43, 45, 51, 155, 216, 217, 219, 220, 221
Dolichos biflorus, 109
drought, 75, 79, 87, 193, 207, 212, 214-218, 248, 272
dry deposition, 67-69, 80, 81

ecosystem(s), 66, 69, 77, 8, 232
ectomycorrhizal (EM) fungi / communities, 73-78, 82
efflux, 5, 7, 8, 26, 137, 138, 189, 192, 267
Eleagnaceae, 102
encytraeids, 73
endodermis, 3, 4, 114, 115
endosymbionts, 100, 101, 265
environment(al), 2, 22, 40, 46, 52, 65, 68, 74, 80, 83, 88, 99, 136, 139, 141-145, 150, 172, 175, 201, 207, 232, 248, 251, 266, 274
environmental stress, 208, 215, 216, 275
Epehdra, 238
epidermal cells, 3, 43, 108, 113, 265
epidermis, 3, 5, 10, 37, 41, 44, 54, 108, 112, 113, 152
epitope-tagging, 220-222
Erythroxylum coca, 239

Escherichia coli, 42, 51, 154, 161, 169, 173, 176
ethylene (biosynthesis / production), 24, 108, 174, 200, 209, 215, 200
etiolated leaves / plants, 150, 156, 160, 164, 166, 167, 171, 175
Eucalyptus, 243, 244
eukaryotes, 9, 153, 222
eutrophication, 65, 69
exodermis, 37, 41, 44, 46, 54
exopolysaccharide, 113
extracellular polysaccharides, 106
extramatrical mycelia, 77

far-red light, 151, 156, 160
fatty acyl chain, 106
FeMo cofactor, 118
ferredoxin-thioredoxin system, 153, 170, 176
ferrosulphoproteins, 118
fertiliser(s), 23, 65, 66, 69, 72, 75, 85, 86, 89, 99, 239, 250, 261-264, 274
fibrous root, 2
flavin, 43, 196
flavonoid(s), 106, 172, 176, 222, 233, 264
flavoproteins, 151
flowering, 19, 149-151, 177, 252, 262
foliar uptake, 80, 85, 88
food, 174, 215, 231-235, 237, 242, 245, 252
forest(s), 35, 65, 66, 68-75, 77-79, 81, 82, 84-86, 88, 89
forestry, 102, 207
Frankia, 100-103, 115
free-living bacteria / rhizobia, 100, 117
freezing, 75, 87, 207
frost, 67, 87, 213
fungal attacks / pathogens, 87, 88
fungi / fungus, 38, 75-78, 102, 111, 112, 154, 159, 168, 234, 237, 240
fungicides, 240

GABA, 210, 213
galactose, 208
garlic, 246, 251
gene(s), 5, 9-12, 14, 16, 18, 20, 21, 26, 35-37, 39-43, 45-47, 50-53, 82, 83, 102, 105-108, 110, 112, 115, 117-119, 121, 145, 149, 150, 152-156, 159, 162-164, 166-170, 173, 174, 176, 177, 207, 208, 212, 214-220, 223, 264-267, 270, 273, 274
genetic screens, 102, 121, 122
genetically engineered foods, 233, 245
genome(s), 50, 52, 104, 105, 158, 163, 177, 215-217, 219, 220, 222, 223, 266, 274, 275
genomic(s), 45, 208, 215, 219, 220, 223, 224
germination, 151, 175, 200, 236
gibberellin(s), 23, 155, 264
glucobrassicanapin, 241
glucobrassicin, 240
gluconapin, 240, 241
glucosamine, 106
glucose, 16, 86, 87, 167, 169, 171, 190, 240, 242
glucosinolate(s), 231, 233-235, 240, 241, 251
glucotropaeolin, 251, 252
glutamate (Glu), 17, 18, 35, 38, 47-51, 53, 55, 118, 119, 153, 154, 157, 167-170, 172, 173, 210, 211, 270, 272, 273
glutamate catabolism, 49
glutamate decarboxylase, 210, 211
glutamate dehydrogenase (GDH), 38, 39, 47-51, 53-55, 270
glutamate dehydrogenase (GDH) mutants, 48, 49
glutamate dehydrogenase (NADH-GDH), 3, 47, 48, 54
glutamate dehydrogenase (NADPH-GDH), 47, 51
glutamate dehydrogenase (NADPH-GDH) gene, 47
glutamate synthase (Fd-GOGAT), 38, 39, 43-46, 48, 50, 54, 170-172
glutamate synthase (Fd-GOGAT) genes, 171, 172
glutamate synthase (GOGAT), 17, 38, 39, 41, 43-48, 50-54, 118, 119, 157, 161, 170, 171, 211
glutamate synthase (NADH-GOGAT), 37, 43-46, 51, 52, 54, 170-172

glutamate synthase (NADH-GOGAT) gene, 46, 51
glutamate synthase (NADH-GOGAT) knockout mutant, 47
glutamine (Gln), 17, 18, 35, 37-42, 44-46, 50, 51, 54, 83, 101, 118, 119, 153-155, 157, 160, 167-170, 172, 173, 176, 211, 244, 268-273, 275
glutamine synthetase (GS), 37-39, 41, 42, 46-48, 50-54, 83, 118, 119, 153, 154, 157, 168-170, 172, 176, 211, 269, 270, 271, 273, 275
glutamine synthetase (GS) gene(s), 40, 42, 154, 168-170, 172
glutamine synthetase (GS) overexpressors, 271, 272
glutamine synthetase (GS) phosphorylation / dephosphorylation, 42
glutamine synthetase (GS1), 39-42, 51-54, 169, 170, 264, 270
glutamine synthetase (GS1) genes, 40, 41, 50-52
glutamine synthetase (GS1) overexpression, 264, 270
glutamine synthetase (GS2), 39-43, 51, 54, 169, 170, 173, 270-272
glutamine synthetase (GS2) genes, 51
glutamine synthetase (GS2) overexpression, 270, 272
glutamine synthetase (GSr), 40, 41, 46, 54
glutamine synthetase / glutamate synthase (GS /GOGAT), 38, 39, 47, 48, 53, 119
glutathione (GSH), 233, 246, 247, 251
glutathione peroxidases, 247
glutathione reductase, 247
glutathione-S-transferase, 247
glycine, 36, 168, 176, 177, 196, 271
glycine decarboxylase (GDC), 168, 177, 196
glycolipid(s), 101, 103
glycolysis, 77, 176
glycoprotein, 113, 120
glycoside(s), 242-245
glyphosate, 175

grain(s), 44, 47, 51-53, 261, 262, 264, 270, 274
Gramineae, 100
grapevine, 48, 49
grasslands, 137
Griffonia simplicifolia, 245
growing season, 74, 79, 87, 141
growth, 2, 5, 10, 13, 14, 16, 19, 20, 22-25, 35, 40, 49-51, 66, 70-80, 82-86, 88, 89, 102, 104-106, 110, 111, 113, 133, 136, 139-141, 149, 150, 155, 165, 167, 169, 174, 186, 200, 207, 218, 231, 238, 240, 242, 246, 261, 268-271, 273, 275
growth rate, 2, 13, 14, 16, 20, 22, 24, 50, 66, 67, 84, 89, 104, 105, 140, 231, 238, 271
guanidine, 196
guanylate cyclase, 199
guanylhydrazone, 217
guard cells, 40
Gunnera, 100, 101
GUS reporter gene, 41, 46, 161, 162, 168, 175
gymnosperms, 101

Halliwell cycle, 235
haploid genome, 159, 166, 167
health, 174, 232, 233, 235, 242, 245, 248
heat stress, 213, 214
heavy metal(s), 79, 210
herbicide(s), 49, 175, 196
herbivores, 88, 234-236, 242, 244
heterocysts, 101
heterotrophic bacteria, 73
hexose phosphates, 190, 194
hexoses, 190
high affinity ammonium transport (HATS), 36
high affinity nitrate transport (constitutive, cHATS), 6, 7
high affinity nitrate transport (inducible, iHATS), 6-8, 10, 11-14, 36, 82, 156
high affinity nitrate transport(ers) (HATS), 6-8, 10-13, 36, 82, 156
histamine, 241, 242

homeostasis, 65, 74, 239
homocysteine, 174
homoserine dehydrogenase, 175
hormone(s), 19, 23, 27, 140, 232, 241, 263, 264
horseradish, 251
host, 50, 75-77, 82, 100, 101, 103-107, 114, 116, 119, 121, 122, 218, 237, 239, 263, 264, 266
human(s), 65-67, 99, 174, 232, 234, 235, 237, 241-243, 246, 249, 252
Hydrilla verticillata, 273
hydroxyproline-rich protein, 120
Hyoscyamus, 238
hyphae, 78, 101-103
hypocotyl(s), 151, 160, 200

IAA, 23, 25, 263
Ilex bonplandiana, 238
immune system, 244, 246, 247
immuno(gold) labelling, 39, 41, 43, 45
immunolocalisation, 40, 48
immunological methods, 196, 220, 221,
in situ mRNA detection, 37
indeterminate nodule(s), 114-116, 119
indole glucosinolates, 240, 247
indolizidine alkaloid, 249
Induced Systemic Resistance, 23
infection, 102, 108, 110, 112-114, 116, 121, 236, 265
infection thread(s), 102, 108, 109, 113, 114, 116, 265
influx, 5-8, 10-14, 17, 19, 108, 120, 136, 137, 267-269
inorganic nitrogen (N), 1, 35, 49, 71, 74, 77, 78, 270
insecticides, 234, 240
internode elongation, 200
intron-tagged epitope technology, 221
iso-alliin, 246, 247
isocitrate, 200
isoleucine, 174, 242
isoquinoline, 239
isothiocyanates, 235, 241, 245, 247, 252

kinase(s), 111, 117, 157, 164, 187, 189

knockout mutant(s) / alleles, 26, 37, 42, 47, 49, 53, 217

Lactarius, 76
lactate, 194, 195
lateral root(s), 4, 20-24, 41, 102, 103, 113, 265
leaching, 75, 99, 232
lectin(s), 109, 113, 120
leghemoglobin (Lb), 120, 266
lettuce, 169
Leucaena, 105
leucine (Leu), 18, 242
leucine rich repeats (LLRs), 111, 117
lichens, 101
life cycle, 66, 151, 208
life span, 71, 168
light, 14, 16, 35, 39, 40, 43, 46, 50, 51, 66, 74, 83, 88, 139, 141-144, 149-154, 156-158, 160-177, 185, 187, 188, 190, 192, 193, 197, 198, 238, 244, 251, 271
light perception, 152, 171
light receptors, 149, 151
light responsive elements (LRE), 152, 161, 163, 168
light-induced genes, 152, 169
lignin(s), 40, 73, 75, 172, 175, 176
Liliaceae, 238
lipid, 112, 213, 215, 248
lipid peroxidation, 213, 215, 248
litter(fall), 71-73, 88, 145
Loganiaceae, 238
long distance transport, 36, 83
Lotus, 104, 105, 111, 112, 117, 119, 121, 270
low affinity ammonium transport (LATS), 36
low affinity nitrate transport(ers) (LATS), 7, 8, 10-13, 36, 156
low light, 84, 200
Lunaria, 237, 238
lupine(s), 104, 237, 239, 242
Lupinus, 134, 238, 239
Lycopersicon, 238
lysine, 174, 211, 237, 238
lysine decarboxylase, 211

MADS-box, 21
magnesium (Mg), 38, 72, 86, 164, 189, 248
maize, 39-43, 45, 47-49, 52, 53, 140, 158, 159, 171, 193, 196, 199, 218, 248, 272, 274
malate, 15, 47, 118, 153, 272, 273
malate dehydrogenase, 153
malic acids, 49
marine environmental pollution, 232
mass flow, 137
Medicago, 104, 105, 109, 111, 112, 117, 121
Medicago genes, 112
Melilotus, 104, 105
membrane, 6, 8, 9, 37, 82, 87, 100, 108, 112, 113, 116, 119, 156, 210, 248
meristem, 22, 44, 115, 116
mesophyll, 40, 43, 50, 51, 54, 152
Mesorhizobium loti, 104, 105
mestome sheath, 46, 54
metabolic fluxes, 267
metabolic inhibitors, 8, 39, 50
metabolic pathway, 168, 172, 208, 215, 223, 224, 235, 270
metabolome, 53
methionine, 174, 269
methionine sulfoximine, 269
microarray(s), 208, 216, 223
microbe(s), 100-102, 106, 107
microbial activity, 165
microbial infections, 237
microbial population, 22, 74
microcolonies, 113
microcosms, 78
microfaunal decomposers, 73
microflora, 79
micronutrient(s), 66, 70, 74
microorganisms, 2, 22, 70, 73, 75, 100, 117, 121, 199, 243
microsymbiont, 100, 105, 113, 121, 265
Mimosoideae, 103
mineral nitrogen, 75, 137, 139, 140, 232
mineralisation, 74, 75
mitochondria, 36, 47, 49, 50, 100, 118, 120, 177, 235

mobile anion effect, 72, 73, 89
modelling, 134-137, 140, 141, 143-145
molecular breeding program, 267
molecular mass fingerprints, 223
molecular mechanisms, 42, 46, 120, 223, 224
molecular oxygen, 185, 186, 198, 201
molybdenum (Mo), 118
molybdenum co-factor (MoCo), 163, 196
monocots / monocotyledons, 2, 7, 40, 156, 159, 160
Monoraphidium braunii, 157
monosaccharides, 118
monoterpenes, 233
morphological response(s), 111, 265
mRNA(s), 41, 42, 47, 154, 158, 161, 171, 173, 176, 185, 209, 216, 267
Mucuna mutisiana, 245
multiproteic complex associations, 221, 222
mustard, 169, 234, 240, 241, 251
mutant(s), 9, 10, 12, 21, 23, 37, 39, 40, 44, 48, 49, 104, 110, 111, 117, 155, 163, 166, 198, 217, 218
mutations, 119, 165, 215-218
mycelia, 75-78
mycorrhiza(e), 76, 138
mycorrhizal association, 112, 265
mycorrhizal fungi, 77, 112
mycorrhizal roots, 76, 78
mycotoxins, 231, 248
Myrica, 102, 103
myrosinase, 235, 240, 251

^{13}N, 5
^{15}N, 4, 5, 18, 19, 49, 50, 79, 158, 196
necrosis, 231
necrotic lesions / tissues, 213, 214
needle(s), 71, 79, 85, 86, 87
Neptunia natans, 105
Neurospora, 155, 168
Nicotiana, 159, 162-164, 166, 167, 238
nicotine, 209, 234, 237-239
nitrate (NO_3^-), 1-16, 18-22, 24-27, 35, 36, 39, 41, 44, 46-51, 52, 66, 68, 72, 73, 79, 82, 83, 99, 137, 139, 140, 149, 150, 152, 154, 155-161, 163,

165-168, 170-172, 185-187, 189-194, 196-198, 201, 211, 232, 262, 266, 267, 269, 270
nitrate acquisition, 1, 8, 14, 20, 24, 25
nitrate assimilation, 15, 44, 83, 187, 190-193, 269
nitrate efflux, 6, 8, 27
nitrate influx, 5-8, 10-14, 16-19, 158, 269
nitrate leaching, 72
nitrate reductase (NR), 4, 7, 15, 21, 22, 49, 83, 136, 152, 155, 157, 159-161, 163-166, 170, 171, 185-211, 269, 270, 274
nitrate reductase activation / inactivation, 160, 164, 186-194, 197
nitrate reductase deficient transformant, 194
nitrate reductase gene(s), 155, 159-163, 166, 167
nitrate reductase kinase(s), 164, 190, 192, 195
nitrate reductase modulation, 159, 186, 189, 191, 195
nitrate reductase mutant, 160
nitrate reductase phosphorylation / dephosphorylation, 164, 186-189, 194, 201
nitrate reductase P-NR-14-3-3, 188, 189, 191, 201
nitrate reduction, 15, 16, 36, 48, 50, 83, 140, 192-194, 201, 269
nitrate sensing, 27
nitrate transport, 3, 5-7, 9, 11-16, 21, 26, 27, 267, 269
nitrate transporter genes, 4, 9-12, 14-19, 26, 155, 156, 158, 159
nitrate transporters, 4, 5, 9-12, 15-17, 19, 26, 155-159
nitrate uptake (rate), 1-10, 12-16, 18-20, 25, 26, 80, 82, 149, 154, 156-158
nitrate uptake mutant *chl1*, 10
nitrate-HCO_3^- ions exchange, 15
nitrate-inducible root-specific gene (*ANR1*) gene, 20, 21, 24
nitrate-selective microelectrodes, 4

nitric oxide (NO), 165, 185, 186, 196-201, 211
nitric oxide in plant pathogen interactions, 200
nitric oxide oxidoreductase, 197
nitric oxide synthases (NOS), 196, 197, 199, 200, 211
nitrification, 73, 74, 89, 262
nitrifying bacteria, 35
nitrile(s), 235, 242
nitrite (NO_2^-), 6, 17, 18, 35, 154, 157, 159, 165, 168, 185, 186, 193, 194, 196-199, 201, 211, 232, 242
nitrite accumulation, 197, 198
nitrite reductase (NiR), 157, 165-168, 198, 211
nitrite reductase gene(s), 166-168
nitrite reduction, 194, 197, 198
nitrite:NO oxidoreductase, 197
nitrogen (N), 1-3, 5, 7, 8, 10-16, 18-27, 35-42, 45, 46, 50-54, 65-100, 106, 119, 121, 133-145, 158, 161, 163, 166, 167, 174, 175, 187-189, 192, 208-218, 231, 232, 234-239, 241-246, 248-251, 261-270, 273, 274
nitrogen (N_2) fixation, 35, 36, 99, 101-105, 115, 118, 119, 121, 262, 264-266
nitrogen acquisition, 66, 82, 266, 267, 274, 268
nitrogen assimilation / metabolism, 15, 16, 36, 45, 52, 53, 76, 80, 83, 84, 114, 149, 150, 154-158, 160, 161, 164, 168, 169, 177, 189, 192, 208, 262, 270
nitrogen availability / supply, 5, 11, 23, 41, 45, 46, 65, 66, 70, 77-79, 89, 139, 149, 232, 235, 238, 239, 241, 243, 247, 248
nitrogen compounds / metabolites, 21, 119, 133, 134, 137-139, 141, 231, 235, 234, 268, 269
nitrogen content, 14, 16, 25, 51, 52, 74, 83, 88, 99, 239, 247, 248, 274
nitrogen cycle / cycling, 65, 66, 72, 74, 78, 83, 99
nitrogen deficiency, 13, 41, 70, 158, 231, 239

nitrogen demand / requirement, 14, 16, 18, 19, 20, 26, 74, 79, 83
nitrogen depletion / deprivation, 13, 14, 16, 18, 19, 37
nitrogen deposition, 65-79, 81, 84, 85, 87, 89
nitrogen eutrophication, 66, 87, 88
nitrogen fertilisation / fertilisers, 66, 76-88, 99, 232, 234, 239, 241-243, 245, 247, 248, 250, 251
nitrogen fluxes, 268, 270
nitrogen input(s), 68, 69, 71, 74, 77-79, 86, 89, 264
nitrogen load(s), 67, 68, 73
nitrogen nutrition, 2, 19, 22, 25, 66, 83, 187, 242, 245
nitrogen signalling, 27
nitrogen source(s), 2, 10, 13, 41, 68, 78, 83, 99, 100, 103, 117, 154, 155, 167, 239, 268
nitrogen status, 12, 14, 16, 18-21, 24, 42, 46, 76, 80, 82, 85, 153, 158, 192, 248, 261, 267
nitrogen transport(ers), 266, 267
nitrogen uptake, 54, 66, 75, 78-80, 82, 86, 88, 137-140, 262
nitrogen use efficiency (NUE), 36, 49, 50-54, 248, 262, 263, 273, 274
nitrogenase, 100, 101, 103, 115, 117-121, 211, 266
nitrogenase *nif* gene(s), 115, 118, 265, 266
nitrophilic species, 66
nitrosamines, 242
nodulation (Nod factors), 102, 104, 106-113, 109, 111, 112, 117, 121, 122, 264, 265
nodulation genes (*nod*), 106, 107, 112, 264, 265
nodule(s), 22, 45, 101-106, 108, 113-121, 172, 264, 265, 266
nodule cortex, 114, 115
nodule metabolism, 116, 118
nodule parenchyma, 114, 115, 120
nodule primordium, 113, 114
non-cyclic photosynthetic electron transfer, 38

non-protein amino acids (NPAAs), 231, 234, 236, 242, 244, 245
Norway spruce, 70, 71, 78, 83-86
Nostoc, 100, 101
nuclear genes, 152
nuclear proteins, 46
nucleic acids, 35, 66, 133, 231, 242
nucleotide(s), 152, 231
nucleus / nuclei, 68, 151-153, 161, 219
nutraceuticals, 233-246
nutrient(s), 2-5, 8, 13, 16-20, 23-26, 66, 71-75, 77, 78, 84-86, 88, 89, 113, 137, 140, 142, 143, 145, 165, 171, 210, 212, 232, 233, 238, 245, 248, 251, 263, 268
nutrient acquisition, 25, 75
nutrient cycle / cycling, 72, 77
nutrient deficiency, 84, 86, 210, 212
nutrient demand, 73
nutrient imbalance, 84, 88
nutrient uptake, 2, 25, 26, 75, 77, 84, 88
nutrient-rich patches, 19, 137, 143
nutrition(al), 1, 14, 18, 22, 24, 26, 133, 232, 245
nutritives, 231

oat, 212, 215, 218
oilseed rape, 36, 51, 241
oligonucleotide(s), 156, 216
onions, 246, 247
organic acid(s), 8, 15, 16
organic farming, 232, 233, 241
organic nitrogen, 77, 78, 103
organic nutrients, 22
organic soils, 72, 74
organisms, 74, 99, 100, 104, 121, 154, 186, 231, 232, 241
organosulphur compounds, 246
ornithine (Orn), 209, 238
ornithine aminotransferase, 211
ornithine carbamoyl-transferase, 211
ornithine decarboxylase, 209, 211
orographic cloud, 68
Oryza sativa, 52
osmiophilic vacuolar bodies, 76
osmolyte biosynthesis, 208
osmotic agents, 212
osmotic compound, 2

osmotic regulation, 245
osmotic stress, 207, 208, 210, 212, 213, 215, 218
oxaloacetate, 118, 273
oxidative attacks, 212
oxidative damage, 165
oxidative decarboxylation, 177
oxidative pentose phosphate pathway / cycle (OPPC), 44, 166, 194
oxygen (O_2), 47, 50, 73, 100, 101, 103, 115, 118-121, 165, 194, 197-200, 262, 266, 271
oxygen diffusion barrier, 120
oxygen-tolerant nitrogenase, 266
ozone, 213, 216, 217
ozone sensitive and tolerant cultivars, 213, 214

paddy rice, 35, 39, 265, 266
Papaver somniferum, 236
Papilionoideae, 103
Parasponia, 102, 104, 105, 120
parenchyma cells, 4, 36, 43
pathogen(s), 22, 23, 66, 86, 88, 89, 175, 196, 200, 232, 235, 237, 243
Paxillus involutus, 78
pea, 39, 40, 43, 45, 104, 111, 112, 117, 119, 156, 169, 173, 174, 214
peanut(s), 49, 120, 104
pentose phosphate, 39, 194
peptide transporter, 9
peptide(s), 220, 221
perennial ryegrass, 14
peribacteroid membranes, 115
pericarp cross-cells, 44
pericycle, 108
periderm, 5, 115
peroxisome(s), 38, 39, 177
peroxynitrite, 165, 185, 186, 199
Persea gratissima, 247
pest(s), 66, 67, 71, 86, 88, 89, 244
pesticides, 232
petioles, 101, 144, 152
pH, 8, 15, 35, 47, 72, 73, 81, 88, 191, 194, 242
Phaseolus vulgaris, 161, 169
phenolic acids, 209, 233
phenolics, 88, 248

phenotype, 50, 160, 217, 218, 273
phenylalanine, 36, 40, 154, 168, 172, 175, 176, 237, 238, 242
phenylalanine ammonia lyase (PAL), 36, 40, 54, 168, 172, 176
phenylalkylamine, 237, 238
phenylpropanoid(s), 36, 175, 176
phloem, 1, 3, 15-19, 22, 36, 40, 46, 54, 118, 138, 140, 173, 210
phloem-translocated amino acids, 17, 19
phloem-translocated signal(s), 15, 17, 18, 22
phosphatases, 164, 189
phosphoenol pyruvate carboxylase (PEPC), 48, 18, 264, 271, 272, 273
phosphoenolpyruvate, 118, 175, 271, 273
phosphoenolpyruvate carboxykinase, 273
phospholipids, 210, 215, 241
phosphorylation, 42, 83, 154, 157, 161, 163, 164, 169, 176, 188-192, 194
photomorphogenetic pigments, 150
photooxidants, 165
photoreceptor(s), 151, 152
photorespiration, 36, 38, 39, 44, 48, 50, 51, 168, 193, 270, 271, 273
photorespiratory ammonium, 51, 54
photorespiratory enzymes, 248
photorespiratory pathway / cycle, 176, 177, 270-272
photosynthesis, 13, 16, 36, 49, 54, 82, 83, 87, 99, 101, 118, 136, 137, 141, 149, 150, 152, 157, 158, 160, 163-166, 174, 175, 177, 192, 193, 262, 272, 274
photosynthetic apparatus, 149, 177, 243, 272
photosynthetic electron transport, 143, 170
phototropins, 151
Phyllobacterium, 24
phylogenetic studies, 102
phylogeny, 107, 234
phytoalexins, 175, 236
phytochelatins, 247
phytochemicals, 248

phytochrome(s), 149, 151, 152, 154, 156, 160, 164, 166, 168-171, 173-175
phytochrome deficient mutant, 166
phytochrome genes, 151, 152, 156, 174
phytoglobines, 200
phytohormones, 23, 24
phytopharmaceuticals, 234, 241, 250, 251
phytosterols, 233
pigment/protein complexes, 248
pigmentation, 150, 176
pigments, 133, 150, 151, 175
PII protein 42, 153, 154, 164, 169
Pilocarpus, 238
pine, 43, 86, 87, 171
Pinus ponderosa, 239
pioneer species, 102
plant development, 19, 208, 216
plant growth, 1, 2, 13, 22, 23, 25, 133, 134, 138, 139, 141, 149, 176, 193, 207, 234, 239, 265, 267, 269
plant growth-promoting bacteria , 23
plant growth-promoting rhizobacteria (PGPR), 2, 22-27
plant nutrition, 12, 23, 133, 210
plant pathogen interactions, 210, 251
plant physiology, 3, 136, 137, 145, 208
plant resistance mechanisms, 23
plant species, 1, 2, 4, 6, 7, 19, 41- 43, 53, 102, 166, 169, 234, 239, 267, 270
plant tissue(s), 16, 48, 133, 134, 139, 140, 197, 209, 212, 213, 231, 234, 243, 248
plasma membrane(s), 1, 3, 5, 6, 8, 9, 16, 103, 108, 113, 116, 267, 268
plasmalemma, 82
plasmodesmata, 1, 3
plastid(s), 39, 41, 43-45, 48, 165, 174, 175, 194, 266
pollen, 213
pollination, 45, 234
pollutant(s), 65, 67, 69, 72, 80, 83, 89, 207
pollution, 136, 145, 232
polyacetylenes, 234

polyamine biosynthesis / metabolism, 207-210, 213-217, 209, 221, 222, 224
polyamine biosynthesis / metabolism genes, 214, 217, 218
polyamine catabolism, 210
polyamine oxidase, 210, 211
polyamine(s) (Pas), 164, 174, 187, 191, 207-214, 218, 223, 239, 242
polyketides, 238, 234
polypeptides, 41, 47, 118, 222
polysaccharides, 101, 113
poplar, 50, 270
post-genomic research, 35
post-transcriptional regulation, 149, 212
post-translational modifications, 154, 177, 219
post-translational modulation, 185
post-translational regulation, 42, 163, 164, 201
potassium (K), 18, 26, 66, 70, 72, 84, 86-88, 115, 145, 212, 231, 267, 272
potassium channels, 267
potato, 214, 237
primordium / primordia, 24, 108, 109, 115
process-based models, 133, 136
programmed cell death, 200
prokaryotes, 121, 153, 164
proline, 40, 212, 213
proteasome, 222
proteids, 231
protein(s), 6, 9, 12, 15, 26, 35, 39, 40, 42-48, 50, 52, 65, 66, 76, 83, 86, 87, 103, 106, 111, 112, 116-118, 120, 139, 140, 144, 145, 152-157, 160, 161, 163, 164, 166-168, 170, 174, 175, 177, 187-190, 191, 193, 195, 196, 208, 209, 212, 213, 217-223, 231, 232, 244, 245, 248, 273
protein kinase(s), 157, 187, 189-191, 195
protein phosphatases, 46, 189, 190, 195, 197
protein synthesis, 39, 86, 87, 144, 175, 213, 231
proteogenesis, 17

proteolysis, 17, 43, 152, 153, 188, 273
proteome, 52, 53, 219, 220
proteomic(s), 208, 215, 219, 222, 223, 224
proton influx, 191
proton pump, 26, 194
protoxylem, 43, 108
Pseudomonas rhizobacteria, 23
purine, 238
putrescine (Put), 208, 209, 211, 213, 218, 242
putrescine hydroxycinnamoyl transferase, 211
pyruvate carboxylase, 48
pyruvate orthophosphate dikinase, 272, 273

quantitative trait loci (QTLs), 1, 52-54
quarternary amines, 241
quinolizidine alkaloid(s), 239

radial root, 1
radial symplasmic transport, 4
radical / reactive oxygen species (ROS), 120, 197-200, 271
radical scavenging, 208
radish, 241
Rauvolfia, 238
receptor(s), 108, 109, 111, 112, 117, 149, 151, 171, 174, 177
receptor kinases, 111, 112
recombinant-DNA breeding methods, 262
red and blue-light receptors, 150, 169, 177
red light, 151, 156, 160, 162, 164, 169, 171
reduced nitrogen, 10, 38, 39, 68, 69, 74, 155, 161, 244
regulatory protein(s), 155, 168
reporter gene, 12, 40, 152, 159, 161, 163, 167, 175, 219
reproductive stage, 159
respiration, 8, 47, 78, 120, 194, 200
reverse genetic screens, 217
rhizobacteria, 23-25
rhizobia, 100, 102, 104-109, 111- 113, 116, 119, 121, 265

Rhizobiacea, 104
rhizobial genes, 106
rhizobial mutants, 108, 113
rhizobial surface polysaccharides, 113
rhizobial taxonomy, 105
Rhizobium, 22, 103-106, 109, 111, 112, 114, 119, 121, 122, 263, 264
rhizomes, 239, 245
rhizosphere, 2, 22-24, 76, 100, 106, 117
rice (plants), 36, 37, 40-46, 48, 51-54, 141, 157, 159, 169, 212, 213, 215, 217, 218, 243, 245, 261-268, 270-272, 274, 275
rice/clover-*Rhizobium* rotation system, 264
rice cultivars, 52
rice globulin, 245
rice root, 37, 41, 44-46, 262, 263, 265, 271
Ricinus communis, 42, 154, 238
root(s), 1-8, 11-27, 36, 37, 39-50, 53, 75-85, 88, 100-104, 106, 108-114, 116, 117, 121, 136-145, 152, 156, 158, 159, 161, 162, 167, 168, 171, 175, 189, 191-195, 197, 207, 253, 263- 268
root amino acid content, 17
root apical meristem, 3
root branching, 20, 22, 24
root cell(s), 1, 6, 7, 8, 12, 13, 15, 17, 21, 268
root cortex, 3, 102, 112
root development, 1, 2, 19, 20-24, 25, 27, 44
root elongation, 21
root epidermis, 3, 46, 102, 108
root exudate(s), 22, 106, 138, 263
root growth, 2, 19, 21, 24, 25, 79, 83, 88, 137, 140, 141, 143
root hair(s), 3, 23, 26, 102, 108-114, 117, 265
root hair curling, 108, 112
root hair deformation, 108, 109, 111, 112, 265
root morphology, 23, 24
root plasmalemma / PM, 82, 197
root plastid(s), 39, 42, 266
root primordia, 103

root surface, 2, 4, 5, 24, 25, 44, 100, 111, 137
root symplasm, 1, 3, 5, 8
root system, 2, 14, 18-20, 22, 25, 82, 102, 117, 137, 140, 142, 143, 197, 207
root tips, 37, 44, 75, 77, 78
root vascular system / tissue, 42, 264
root xylem, 193
Rubiaceae, 238
Rubisco, 83, 133, 139, 143, 144, 177, 271, 273, 274
Rutaceae, 238

Saccharomyces, 36, 155
S-adenosylmethionine, 174
salicylate derivatives, 196
salicylic acid, 176
saline conditions, 271
salinisation, 207, 215
salinity, 193, 215, 216, 272
Salmonella, 231
salt, 187, 193, 207, 208, 212, 218, 272
salt stress, 187, 193, 212, 272
salt-resistant cultivars, 212
Sanguinaria canadensis, 239
saprophytic fungi, 73
saprotrophs, 77
sclerenchyma, 40, 41, 44, 54
Scots pine, 71, 79, 81, 82, 84, 85, 87, 169
secondary compounds / metabolites, 133, 157, 172, 175, 231-234, 236, 238, 244-246, 249, 251, 252
secondary metabolism, 54, 168, 208, 222
seed(s), 52, 104, 150, 234, 235, 239, 241, 251
seed dispersal, 234
seedling, 71, 143, 150, 157, 160, 166, 177, 270
senescence, 40, 42, 48, 115, 138, 200, 208, 273
senescent nodule, 121
senescent zone, 115, 116
senescing, 40, 52, 54, 85, 144
sensing, 25, 42, 141
sensitive species, 76

sensor, 16, 42, 106, 161
serine (Ser), 36, 38, 39, 111, 117, 163, 176, 187-189, 191, 163
Sesbania rostrata, 105
shikimate pathway, 172, 175, 176
shoot(s), 1-4, 10, 11, 13, 15, 17-22, 25, 37, 41, 45, 47, 51, 82-84, 88, 89, 117, 136, 138, 140, 141, 142, 145, 159, 193, 194, 196, 207, 218, 267
shoot/root ratio, 21, 26, 88
signal(s), 20, 26, 42, 46, 106, 111, 140, 153, 166, 177, 189-191, 207, 220, 266
signal transduction, 21, 42, 141, 152, 153, 157, 158, 161, 166, 167, 174, 177, 236
signalling, 14, 15, 16, 18-20, 24, 82, 83, 109, 110, 111, 141, 156, 176, 185, 200, 201, 210, 216
Sinorhizobium meliloti, 104, 105, 107
Sitka spruce, 79, 85, 88
sodium chloride stress, 218
soil(s), 1, 2, 5, 19, 22, 24, 25, 67, 69-75, 77-81, 83-86, 89, 99, 102, 103, 106, 137-140, 143, 145, 150, 165, 193, 197, 199, 207, 263, 269, 274
soil acidity / acidification, 72, 73
soil chemistry, 70, 138
soil microbes / microorganisms, 74, 75, 89, 197
soil nitrogen, 2, 71, 74, 78, 99, 138
soil pH, 73, 74
soil water, 71, 75, 79, 86, 138, 139
Solanaceae, 159
Solanum, 238, 243
source leaves, 2, 17, 53
soybean, 4, 15-17, 50, 104, 105, 117, 119, 159, 167, 169, 196, 242, 265
spermidine (Spd), 187, 208-214, 218, 221, 242
spermidine synthase, 209, 211
spermine (Spm), 208-214, 217, 218, 241, 242
spermine synthase, 209, 211, 221
Sphaeropsis sapinea, 88
spinach, 43, 164, 167, 168, 175, 188, 193
Spirodela, 156, 169, 171

split-root experiments, 14, 18, 19, 140
spores, 106, 240
sporocarp, 76, 78
sporophore producing fungi, 76
spruce, 78, 88, 268, 269
stele, 3, 4, 41, 44
stem(s), 3, 16, 40, 53, 86, 86, 103, 105, 144, 150, 152, 251, 252
stem girdling, 16
stemflow, 74
stomata, 80, 81, 83, 137, 141, 151, 152, 193, 207, 271
stomatal conductance, 136, 272
stomatal control, 272, 273
stomatal uptake, 137
Streptomyces thermoautotrophicus, 266
stress responses, 207, 208, 216, 223
stress signalling, 208, 217
stress tolerance, 207, 208, 214, 218, 272
stress(es), 49, 53, 67, 76, 84, 86-88, 176, 207, 208, 210, 212-218, 223, 236, 237, 239, 245, 272
stress-induced genes, 208
sub stomatal cavity, 81, 136
subcellular compartment, 219, 273
subcellular fluxes, 267
subcellular localisation, 37, 219, 220, 221, 223
succinic semialdehyde dehydrogenase, 211
sucrose, 16, 35, 141, 154, 158, 160, 161, 171, 173, 189, 190, 212, 273
sucrose phosphate synthase, 189, 190
sucrose synthase, 189
sugar(s), 106, 107, 149, 164, 169-171, 190, 191, 192, 242
sugarcane, 100
Suillus, 76, 78
sulfoxides, 246
sulphate transporter gene, 18
sulphides, 233, 247
sulphur based pollutants, 69
sulphur containing volatiles, 246
sulphur deficiency, 240
sulphur fertilisation, 241, 247, 248, 251
superoxide, 165, 185, 186, 198, 248
superoxide anion scavengers, 248

superoxide dismutase (SOD), 199
symbionts, 105, 116, 119, 121, 210
symbiosis(es), 76, 100-105, 107, 110, 113, 116, 119, 121
symbiosome (SM), 100, 116
symbiotic associations, 22, 99, 100
symplasm, 1, 3, 8
symplasmic domains, 3
Synechococcus, 154
synthetic peptides, 188, 189
synthetic pesticides, 232

Tabernaemontana pachysiphon, 239
Taxus, 238, 249
temperature(s), 8, 67, 70, 71, 73, 74, 75, 82-84, 138, 213, 248, 251, 271
terpene(s), 234, 237, 239
Thalictrum rugosum, 236
Theobroma cacao, 238
Thermus thermophilus, 213
thiocyanate(s), 235, 237, 241
thiopropanal S-oxide, 246
thioredoxin(s), 153, 170, 176
threonine, 174
thylakoid(s) membranes, 83, 215, 248
tobacco, 21, 40, 42, 43, 48, 51, 154, 155, 161, 167, 169, 175, 194, 195, 197, 198, 213, 215, 239, 242, 270
tomato, 36, 43, 155, 159, 160, 166, 169, 171, 174, 176, 212, 213
tomato *aurea* mutant, 167, 174
transaminase(s), 157, 172
transamination, 47, 177, 210
transamination inhibitor, 47
transamination mutants, 48
transcription, 162, 168-170, 174, 219
transcriptome, 53, 208
transcripts, 37, 42, 216
trans-cuticular route, 83
transduction pathways, 26
transformant(s), 12, 50, 51, 194, 273
transgenic plants, 26, 50, 51, 161, 163, 167, 218, 221, 272
transgenic rice, 40, 46, 51, 218
transgenic tobacco, 40, 51, 154, 161, 215
transglutaminases, 209
transit peptide, 43, 51

translation, 42, 165, 185, 192, 209, 266
translational inhibitors, 6
transmembrane domain(s), 9, 111
transmethylation reactions, 209
transpiration, 3, 80, 138, 139, 207
transport processes, 136, 139
transport proteins, 3, 12
transport systems, 2, 6, 7, 8, 13, 14, 26, 36, 138, 139, 267
transporters, 1, 4, 6, 9, 10, 15, 16, 26, 36, 82, 119, 155-158, 267
tree(s), 66, 74, 75, 77-82, 85-89, 102, 207, 250
tree growth, 70, 72, 77, 78, 85
tree roots, 75, 76
Trichocereus pachanoi, 238
Tricholoma, 76
Trifolium alexandrinum, 263
Trigonella, 104, 105
Tritordeum, 218
Tropaeolum majus, 251, 252
tryptophane, 172, 175, 237, 238
tyramine, 242
tyrosine, 172, 175, 238, 242

Ulmacea, 105
Ulva rigida, 156, 157
ureides, 119
uridylylation, 154
UV, 152, 166, 171, 176

vacuolar arginine, 84
vacuolar defence compounds, 240
vacuolar nitrate, 192
vacuole, 2, 36, 156, 242, 246, 267, 269
valine, 242
vascular bundles, 46, 48, 54, 114, 115
vascular parenchyma, 41, 54
vascular tissue(s), 40, 46, 51, 53, 103, 152
vegetative stage, 149
vegetative tissue, 274
vesicle(s), 103, 115
Vicia, 104, 200
Vinca, 238

water, 2, 87, 145, 181, 193, 197, 207, 208, 271
waterlogged areas / plants, 105, 194
wet deposition, 68, 87
wheat, 51, 160, 213, 264, 270, 274
woody species, 7, 101

xanthine oxidase/dehydrogenase, 196, 197
Xantobacter, 105
Xenopus oocytes, 10
xylem, 1-4, 8, 17, 39, 41, 43, 46, 54, 81, 119, 138, 139, 156, 173, 210, 265

yeast(s), 36, 154, 155, 173, 189, 190, 208, 219-222
yield(s), 51, 52, 66, 70, 71, 186, 196, 207, 210, 215, 220, 231, 235, 242, 246, 261-264, 267, 269, 270, 272-275

Zea mays, 250

Plant Ecophysiology

1. M.J. Hawkesford and P. Buchner (eds.): *Molecular Analysis of Plant Adaptation to the Environment.* 2001   ISBN 1-4020-0016-2
2. D. Grill, M. Tausz and L.J. De Kok (eds.): *Significance of Glutathione to Plant Adaptation to the Environment.* 2001   ISBN 1-4020-0178-9
3. S. Amâncio and I. Stulen (eds.): *Nitrogen Acquisition and Assimilation in Higher Plants.* 2004   ISBN 1-4020-2727-3

KLUWER ACADEMIC PUBLISHERS – DORDRECHT / BOSTON / LONDON